Praise from the Experts

"I enjoyed this book because it is easy to read and understand. Using examples that are faced in real-life situations, it explains important statistical concepts clearly enough for anyone to grasp. A must-read for anyone interested in statistics applied to management."

Issa Bass
Senior Consultant
Manor House and Associates

"Practicing statisticians in business and industry often have to deal with an unfortunate reality: many of our technical and managerial colleagues have a narrow view of how statistics applies to what they do. This is often a result of their negative experience with the introductory-level college course on statistics. In *JMP Means Business,* Schmee and Oppenlander strive successfully to change the perception of statistics as a dull topic. They emphasize applications and focus on the problem to be solved, rather than the tools. The book is based on the premise that real-life applications almost always involve an iterative learning process as opposed to a single-shot solution to a well-defined technical problem. And, as the authors convincingly illustrate, there is much that statisticians can contribute to up-front planning of studies instead of analyzing data thrown at them after the fact.

"*JMP Means Business* makes an exceptional textbook for business majors and MBA students. It can also be used for self-study by practitioners wishing to learn how statistics applies to real problems in business and industry."

Necip Doganaksoy, Ph.D.
Principal Technologist
GE Global Research

JMP® Means Business
Statistical Models for Management

Josef Schmee

Jane E. Oppenlander

The correct bibliographic citation for this manual is as follows: Schmee, Josef, and Jane E. Oppenlander. 2010. *JMP*® *Means Business: Statistical Models for Management.* Cary, NC: SAS Institute Inc.

JMP® **Means Business: Statistical Models for Management**

Copyright © 2010, SAS Institute Inc., Cary, NC, USA

ISBN 978-1-59994-299-5
ISBN 978-1-60764-427-9 (electronic book)

SAS® Publishing provides a complete selection of books and electronic products to help customers use SAS software to its fullest potential. For more information about our e-books, e-learning products, CDs, and hard-copy books, visit the SAS Publishing Web site at **support.sas.com/publishing** or call 1-800-727-3228.

Contents

Preface   ix

Acknowledgments   xi

Chapter 1   About Data   1

1.1   Introduction   2
1.2   Why Data Are Needed   4
1.3   Sources of Data   5
1.4   Data Scales   11
1.5   Summary   15
1.6   Problems   16
1.7   Case Study: Green's Gym—Part 1   17

Chapter 2   Data Collection in Surveys   19

2.1   Introduction   20
2.2   Questionnaires   21
2.3   Sampling   33
2.4   Summary   46
2.5   Problems   47
2.6   Case Study: Green's Gym—Part 2   51
2.7   References   51

Chapter 3   Describing Data from a Single Variable   53

3.1   Introduction   54
3.2   Example: Order Processing in an Herbal Tea Mail Order Business   57
3.3   Descriptive Statistics with the JMP Distribution Platform   58
3.4   Interpretation of Descriptive Statistics   64
3.5   Practical Advice and Potential Problems   71
3.6   Summary   76

3.7 Problems 77
3.8 Case Study: New Web Software Testing 79

Chapter 4 Statistical Models 81

4.1 Introduction 82
4.2 Classification of Statistical Models 90
4.3 Model Validation 97
4.4 Summary 100
4.5 Problems 101
4.6 Case Study: Models of Advertising Effectiveness 102

Chapter 5 Discrete Probability Distributions 105

5.1 Introduction to Distributions 106
5.2 Discrete Distributions 111
5.3 Binomial Distribution 113
5.4 Distributions of Two Discrete Random Variables (Y_1, Y_2) 118
5.5 Summary 124
5.6 Problems 126
5.7 Case Study: Assessing Financial Investments 128

Chapter 6 Continuous Probability Distributions 131

6.1 Introduction to Continuous Distributions 132
6.2 Characteristics of Continuous Distributions 132
6.3 Uniform Distribution 135
6.4 The Normal Distribution 136
6.5 Central Limit Theorem 146
6.6 Sampling Distributions 148
6.7 Summary 152
6.8 Problems 154
6.9 Case Study: Julie's Lakeside Candy 156

Chapter 7 Confidence Intervals   159

7.1   Introduction   160
7.2   Point Estimates of Mean and Standard Deviation   160
7.3   Confidence Intervals for Mean and Standard Deviation   165
7.4   Detail Example: Package Delivery Times of Herbal Teas   169
7.5   JMP Analysis of Herbal Tea Package Delivery Times   170
7.6   Prediction and Tolerance Intervals   174
7.7   Summary   180
7.8   Problems   181
7.9   References   183

Chapter 8 Hypothesis Tests for a Single Variable Y   185

8.1   Introduction to Hypothesis Testing   186
8.2   Sample Size Needed to Test H_0: Mean = $Mean_0$ versus H_A: Mean = $Mean_A$ 203
8.3   Summary   216
8.4   Problems   217
8.5   Case Study: Traffic Speed Limit Change   221

Chapter 9 Comparing Two Means   223

9.1   Introduction   224
9.2   Two-Sample *t*-Test   227
9.3   Paired *t*-Test   233
9.4   Paired *t*-Test versus Two-Sample *t*-Test on the Same Data   238
9.5   Summary   241
9.6   Problems   241
9.7   Case Study: Westville Meat Processing Plant   244
9.8   References   246

Chapter 10 Comparing Several Means with One-Way ANOVA   247

10.1 Introduction   248
10.2 Detail Example: Training Method and Time to Learn   253

10.3 One-Way ANOVA in JMP 254
10.4 Checking Assumptions of ANOVA Model 265
10.5 Summary 272
10.6 Problems 273
10.7 Case Study: Carpal Tunnel Release Surgery 276

Chapter 11 Two-Way ANOVA for Comparing Means 277

11.1 Introduction 278
11.2 Two-Way ANOVA without Replications 280
11.3 Two-Way ANOVA with Equally Replicated Data 292
11.4 Two-Way ANOVA with Unequal Replications 306
11.5 Summary 306
11.6 Problems 307
11.7 Case Study: Fish Catch near Oil Rig 312

Chapter 12 Proportions 315

12.1 Introduction 316
12.2 Proportions from a Single Sample 317
12.3 Chi-Square Test for Equality of k Proportions 327
12.4 Summary 334
12.5 Problems 335
12.6 Case Study: Incomplete Rebate Submissions 337

Chapter 13 Tests for Independence 339

13.1 Statistical Independence of Two Nominal Variables 340
13.2 Stratification in Cross-Classified Data 352
13.3 Summary 364
13.4 Problems 364
13.5 Case Study: Financial Management Customer
     Satisfaction Survey 366
13.6 References 367

Chapter 14 Simple Regression Analysis 369

14.1 Introduction 370
14.2 Detail Example: Yield in a Chemical Reactor 374
14.3 JMP Analysis of the Yield in a Chemical Reactor Example 377
14.4 Interpretation of Basic Regression Outputs 381
14.5 How Good Is the Regression Line? 393
14.6 Important Considerations 397
14.7 Summary 401
14.8 Problems 402
14.9 Case Study: Lost Time Occupational Injuries 405

Chapter 15 Simple Regression Extensions 407

15.1 Simple Correlation 408
15.2 Regression and Stock Market Returns 417
15.3 Curvilinear Regression 426
15.4 Summary 438
15.5 Problems 439
15.6 Case Studies 444

Chapter 16 Multiple Regression Analysis 447

16.1 Introduction 448
16.2 Detail Example: Profits of Bank Branches 451
16.3 JMP Analysis of Bank Branch Profits Example 453
16.4 Evaluating Model Assumptions and Goodness of Fit 462
16.5 Model Interpretation 468
16.6 Summary 475
16.7 Problems 476
16.8 Case Study: Forbes Global 2000 High Performers 479
16.9 References 480

Chapter 17 Multiple Regression with Nominal Variables 481

17.1 Introduction 482
17.2 Detail Example: Loan Amount versus Sales Revenues 483

17.3 Difference of Intercepts of Two Parallel Lines 485

17.4 Regression Models Including Nominal Variables with Three or More Levels 490

17.5 Both Intercept and Slope of Two Lines Are Different 495

17.6 Summary 501

17.7 Problems 501

17.8 Case Study: Coffee Sales 505

Chapter 18 Finding a Good Multiple Regression Model 507

18.1 Introduction 508

18.2 Detail Example: Profit of Bank Branches 510

18.3 All Possible Regression Models 512

18.4 Stepwise Regression 516

18.5 Candidate Models 523

18.6 Model Recommendation 530

18.7 Summary 533

18.8 Problems 534

18.9 Case Studies 539

Chapter 19 Exponential Smoothing Models for Time Series Data 541

19.1 Introduction 542

19.2 Detail Example: 10-Year Treasury Note Closing Prices 543

19.3 Smoothing Models 550

19.4 Summary 569

19.5 Problems 569

19.6 Case Study: Lockheed Martin Stock in Changing Times 572

19.7 References 573

Index 575

Preface

This text evolved from a set of class notes for an MBA course and from teaching materials developed as part of in-house training on quantitative analysis tools in several industrial and financial corporations, some as part of Six Sigma training. Our MBA students, who often have limited inclination for formula-based statistical instruction, need statistical tools to perform course work in finance, economics, operations, and quality management. Students in industry often have job-related projects with strong statistical content. We have found that both groups are quite willing to delve into fairly sophisticated statistical methods if they are given a practical and problem-oriented approach that involves the assistance of modern statistical software.

Accordingly, we aim this text at business and management students, both undergraduate and graduate, and at professionals in industry with a need to use and interpret statistics. The majority of the examples and exercises have been gathered from various sources in manufacturing and service industries. Of course, we have modified many problems to suit the introductory level of the text and to make pedagogical points, but have retained the flavor of the originals.

We offer a standard selection of introductory topics in classical statistics. We have chosen them on the basis of their usefulness to business and management and their necessity in other management courses. We have also included questionnaire design because of its importance in surveys, and exponential smoothing because of its importance in operations. We also cover the statistical steps for dealing with the capital asset pricing model and price elasticity of demand. Page limitations required that we omit topics such as quality methods, design of experiments, simulation, or partitioning.

Most of our students had a recent exposure to probability for about half a semester. We offer two probability chapters to review probability and to demonstrate the capabilities of JMP. These two chapters should enable readers to understand the remainder of the text.

We have attempted to avoid statistical formulas as much as possible and to define important concepts in plain English and graphically. Most chapters demonstrate a statistical tool by working through a detailed example according to the following template:

- problem statement
- data requirements
- implementation in JMP

- discussion of the JMP results
- interpretation to address the problem statement

We used JMP 8 to develop the examples. We have found that the JMP platforms used in this text did not undergo major changes from version to version. However, some option menus have been extended. In some instances, such as the case of sample sizes for proportions, a newer method has replaced an older one.

Each chapter includes a problem set and all but one ends with at least one case study. The problems range from simple to intermediate in their level of difficulty, with specific questions intended to reinforce key concepts. The case studies offer more unstructured tasks and are more open-ended than the problems.

The following supportive materials are available from http://support.sas.com/authors:

- data sets for the examples in the text
- data sets for the problems at the end of each chapter
- solutions to most problems
- Microsoft PowerPoint presentations of the main points in each chapter
- supplementary chapters and examples

Our approach to teaching a statistics course for management has been analogous to owning a car. Whoever owned a car around 1900 needed to be a mechanic or had to hire a driver with good mechanical skills. Nowadays, car technology has advanced to a state where a car owner needs to know only how to drive a car and what the rules of the road are. Still, highly trained automotive engineers continue to be engaged in research to improve cars.

Similarly, before the widespread availability of statistical software, statistical analysis required advanced training to perform long-winded calculations. Statistical software has changed that. Complicated calculations can be performed on a laptop in a split second. Complex graphs can be drawn and redrawn with a mouse click. But the laptop owner needs to know the rules of the statistical road. Meanwhile, highly trained statisticians are engaged in researching new tools and in improving existing ones.

Acknowledgments

We had many beneficial discussions with several people. Lesly Regis, Senior Vice President of Business Transformation at HSBC, was of invaluable help. His frequent questions shaped our approach and many detailed suggestions improved our chapters. We are deeply indebted to Zhilan Feng and James Lambrinos of Union Graduate College for reviewing our writing on the capital asset pricing model and price elasticity. We also thank Necip Doganaksoy of GE Global Research for his comments on the text. We both had the benefit of having been taught by Gerald J. Hahn of GE Global Research (retired). We also thank David Bobeck, Senior Vice President of Credit Suisse, for his support in the early phases of the project.

We thank Christopher Gotwalt of SAS for his help with the new version of the sample size platform. We also thank our technical reviewers Mark Bailey, Marissa Langford, Paul Marovich, and Tonya Mauldin for many helpful comments and clarifications.

The staff of SAS Press has been most helpful throughout the project, especially Stephenie Joyner, who responded with speed and courtesy to all our requests. We thank our copyeditor Joel Byrd for many improvements to the text. We gratefully acknowledge the contributions of managing editor Mary Beth Steinbach, production specialist Candy Farrell, designers Patrice Cherry and Jennifer Dilly, and marketing specialists Stacey Hamilton and Shelly Goodin.

Lastly, we want to thank our students of the last decade. They have accepted many trial versions of notes and presentations with great patience and helped us understand their problems in dealing with these subjects.

C h a p t e r 1

About Data

1.1 Introduction 2

1.2 Why Data Are Needed 4

1.3 Sources of Data 5

    1.3.1 Existing Data versus New Data 5

    1.3.2 Existing Data 6

    1.3.3 New Data 7

1.4 Data Scales 11

    1.4.1 Ratio Scale 13

    1.4.2 Interval Scale 13

    1.4.3 Ordinal Scale 13

    1.4.4 Nominal Scale 13

    1.4.5 Likert Scale 14

1.5 Summary 15

1.6 Problems 16

1.7 Case Study: Green's Gym—Part 1 17

1.1 Introduction

Data are an essential ingredient in learning about the business world and solving business problems. The list that follows gives you a flavor of the numerous business situations requiring data:

- In monitoring the delinquency rate of credit card accounts, a bank needs data on delinquencies and the demographic factors that might have contributed to them.

- In evaluating the performance of bank branches, a bank needs data on the branches and the business environment under which each operates.

- In making investment choices, a fund manager needs data to assess the risks associated with investment alternatives.

- In developing a new test for blood sugar levels of diabetics, a company needs to estimate the size of the market from various data sources.

- In making an advertising claim that a dishwasher is superior to functionally equivalent models of several competitors, a company needs to collect data that substantiate that claim.

- In comparing the performance of several suppliers of a plastic material, a car manufacturer needs data on the strength of each supplier's material.

- In new product development, a company requires data on customers' needs and desires.

- In developing an ink for printing on plastic bags, a company needs experimental data that evaluate the adhesiveness of various ink mixtures.

- In pricing a multi-year warranty program for personal computers, a computer manufacturer needs data on the failure rate of components and the cost of replacing them.

- In getting a new cholesterol drug ready for approval, a company needs data that prove that the new drug works and has acceptable risks associated with it.

- In designing a new wastewater treatment facility, a company needs data from a simulation to evaluate the performance of various plant configurations.

Data are measurements of facts that are closely linked to statistical models. Models provide explanations in view of the data. Statistical models are often mathematical expressions. Data are used to estimate model coefficients or parameters of the mathematical expressions. However, the process of statistical model building goes further. Data and statistics help decide which variables to include in the model, as well as how to express variables in combination with other variables.

For example, in building a model for the valuation of medium-size businesses, you might use data to decide whether to include total sales or total assets or both as model variables. Data also help determine whether the value of a business depends linearly on sales and assets or whether a change of scale yields a better model. Last, additional data often suggest modifications to existing models, such as adding new variables, eliminating existing ones, or re-estimating the model coefficients.

Data suggest modifications to existing empirical models. The cycle of data collection and model building reflects the fact that businesses evolve dynamically. They require continuous updating of facts and assumptions as well as continuous adaptation of the model to new realities as suggested by the data. This empirical learning cycle is shown in Figure 1.1.

Figure 1.1 Empirical Learning Cycle

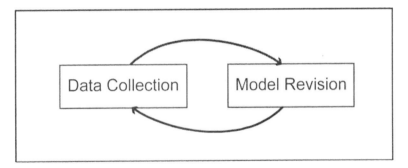

This chapter examines some common sources of data: existing data sources, survey data, data from designed experiments, observational data, and data from simulation experiments. You should recognize from the outset that data reflect reality but are not reality itself. Even though data often seem to give a true picture, there are occasions when data present an incomplete and distorted picture of reality. Whenever data are used, examine the quality and relevance of the data for the purpose to which they are put.

Data are generated or obtained from sources either internal or external to the business. Data might already exist or have to be collected. Sometimes a mixture of internal and external data needs to be used. For example, in benchmarking order processing, a business might compare its own order process to a similar process within the same company, to those of (external) competitors, or even to those of noncompetitors.

1.2 Why Data Are Needed

Management requires data because they encounter variation and uncertainty in decision making. Especially in larger businesses, management has an undiminished need to understand the business and its environment. Global markets have added further complexity and the need to adjust decisions quickly to changes in the business environment. Data provide understanding of the internal business operations as well as of the external environment.

The saying "You cannot manage what you cannot measure!" expresses how essential data are to modern management practice. Data guide strategic, tactical, and operational decision making. Management needs data to determine the future direction of a business, to allocate resources, and to run existing operations smoothly and efficiently. A diversified company needs data to determine which businesses to foster and grow, which to acquire, and which to leave behind. The data for such strategic decisions are not easy to come by. They require the proper context to be useful and informative. A retail chain needs data to determine sites for new retail outlets. Retailers maintain and acquire large databases for such tactical decisions. These data are fed into specialized software to evaluate locations for their suitability. Such software determines the data needs.

Management needs data because business operations, markets, and finance, among others, cannot be predicted with certainty. The outcomes of virtually all business processes are subject to some variation. Markets tend to shift randomly over the short term, often hiding underlying long-term patterns. Therefore, a major reason behind the need for data is to understand variation.

Data play an essential role in managing a business. Data are used in product development, manufacturing, and marketing as well as in support functions such as finance and human resources. Data are needed to

- **demonstrate compliance** with the regulatory requirements of such agencies as the Environmental Protection Agency
- **explain** relationships between variables
- **estimate** a curvilinear relationship that pinpoints a staffing level above which processing times do not increase
- **develop** new products and services by providing quantitative and qualitative information about customer preferences
- **test** the product of one company against similar ones of competitors
- **monitor** the quality of products and customer services
- **predict** the market for various products or services

Data alone would be insufficient for successful management. Data are an aid in decision making. They do not replace business knowledge but increase it. Subject matter knowledge gives management direction in several ways. It guides data collection and is an essential prerequisite for proper data analysis and model building.

Data are needed because of variation. Without variation in customer opinion, one customer would be sufficient to find out what all other customers like and dislike about a product or service. However, this case of customer unanimity is hard to find. Variation occurs when a process yields a finite or infinite variety of uncertain outcomes such as the daily changes in the stock market, varying times to complete a task, or varying dimensions of a part used in the manufacture of a product. Repeated observations of outcomes of the same process under nearly identical conditions yield different values. Process variation is a major cause of concern in improving the quality of products and services. Eliminating variation, or at least taming it by explaining, mitigating, and eliminating some of its causes, is a major effort in continuous improvement efforts. Variation, especially that arising from the uncertainty of the future, can never be eliminated.

1.3 Sources of Data

Data need to be assembled, acquired, or collected. Because data represent factual information, data quality is an important issue. Unfortunately, no matter how they were collected, data often contain errors. In this section, various sources of data and their most common advantages and disadvantages are discussed.

1.3.1 Existing Data versus New Data

The data required to proceed to solve a specific problem might already exist within a business or they might be obtained from commercial and noncommercial sources outside the business. The main task with such existing data is to locate them, verify their quality, and use them judiciously.

Sometimes, existing data are used instead of collecting new data. For example, in estimating the beta risk of a company, you need to rely on recent market prices of a company's stock. Other times, newly collected data are more appropriate. For example, an Internet service provider might contemplate a variety of new services to a market segment. In order to find out which services customers are most likely to accept, new data need to be collected because none exist. New data can be tailored to solve a particular problem, whereas existing data are limited by the available content. We now briefly discuss some data sources that play an important role in business applications.

1.3.2 Existing Data

The first task with existing data is to find the appropriate ones. Sources are varied and often unreliable. When they are the only data available, they might provide better value than trying to obtain new data or not using data at all.

Sources

Sources for existing data are plentiful. Within a company, for example, a business unit providing computer services for databases and Web applications logs data continuously about the usage and update frequency of its servers. Such information is helpful in improving the operation of the servers.

Data might exist but in an unusable format requiring work to make them useful. Good data require some effort to assemble and verify. They usually require considerable resources but often prove themselves as good investments. Examples of existing business internal data are

- delivery times of packages of a package delivery business
- monthly sales by department for the past 10 years of a chain store business
- cost of transplant surgery in a hospital for the most recent 3 years
- results of strength tests of a material for the last year of a material testing lab
- monthly deposits at a bank branch of a bank holding company

Many organizations provide financial or marketing data by subscription or on an individual basis. Benchmarking organizations offer inter- and intra-industry data on a variety of subjects. Financial data on companies are available from organizations such as COMPUSTAT, which is a subsidiary of Standard & Poor's. Another popular financial database is from the Center for Research in Security Prices (CRSP). CRSP is an organization with strong ties to academic researchers. For financial market data, finance.yahoo.com is used in this book. Marketing data are available from many organizations; among the best known is ACNielsen. Specialist firms such as Claritas.com provide data and perform studies on consumer spending, product usage, lifestyle segmentation, and many other applications.

Uses of Existing Data

Existing data can be summarized and analyzed for empirical patterns or relationships that hold true for the past. Existing data are most useful in

- setting standards or targets for formulating future hypotheses
- comparing present or new practices with past practices

Existing data are sometimes used to construct models for predicting the future. In this case, you should use caution because the historical pattern might not predict the future. In using existing data, consider the following points:

- Verify the integrity of the data source. Many data sources, including those that are commercially sold, contain errors. In order to avoid erroneous analyses, always include sanity checks in your preliminary analysis.

- Newly accruing data might change a distribution's shape and the appearance of graphs and numerical summaries.

- An important question in business applications is how long a data history should be considered. Going too far back into the historical record could present information that has little relevance to the current problem. Using a very short history might not reveal important patterns.

- Historical data are often not random and, thus, limit the objectivity of the conclusions you can draw from them.

- Different methods of presentation (for example, histograms) might give different impressions of the data. Use JMP to try many methods and see which one is most useful (see Chapter 3).

1.3.3 New Data

The advantage of new data is that they can be collected with a focus on obtaining problem-specific answers. By using problem-specific variables and optimum data collection methods, you can get results that provide answers that are more specific. However, new data often require considerable expenditures of time and resources and, thus, present the need to trade off the benefits of special-purpose data against the cost and time of acquiring them. Different methods of data collection might be appropriate. Focus groups provide general and highly qualitative insights. Surveys, designed experiments, and simulation often provide very specific quantitative and qualitative data to answer specific questions.

Focus Groups

Focus groups are structured discussion groups of individuals with the aim of obtaining several perspectives on a single problem. Focus groups are useful for exploring personal attitudes and beliefs as well as experiences and reactions from individuals in a group setting. Businesses and politicians use them at an early stage of the problem-solving process. Focus group interviews provide qualitative information. Findings are open-ended and not constrained by finite choices. As a result, they can help generate ideas and develop questions for a follow-up questionnaire. A topic for a focus group might be to investigate the importance of certain features on washing machines. The moderator might give a broad outline of washing machine types (top-loading versus front-loading) and then try to elicit those features that are important to the group members. The data obtained from focus groups often lend themselves to qualitative, rather than quantitative, analysis.

Focus group sessions are administered by a skilled focus group moderator who is neutral, skilled in leading a group, and has good interpersonal skills. Focus group interviews require more planning than other types of interviews. Participants need to meet at specific times and places that are especially equipped for recording the findings. The recommended size for focus groups is 6 to 10 participants per group, although there have been groups as large as 20. Focus group sessions usually last 1 or 2 hours.

Here are some important things to remember:

- Focus groups often help us understand why people attach importance to an issue or a product feature.
- Focus group findings should not be generalized to the entire population without further verification.
- Focus group moderators must not influence the group.
- Focus groups yield results that are open-ended in nature.

Surveys

Surveys consist of a planned and designed collection, analysis, and interpretation of data regarding some aspect of a well-defined population. Populations could be people living in a particular region or having specific characteristics or some other identification of interest to the organization conducting the survey. In business, it is common to conduct satisfaction surveys on past and existing customers. For example, a bank might conduct a quarterly survey of new customers with the aim of monitoring their satisfaction with bank services. Surveys are regularly conducted in certain areas of economic activity, such as housing construction, manufacturing, or household expenditures.

Census

Surveys can collect data from all members of the designated population. In that case, they are called a *census*. The best-known census, prescribed in the U.S. Constitution, is the one conducted every 10 years by the U.S. Census Bureau that covers all people living in the United States. Censuses are also conducted on many smaller subpopulations, such as all the employees of a plant site who are asked about their health care preferences. Censuses of large populations are problematic because it is not always possible to reach all parts of the populations. The missing responses might lead to biased conclusions.

A 100% census can be considered when the cost of sampling is negligible or the population is relatively small and the precision with which results have to be ascertained is fairly high. In stratified audit samples of sales taxes paid, the population is separated into non-overlapping strata by merchandise amount. The stratum containing the high value items often is subject to a census, whereas lower valued strata are sampled.

Sample Surveys

While a census attempts to survey 100% of the members of the designated population, a sample survey aims at surveying only a subpopulation. Depending on how the sample subpopulation is selected, it can be called a *representative sample*, *random sample*, or *judgment sample*, for example. These terms are explained in the next chapter.

In comparing the merits of a sample survey versus a census, several features stand out. Table 1.1 gives a brief comparison of sample survey and census.

Table 1.1 Comparison of Sample and 100% Census

Sample	100% Census
Inexpensive	Expensive
Possible	Often impossible
Allows a more intensive look	Allows a superficial look
Unbiased due to random selection	Biased due to often small number of non-responses

Sample surveys are often cheaper to conduct because fewer population members have to be surveyed. When a large or geographically dispersed population needs to be surveyed, sample surveys can produce results in a more timely fashion than censuses. In a sample survey, you can trade some of the savings due to having to examine fewer population members with a more in-depth look at each member, which also might lead to higher quality data. Last, looking at 100% of a population does not guarantee an unbiased picture, especially when reality shows that some subpopulations might be nearly impossible to examine. Sample surveys carry with them potential disadvantages. Even random selection does not guarantee representativeness. The number of population members surveyed in the sample, the sample size, might be inadequate to draw the desired conclusions.

Designed Experiments

Design of experiments (DOE) is a methodology of systematically varying the inputs, or X-factors, to measure their effect on the output, the Y-response variable, under well-defined experimental conditions. Experiments involve active manipulation of X-factors to study their effects on Y. DOE methodology tells us which factor level combinations to include in the experiment. Consequently, factor effects are measured precisely, accurately, and efficiently. In DOE, you can conduct smaller experiments and augment them by additional experiments. This helps to improve the results, reconcile ambiguities, and bring the experimenter nearer to a cumulative understanding of the problem.

Several points are important with experimental data. Experiments are often performed in laboratories and on a small scale. One needs to be careful when extrapolating results to a larger scale. Experiments are often performed under well-controlled conditions. Extrapolation of controlled conditions to field conditions is often dangerous, because of the additional sources of variation present in the field. Experiments have the ability to manipulate factors (causes) systematically and yield insights unobtainable from historical or field data.

There are many areas where experimentation is impossible or severely restricted (for example, because of ethical concerns).

Observational Studies

Observational studies are occasionally referred to as *quasi-experimental studies* because data are collected using factor combinations similar to those in DOE, but without the benefit of random allocations. Factor level allocations are non-random, because the experimenter cannot manipulate factor level allocations at will. The experimenter may be able to match subject characteristics against desired factor allocations. For example, in studying consumer car buying behavior, personal income is an important X-factor. A subject is selected on the basis of income level. However, the experimenter cannot ask a subject to make more or less money, but can only match subjects to factor levels of their existing income. Because of the difficulty in matching complex factor-level patterns, the number of factors in an observational study is limited.

Computer Simulation

Simulation is a useful tool to understand the behavior of complex systems. With *simulation*, systems can be observed even outside the safe range of operations. Design changes can be evaluated even before the changes are actually constructed. Computer simulation allows the creation of very realistic (although still artificial) models that often match historical data. Computer simulation models also allow for a manipulation of factors in the model so that they can be evaluated regarding their importance on the output variable.

A simulation model is only as good as the assumptions on which it is based. Different people often disagree over those assumptions and the results. The controversy regarding predictions from simulation models concerning climate change is an example. Different experimenters make different assumptions and so predict a different future.

1.4 Data Scales

Data are recorded observations about the physical or perceptual world. In the physical sciences, these data often represent measurements on physical objects. These might be measurements of naturally occurring phenomena or the result of carefully designed experiments. In business, data often occur quite regularly. For example, they occur as regular records of daily sales volume or as information specifically gathered through surveys and experiments.

Data are characteristics measured on elements. *Elements* can be physical objects like customers or parts of a product. They can also be organizationally complex entities such as bank branches, business divisions, or entire businesses. A characteristic that can take on more than one value is called a *variable*. Otherwise, it is called a *constant*. In passenger cars, for example, the number of cylinders in the engine is a variable because it can take on values like 2, 4, 5, 6, 8, 12, and even 16. The number of axles is constant at 2, excepting some special designs that might have more than 2.

In JMP as in other statistical software, data are represented in tables. Columns represent variables or characteristics. Rows represent the elements on which these characteristics have been measured. Figure 1.2 shows an excerpt of the JMP data file *MutualFundInvestment.jmp* containing a record of mutual fund payroll deductions. The following variables are in six columns:

- Column 1: Year (only 1999 is shown)
- Column 2: Week (weeks 14 to 36 are shown)
- Column 3: Dollars (amount invested per week)
- Column 4: Price (per unit of mutual fund)
- Column 5: Units Purchased (calculated from Columns 3 and 4)
- Column 6: Cum Units (cumulative units calculated from Column 5)

Figure 1.2 Data File Excerpt for Mutual Fund Payroll Deduction

	Year	Week	Dollars	Price	Units Purchased	Cum Units
1	1999	14	100	14.5	6.897	6.897
2	1999	16	100	14.57	6.863	13.76
3	1999	18	100	14.72	6.793	20.553
4	1999	20	100	14.78	6.766	27.319
5	1999	22	100	14.19	7.047	34.366
6	1999	24	100	14.29	6.998	41.364
7	1999	26	100	15.51	6.447	47.811
8	1999	28	100	15.83	6.317	54.128
9	1999	30	100	14.81	6.752	60.88
10	1999	32	100	14.76	6.775	67.655
11	1999	34	100	15.09	6.627	74.282
12	1999	36	100	15.14	6.605	80.887

MutualFundInvestment

MutualFundInvestme

Columns (6/0)
- Year
- Week
- Dollars
- Price
- Units Purchased
- Cum Units

Rows
- All rows   96
- Selected   0
- Excluded   0
- Hidden   0

Each row of the data table represents a purchase action of mutual fund units. The weeks are ordered from earliest (Year 1999, Week 14) to latest (Year 2002, Week 46). The payroll deduction amount remained constant at $100. However, there are other amounts at year-end representing dividend disbursements.

The variable Price is measured on a continuous scale because any value greater than zero makes sense. Is $14.5 a high or a low price? That depends entirely on the comparison with other prices within the time frame of the mutual fund. Since the data file is from Week 14 in 1999 to Week 46 in 2002, use this period to judge a high or a low unit price. Suppose that a unit price of $10 or more is considered high and less than $10 is low. The variable Price is now simplified into two categories, high and low. Such a simplification of continuous variables often communicates concepts more effectively; at other times, they are the only form in which data are available. A classification of the unit price of a mutual fund into high and low is a different type of scale.

Numeric variables are often, but not always, measurements on a continuous scale. Such data are usually on the ratio scale or the interval scale. Variables might also record rankings of performance. These data are said to be on an ordinal scale. Last, variables might represent simple categories, such as left and right. These are on the nominal scale.

1.4.1 Ratio Scale

Ratio scales are the highest level of scales. They are called *ratio scales* because when two measurements differ by a multiplicative factor *r*, then the larger value is said to be *r* times larger than the smaller. For example, if the price of a barrel of oil is $100 and subsequently increases to $110, you can say that the price increased by 10%. The ratio factor is *r*=1.1. Likewise, a rise from $100 to $200 would represent a doubling in price, because twice the monetary exchange units are required to buy a barrel of oil. In this case, the ratio factor is *r*=2. Ratio scales have a natural 0. If crude oil costs $0, then no money is needed to buy a barrel of it.

1.4.2 Interval Scale

Interval scales are so called because when two measurements differ by an additive factor d, then the larger is said to be d units higher than the lower. For example, if the temperature in London is 50 degrees and the temperature in New York is 40 degrees, then London is 10 degrees warmer, but you cannot and should not say that London is 25% warmer (as for data on a ratio scale). The difference between the price of oil and the temperature in degrees Fahrenheit is that the Fahrenheit scale has an arbitrary 0 (at approximately –18 degrees Celsius), just as the Celsius scale has an arbitrary 0 at 32 degrees F.

1.4.3 Ordinal Scale

Data on an ordinal scale rank performance measurements. For example, an issue of *Consumer Reports* ranked DVD players from number 1 to number 13, suggesting that number 1 was the best performer and number 13 was the poorest performer of the 13 investigated. This scale did not suggest, however, that the number 1 DVD player was twice as good as the number 2 player nor that the difference between the player ranked 1 and the player ranked 2 was the same as the difference between players ranked 2 and 3. Other examples of variables with ordinal scales involve Likert scale responses discussed in Section 1.4.5.

1.4.4 Nominal Scale

Data on a nominal scale represent a set of categories that differ in some characteristic that does not necessarily have a measurable magnitude. Observations on a nominal scale are classified in one of several non-overlapping categories, also called the levels of the variables. Gender, Type of Bank, and Country of Origin are on a nominal scale.

Table 1.2 compares these four scales, relates them to JMP terminology, and identifies whether a mean or a standard deviation make sense. In JMP, the modeling type of each variable has a specific marker next to its name in the data window.

You can always collapse a higher order scale into a lower scale. For example, categorize days above 80 degrees Fahrenheit as hot, days between 40 and 80 degrees as temperate, and days below 40 degrees Fahrenheit as cold. The collapse is arbitrary because other temperature cutoff points could have been used.

Table 1.2 Measurement Scales

Scale	Modeling Type in JMP	Mean, Std. Dev.	Other Names
Ratio	Continuous ◢	Yes	Quantitative (may be both continuous and discrete=integer)
Interval	Continuous ◢	Yes	Quantitative (may be both continuous and discrete=integer)
Ordinal	Ordinal ◢	No	Qualitative (cannot be continuous)
Nominal	Nominal ▮.	No	Qualitative (cannot be continuous)

1.4.5 Likert Scale

In survey questionnaires, the Likert scale is used for measuring attitudes in marketing research. Likert scales measure the strength of a respondent's perceived agreement or disagreement to statements such as "ABC bank employees are always friendly when doing business with their customers." A typical form of a response on a five-point Likert scale is

Strongly Disagree	Disagree	Neutral Opinion	Agree	Strongly Agree
1	2	3	4	5

In a five-point Likert scale, the responses are scored 1, 2, 3, 4, and 5. The value 3 is the neutral position, while 1 and 5 are the two extremes. The very common seven-point Likert scale augments the responses to Strongly Disagree (1), Disagree (2), Slightly Disagree (3), Neither Agree nor Disagree (4), Slightly Agree (5), Agree (6), and Strongly Agree (7). Here the neutral position is scored with 4. Likert scales are often treated as interval scales because you calculate average responses to statements. Other times they are treated as ordinal, such as when the ordering of the responses needs to be stressed.

1.5 Summary

- Data are needed because variation is unavoidable.

- Data collection and model revision are iterative approaches that are helpful in learning about and explaining business situations.

- In business, data come from many sources. Existing data are plentiful, but they may not always suit the purpose at hand. New data serve their purpose, but they require time and resources to collect and organize.

- Existing data need to be validated and checked for errors. Verify the assumptions under which they were collected. Make sure they describe situations that will continue into the future.

- The Internet is a rich source of data. Internet data need to be checked for quality.

- New data come from focus groups, surveys, designed experiments, observational studies, and computer simulation.

- Focus groups yield qualitative data that is difficult to generalize.

- Surveys are useful in collecting data about opinions and attitudes of large and widely dispersed populations.

- A census, although recording data on 100% of the population, does not necessarily provide better information than a well-selected sample. Censuses are useful for examining small populations or small subpopulations of larger populations.

- Samples, when collected following specific rules, yield unbiased information efficiently and with known precision.

- Designed experiments are useful when the X-factor levels can be easily manipulated. Designed experiments often yield results that are easy to interpret.

- Observational studies collect data in patterns similar to designed experiments. Subjects are chosen because their characteristics match desired X-factor level combinations.

- Data from computer simulations are used when it is impossible or impractical to identify realistic observations or to set-up realistic experiments.

- Data are usually organized by columns and rows. Columns represent variables or factors. Rows represent observations.

- In JMP, variables are continuous, ordinal, or nominal. Continuous variables allow calculation of means and standard deviations. Ordinal and nominal variables allow only counts of occurrences by nominal category.

- Surveys of attitudes often use Likert and similar scales. These scales, although ordinal or nominal in appearance, are treated as continuous in many applications.

1.6 Problems

1. Consider this question: "What is a good apple?" Identify variables that would address this question from several different perspectives:

   a. the apple grower's perspective

   b. the retailer's perspective

   c. the consumer's perspective. Discuss how these different perspectives influence the determination of the quality-defining characteristics.

2. Consider your daily trip to work or school:

   a. Describe the variability in your arrival time.

   b. Identify those factors that are in your control and those that are not within your control.

   c. Describe ways in which your travel process could be changed to reduce the variability.

3. A distribution center for a regional grocery chain is concerned about the losses they are incurring from spoilage among their produce items. This requires labor to inspect and remove spoiled items prior to shipment to a grocery store. In addition, the storage bins must be cleaned more frequently.

   a. Choose an appropriate performance variable and measurement scale for that variable.

   b. Identify important X-factors that characterize produce inventory. For each factor, identify the measurement scale selected (ratio, interval, ordinal, or nominal).

4. A human resources (HR) department is charged with evaluating worker satisfaction with the employee assistance program. The employee assistance program provides resources to help workers in the following areas: childcare, eldercare, care for disabled or chronically ill family members, financial planning, and funding for college.

   The HR department would like to collect data from their employees. Options for collecting data include a focus group, census, or sample survey. Prepare a one-page summary that discusses the advantages and disadvantages of each data collection method for this situation. Give your recommendation for the method that should be used. Provide reasons for your recommendation.

1.7 Case Study: Green's Gym—Part 1

Green's Gym is an independent, locally owned fitness center that serves a suburban area of approximately 65,000. A national fitness center chain is expanding into the region, threatening to reduce Green's market share. Green's offers individual, family, seasonal, and senior citizen memberships. Senior citizens receive a 15% discount. Seasonal memberships are either summer (May to August, targeting college students who have returned home for the summer) or winter (November to April, targeting those who prefer outdoor exercise during the warm months).

The facility includes a gymnasium, weight room, dance studio, fitness room (containing exercise equipment such as stationary bicycles and rowing machines), and a childcare facility. The following services are provided:

- Childcare for up to 25 children under the age of six while parents are using the facility. The childcare is not intended to provide daycare for working parents. Childcare hours are from 8 a.m. to 5 p.m. Monday–Friday.

- Adult evening volleyball and basketball leagues that meet once a week.

- The fitness center and weight room are open from 7 a.m. to 9 p.m., seven days a week.

- Saturday morning children's programs utilize the gym from 9 a.m. to 11 a.m. The balance of the weekend time is available for patron use.

- Aerobics classes are held in the dance studio during the week at 6:30 a.m., 10 a.m., and 7 p.m., and on Saturdays at 9 a.m. and 11 a.m. Senior citizen exercise classes are held on weekdays at 9 a.m. and 2 p.m.

Business Goals

Green's is interested in maintaining customer loyalty in the face of new competition.

Strategy

Green's would like to identify those factors that are important in retaining their members in order to prepare a customer satisfaction survey. The information collected in the survey will enable Green's to adjust their services to meet their customers' needs better, thereby retaining members.

Tasks

Perform the following tasks:

1. Identify three important factors that contribute to satisfied members at Green's Gym. Rank those factors in order of importance.

2. Discuss how each of these factors can be effectively measured. Recommend an appropriate measurement scale for each.

3. Are there any additional data that should be collected to help Green's assess their customer satisfaction?

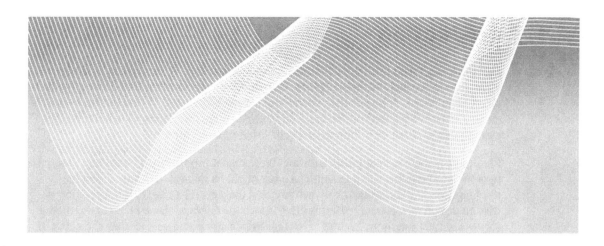

C h a p t e r 2

Data Collection in Surveys

2.1 Introduction 20

2.2 Questionnaires 21

    2.2.1 Layers of a Questionnaire 21

    2.2.2 How to Ask Questions 22

    2.2.3 Questionnaire Design 24

    2.2.4 Guidelines for Writing Questions 24

2.3 Sampling 33

    2.3.1 Sampling Concepts 33

    2.3.2 Probability Samples 36

    2.3.3 Probability versus Non-Probability Samples 37

    2.3.4 Common Types of Probability Sampling 38

    2.3.5 Other Sampling Methods 45

2.4 Summary 46

2.5 Problems 47

2.6 Case Study: Green's Gym—Part 2 51

2.7 References 51

2.1 Introduction

Surveys collect data from populations using questionnaires. A survey can be designed to yield descriptive and predictive results. Government surveys often find descriptive facts about populations. A household expenditure survey is designed to find out how much money households spend on food and other expenditure categories. Other surveys ask manufacturers about their current hiring or firing of employees.

Predictive questionnaires gather information to explain some broad issues and to test hypotheses about them. For example, a bank might be interested in explaining the mortgage lending differences between regional markets. The resulting information is used to design new product offerings that match the needs of those markets. Predictive surveys answer questions such as "How can we design mortgages to meet the needs of different markets?"

The survey process shown in Figure 2.1 begins with designing the survey in its general outline, including writing a problem statement that makes clear the reason why the data are collected, identifying the target population and the sampling frames, and specifying the sample design in terms of sample type and size. It is followed with developing and testing the questionnaire, including specifying the method of administrating the questionnaire (personal interviews, for example), writing the questions and ordering them, and deciding on any measures to minimize non-response.

After the questionnaire is complete, you can begin the selection of the sample as well as the training of the interviewers or observers. Many surveys require a pilot test of the questionnaire on a small sample to fine-tune the language and evaluate whether the questionnaire will yield the desired information. The questionnaire needs to be distributed to and collected from all respondents. The analysis of results includes both data management and statistical analysis. The last step is to present the findings in such a way that the intended audience can easily grasp them.

Figure 2.1 Survey Design Process

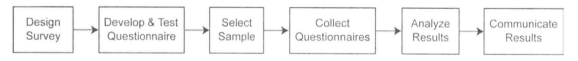

This chapter discusses simple guidelines for writing questions and selecting random samples.

2.2 Questionnaires

A *questionnaire* is an integrated set of questions that are addressed to an intended population to gather information regarding some problem or hypothesis. Questionnaires are more than a series of unrelated questions.

2.2.1 Layers of a Questionnaire

Patricia J. Labaw, in her monograph *Advanced Questionnaire Design* (1980), outlines some conceptual principles about questionnaires. According to Labaw, a questionnaire consists of four layers. Each layer contributes to the success or failure of a questionnaire. The four layers are words, questions, format, and hypothesis.

Words

Words need to be checked for ambiguity. Words can have multiple meanings or be closely related to and confused with similar sounding words. Some words are derived from professional jargon and are too difficult for lay respondents to understand. Word choices may bias responses. It is therefore important to clarify confusing terms. For example, Medicare and Medicaid are two different government programs that are often confused by respondents. Before asking a question about either one of them, the questionnaire should explain the differences between them and ask questions that eliminate one or the other alternative. Another important practice is to use neutral words that do not bias the response. For example, the news organization CNN switched from the term "foreign correspondent" to "international correspondent" to maintain neutrality with respect to its worldwide audience.

Questions

Open-ended questions often provide unanticipated insights, but they are difficult to summarize and quantify. Closed questions provide a fixed choice of answers and limit the responses to those pre-specified. Ask yourself whether respondents are able to answer the questions. Questions that are too general or complex or that are ambiguous should be avoided. Establish an adequate frame of reference for your question. A question like "What is your financial situation?" is too broad because it does not take into account the differences in financial goals individuals may have.

Yet another consideration is the order in which questions are asked. Question order should avoid leading respondents to a response that is a conclusion of previously asked questions. Question order can be used to suggest certain answers, such as when an open-ended question regarding the most important issues facing the nation is immediately preceded by a question about national security.

Format

Internet questionnaires offer few questions on many screens. Paper questionnaires are often filled with many questions on a single page. The layout should have a nice flow that does not confuse the respondent and is easy to follow.

Much research has reviewed questionnaire layout both in connection with paper and with online questionnaires. Questionnaires should be easy to follow from question to question, especially when branching questions are required. Different questions or blocks of questions should be easily identifiable, possibly by using color or other visible identifications for the questions. Fields that respondents need to fill in should be easily identifiable and linked to the questions. Questionnaires should avoid a clutter of questions on a page or screen.

Hypothesis

With predictive surveys, a questionnaire becomes a problem-solving instrument. The questionnaire should be focused on probing specific questions. It should not give the impression of a fishing expedition.

2.2.2 How to Ask Questions

A key question that needs to be answered before designing the questionnaire is how the questionnaire should be administered. Different modes of administration require different safeguards. Survey data are collected from personal, face-to-face interviews of the respondents, from telephone interviews, from self-administered mailed questionnaires, from online questionnaires, and by direct observation.

Personal Interviews

In personal interviews, interviewer and respondent meet face to face. They have the advantage that respondents usually answer questions when they are confronted by the interviewer, assuming that they initially agreed to the interview. While interviewers can clarify questions that respondents have difficulty understanding, they might bias results by influencing respondents either by their body language, by the way the questions are read, or by any comments tagged on to the carefully worded questions. Consequently, interviewers need to be trained in interviewing skills. Personal interviews are expensive because of the time involved in training interviewers and the interviewer time needed to conduct the interviews.

Telephone Interviews

In telephone interviews, the interviewer asks questions from respondents over the telephone. They are usually cheaper than personal interviews but are usually of a limited duration. Interviewer quality can be monitored, but refusal to answer certain questions or hang-ups during the interview increase non-response. Problems with unlisted numbers can be surmounted with random digit dialing. In households with more than one person, a specific person can be identified by asking for the person with the most recent birthday. Households that use only cell phones without land-based phones present new problems. For example, in a pre-election poll, area codes for cell phones do not necessarily predict where people are going to vote. Another concern is that cell phone calls increase the cost of the interview process.

Self-Administered Questionnaires

Self-administered questionnaires are mailed to and returned completed by respondents without the aid of an interviewer. Alternatively, questionnaires can be administered online. Respondents can complete these questionnaires at their leisure and therefore they can be longer than telephone questionnaires. However, they have higher non-response rates. They are useful in market and industry studies, especially when detailed attitudes and their change over time are sought. In business trend research mailed to experts in business, the length and frequency of questionnaires must take into account the job responsibilities of those asked to respond.

Mailed Questionnaires

Mailing questionnaires to respondents is cheap and can be done on a large scale. Mailed questionnaires may have a large percentage of non-respondents. They are useful when they are sent to groups that have an interest in the outcome. For example, a questionnaire mailed to a professional group about their views on new legislation concerning the profession can be expected to receive much interest and a high response rate. Mailed questionnaires in which some subgroup (for example, trial lawyers) is especially interested in the outcome, while other subgroups (for example, all non-trial lawyers) are not often lead to biased results because of different response rates.

Online Surveys

In online surveys, questionnaires are presented to the respondents over the Internet in a series of screens. Online surveys are very popular because they usually have a lower cost per questionnaire, they eliminate interviewer bias but rely on respondents being able to understand the questions, and they can be sent cheaply and quickly to a target audience over a wide geographic reach. Such surveys can be conducted very rapidly, because multiple respondents can respond at any given time. Video clips, sound tracks, advertising layouts, and similar objects can be included. Respondents can take their time completing these surveys, providing more accurate responses. The downside is an uncertain selection method and the difficulty in verifying the identity of the respondent.

Direct Observation

Direct observation does not require a questionnaire per se, but it requires a data collection instrument to record the responses. It is useful because certain information can be observed directly, such as the number of shoppers at shopping centers or non-verbal responses that would be omitted from a standard questionnaire. It can include audio or video recording as part of the data collection. Direct observations may require measurement or counting technologies to produce data useful in both scope and quality. For example, traffic counters are used to record vehicle volumes passing through an intersection.

2.2.3 Questionnaire Design

Questionnaires are data collection instruments. They are structured and coherent collections of questions designed to shed light on specific issues. A good questionnaire will yield valid and reliable information. *Valid* means that the information obtained is relevant for the purpose of the survey and can be used to solve the underlying problem. *Reliable* means that you would obtain the same answers to the questions from the same respondents if asked again.

The typical questionnaire follows a simple outline. The introduction contains the purpose of the survey with a statement about who is conducting it and for whom. It gives assurances about privacy and offers the possibility to terminate the interview at any time. The purpose should be stated in such a way that, together with the opening questions, it motivates respondents to participate and increases their confidence in and comfort with the survey process. The body of the questionnaire should reveal the purpose of the survey. Sensitive questions, such as questions about personal income or private habits, should be asked only if they are necessary to understand the underlying problem. Respondents are more likely to give inaccurate or incomplete answers to sensitive questions. Non-response becomes a way to protect their privacy. Sensitive questions should never be asked upfront but should be placed after lead-in questions.

2.2.4 Guidelines for Writing Questions

The classic book on writing questions for questionnaires is S. L. Payne's *The Art of Asking Questions* (1951). Next is a brief list of principles to keep in mind when writing questions for a questionnaire:

- Use simple words that are familiar to all respondents.
- If necessary, clarify confusing terms.
- Use clear and specific questions.
- Use questions that relate directly to some potential or anticipated action.
- Avoid double-barreled or compound questions.

- Avoid leading or loaded questions.

- Make sure questions are applicable to all respondents or provide branching questions to skip non-applicable questions.

Questions to Avoid

Double-barreled or compound questions ask two or more issues in one question. "Do you plan to sell your house and buy another one?" requires two separate responses. The first response is to the question of selling the house, the second to buying another house.

Leading questions push the respondent toward a certain answer, either directly or indirectly. A question starting with "Don't you agree that …?" effectively asks for a certain response and might lead to biased answers. At times, some pushing might be required to elicit a response, especially on sensitive questions. "By how much do you under-report your income?" might yield better answers than "Do you cheat on your taxes?"

Loaded questions have characteristics such as the following:

Partial mention of alternatives: For example, the question "What type of financial planning, such as a retirement planning, did you conduct within the last year?" might lead to more responses about retirement planning than other categories of financial planning.

Using emotionally charged words: Emotionally charged words such as *fair profits* or *unfair profits* will bias results.

Appealing to or threatening the self-esteem of respondent: Certain questions, though benign in the eye of the interviewer may nevertheless be perceived as threatening. For example in a question with respect to employment status, a phrase like "Do you work?" may intimidate certain respondents and therefore tend to produce false responses. Instead, one could ask the more neutral "What is your current employment status?"

Using context to influence answers: Context also influences answers, especially in closed questions. For example, in eliciting reasons for choosing a bank, you could use a checklist of alternatives such as *online capabilities*, *extent of branch network*, and *banking conditions and terms*. Questions concerning each of these alternatives should be equally distributed since different frequencies can bias the results.

Personalization of questions also leads to bias.

- Rather than asking "Do you think that gambling should be legalized?" simply ask, "Should gambling be legalized?"

- Respondents are steered toward a specific answer by implication—"How much do you cheat on your income taxes?" implies that the respondent cheats—or by outright suggestion—"You did cheat on your income taxes, didn't you?"

Sometimes such suggestive methods are needed to elicit a truthful response, but mostly they are avoided because of their potential for biased responses.

Open versus Closed Responses

An *open response* allows the respondent to answer a question as they see fit. A *closed response* restricts the respondent to a fixed set of alternative responses. Both open and closed responses to questions have their advocates. Open responses do not suggest alternative answers. They ask for a free-form answer to get spontaneous responses and quotable quotes. Open responses are useful in exploring qualitative aspects of a problem. The range of responses can be quite large and can produce responses that nobody thought of beforehand. However, open responses are poor for coding and analysis and require more time and energy by the respondent, resulting in a lower response rate. The saying about open responses is "How can anything so good be so bad?" Use them judiciously at the beginning of a project, when response categories are still being developed.

Closed response questions offer a set of possible answers and, therefore, are easier to answer. Respondents preferably answer with the response alternatives offered by the question. Closed responses yield fixed response sets that are easy to tabulate but limit the response range. Take, for example, a closed response set to a question about Internet security lapses:

- ☐ Computer records infected by virus.
- ☐ Communications with bank misdirected to wrong account.
- ☐ Hacker retrieved account information.
- ☐ Other (specify):_____.

The set clearly limits the response. Even though the Other (specify) response is possible, respondents might be unwilling to take advantage of this category.

Question by Type of Response

Questions can ask for different closed type responses. Some important examples are checklists, questions with quantity and intensity scales (Likert scales), semantic differentials, and rankings.

Checklists: Checklist questions (or statements) are cafeteria style in which the respondent is shown a list of possible answers. Respondents may check as many questions or statements as apply. A concern with checklists is that an appropriate range of response categories is offered to the respondent. In a pre-election poll, a balanced question regarding the reasons to vote for or against a candidate needs to include a sufficient and balanced number of response categories for and against the specific candidate.

Quantity or Intensity Scale Responses: In quantity scales, respondents are asked to rate a concept, product, or service ranging from high to low, from more to less. For example, a question concerning alcohol usage could be listed as:

How often do you drink alcohol?

- ☐ Never
- ☐ 1 or 2 drinks per week
- ☐ 3 to 7 drinks per week
- ☐ 8 to 14 drinks per week
- ☐ More than 14 drinks per week

A question about Internet banking could be phrased as follows:

For what proportion of your banking needs do you use Internet banking?

- ☐ For almost all of my banking needs
- ☐ For most of my banking needs
- ☐ For about half of my banking needs
- ☐ For very few of my banking needs
- ☐ For almost NONE of my banking needs

Intensity scales ask respondents to select a value that expresses the strength of their attitude or preference toward a subject. The response categories range from very negative to very positive or from strong dislike to strong like. These scales are also written in the form of Likert scales, where respondents are presented with an attitude statement and they are asked to mark the strength of their agreement or disagreement. An example of a Likert scale response to a statement about Internet banking could be as follows:

Internet banking is a safe way to conduct banking business.

- ☐ Strongly disagree
- ☐ Disagree
- ☐ Uncertain
- ☐ Agree
- ☐ Strongly agree

The difference between the earlier quantity scale and the Likert scale is quite striking. Both give five response alternatives. However, Likert scale questions ask for attitudes rather than frequency of usage and so are more likely to lead to insights.

The typical number of response values is 3, 5, and 7 for odd-numbered responses. With even responses, 4 response values are common, but 6 are also used. Higher numbers can be found but are less common, because the finer scale often introduces artificial differences. The two schools of thought concerning the number of alternatives advocate either an even number of response values (avoiding the neutral response) or an odd number of responses (including the neutral or midpoint response). Even-numbered response values avoid neutral responses and force a preference. Those preferring odd-numbered response values argue that to force an opinion generates artificial choices and can lead the respondent.

Semantic Differentials: Semantic differentials measure a respondent's reactions to stimulus words. As with Likert scales, there are two opposite scale endpoints, but in the semantic differential the bipolar scale is defined with contrasting words. An example of a response scale between bad and good is shown in Figure 2.2.

The middle category 0 is neutral, 1 is slightly, 2 is very, and 3 is extremely good or bad, depending on the sign. A semantic differential scale measures both the type and direction of the reaction (for example, good versus bad) and the intensity of the reaction (slight, strong, extreme). Semantic differentials are most effective when respondents are asked to evaluate some concept, for example a soft drink brand. By presenting a series of semantic differentials, the respondent can rate the brand on each of these semantic scales. From the ratings of many respondents, a composite brand image emerges.

Figure 2.2 Scale of Semantic Differential

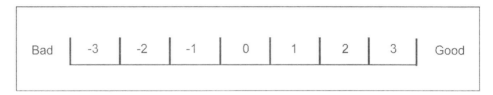

Semantic differentials use three major factors of judgment:

- Evaluative factors such as good–bad, comfortable–uncomfortable, cheap–expensive
- Potency factors such as strong–weak, slow–fast, heavy–light
- Activity factors such as active–passive, work–vacation, tense–relaxed

Semantic differentials are not without problems. They assume that selected opposites have the same meaning to all respondents. However, it is difficult to ascertain that, for example, words like comfortable and uncomfortable have the same meaning to all respondents. Semantic differentials use language to get a sense of how respondents would describe an object or a concept. It is related to the Likert scale in that it preferentially uses a 7-point scale and that the ratings are along opposite semantic differences.

As an example of using semantic differentials in evaluating a car, the following opposites might be useful: slow–fast, work–vacation (association), happy–sad, comfortable–uncomfortable. Of course, there are many more ways to describe a car. Figure 2.3 shows that the response of an individual respondent can be traced in a semantic differential profile. The same profile could be constructed for the average of responses.

Figure 2.3 Individual Response Profile to Semantic Differential of a Car Brand

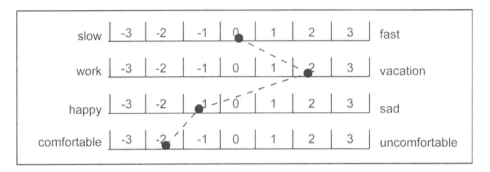

Ranking Questions: These questions request that the respondent rank a number of attributes from high to low.

Example: In choosing a bank for personal banking, rank the following services from (1), representing the most important, to (4), representing the least important:

- ☐ Internet access to my accounts
- ☐ Convenient branch network
- ☐ Debit and credit cards available at no charge
- ☐ Financial planning services

Any number of alternatives can be chosen. The advantage of a ranking question is that you get a sense of which features are important to customers. The problem is that often the choices are forced.

Composing a Questionnaire: A complex questionnaire for multi-issue problems requires experience and expertise. Start from simple principles to compose a useful questionnaire. Test it on a small set of respondents to reveal problems with the questionnaire as a whole or with specific parts. Address concerns about wording, ordering of questions, duration of completing the questionnaire, and responses to sensitive questions at this stage.

Maintain the good will of the respondents during the entire survey and reassure them of the value they contribute to the survey. They should be given the choice to receive the results of the completed survey. Their required effort in completing the survey should be minimized. They should receive thanks for their efforts. Figure 2.4 shows the header of a questionnaire regarding customer satisfaction with service quality at a car dealership. The header text identifies the make of the car, the customer name, and the dealership at which the car was serviced. It contains the following text.

Figure 2.4 Questionnaire Header for Customer Satisfaction Survey

Dear CUSTOMER NAME:

Thank you for having your CAR serviced at DEALERSHIP. Enclosed is a questionnaire that we specifically designed to measure how well CAR DEALERSHIP is meeting your expectations. At DEALERSHIP, we are committed to Customer Satisfaction and look forward to the evaluation of DEALERSHIP service personnel by the person who took your CAR to the dealership.

Please take a few minutes to complete the survey about your most recent service experience and return it to CAR CUSTOMER CENTER in the enclosed postage-paid envelope.

Thanks again for choosing CAR DEALERSHIP for your service needs.

The questionnaire containing 16 questions is shown in Figure 2.5. The questionnaire covers the experience of the service process. In practice, additional questions could ask the customer about several other areas:

- Questions concerning the quality of the service: Was the problem identified early in the service process? Was the problem fixed properly? Did the problem reoccur? How many additional visits to the car dealership were necessary to fix the problem?

- Questions concerning the service: Were the charges reasonable? Did you get value for the charges? Did you sense that you were valued as a customer? Was the service operation mindful of your time constraints?

- Questions concerning your ownership experience: When did you buy the car? Did you buy it from the dealer at which you requested service? Did you buy the car new or used? Do you like driving the car? Would you buy another car of this make? Would you buy again from this dealership? How likely are you to recommend the car maker to others? How likely are you recommend the dealership to others?

Figure 2.5 Questionnaire Concerning Service Experience at Car Dealership

MAKING YOUR APPOINTMENT FOR YOUR SERVICE VISIT

1   On your most recent visit to the dealership, which of the following services did you have performed? (**Mark all that apply!**)

     ◯ Oil change   ◯ Scheduled maintenance   ◯ Car and Body Repairs

2   Were you able to get an appointment for the day of your choice?

     ◯ Yes   ◯ No ⟶   If not, how many days did you have to wait beyond your desired appointment date? [   ]

                                                                                           Enter # of days

3   On a scale from 1 to 7, where 1 is Poor and 7 is Excellent,   Poor   Average   Excellent
     rate the dealership's ability to schedule your visit as desired: ① ② ③ ④ ⑤ ⑥ ⑦

PERFORMANCE OF THE SERVICE REPRESENTATIVE

On the same scale from 1 to 7, where 1 is Poor and 7 is Excellent, rate the service representative on how well they

                                                                 Poor   Average   Excellent

4   Talked to you about your car service needs, present and future   ① ② ③ ④ ⑤ ⑥ ⑦

5   Showed knowledge in servicing cars   ① ② ③ ④ ⑤ ⑥ ⑦

6   Listened to your concerns and needs   ① ② ③ ④ ⑤ ⑥ ⑦

7   Asked you questions to clarify your service needs   ① ② ③ ④ ⑤ ⑥ ⑦

APPEARANCE OF THE SERVICE FACILITIES

8   Did you wait at the service facility, while your car was being serviced?   ◯ Waited   ◯ Left the car

                                                                         Go to next question   Skip to question 12

On the same scale from 1 to 7, where 1 is Poor and 7 is Excellent, rate the service facility on

                                                                  Poor   Average   Excellent

9   How well you were able to spend your time while waiting?   ① ② ③ ④ ⑤ ⑥ ⑦

10   How clean were the waiting areas?   ① ② ③ ④ ⑤ ⑥ ⑦

11   How comfortable were the waiting areas?   ① ② ③ ④ ⑤ ⑥ ⑦

QUALITY OF SERVICE WHEN PICKING UP CAR

12   Was your car ready at the promised time?   ◯ Yes   ◯ No

                                                                        Skip to question 14   Go to next question 13

13   If the car was not ready, what was the reason given for the delay? (**Select only one alternative**)

     ◯ Service found additionals problems   ◯ Parts were not available
     ◯ Service could not work on car in time   ◯ Other reasons (Specify): _____

On the same scale from 1 to 7, where 1 is Poor and 7 is Excellent, rate the following:

                                                           Poor   Average   Excellent

14   Explanation of the work performed on your car?   ① ② ③ ④ ⑤ ⑥ ⑦

15   Explanation of the charges on your bill?   ① ② ③ ④ ⑤ ⑥ ⑦

16   Cleanliness of your car inside and out?   ① ② ③ ④ ⑤ ⑥ ⑦

COMMENTS

Please tell us about any other aspects of your experience at the dealership: _____

_____

_____

The questionnaire in Figure 2.5 serves a narrow purpose. In designing a questionnaire, keep the respondents in mind. A questionnaire should allow respondents to answer questions easily and should have understandable words, with questions presented in a sequence that makes sense and that can be followed by respondents.

2.3 Sampling

This section offers a brief overview of important sampling concepts. Some of the concepts, such as population and sampling frame, apply to both census and sampling. Others, such as those related to random sampling, do not apply to censuses.

Sampling takes time and resources. The purpose of sampling depends on the problem statement or survey objective. Having defined the problem, you need to specify what needs to be known in order to solve the problem. As a related topic, the chapter on statistical models discusses the fact that specifying statistical models includes specifying the variables that are needed. This leads to specifying data requirements. You need to determine where to collect the data and how many observations are adequate to achieve useful answers.

2.3.1 Sampling Concepts

In sampling, the basic unit of observation is called an *element*. Elements are objects, records, or persons on which measurements are taken. A *population* is a collection of all elements of interest in the study. The unambiguous definition of the target population can be quite difficult in practice. For example, in an election poll, the population should be all those individuals who are going to vote in the upcoming election. However, it is impossible to predict with certainty the actual voters in a future election. Instead, a substitute population has to be chosen. Is it going to be all registered voters? Not all registered voters actually vote. Can we select from registered voters those who are likely to vote? What if some voters have yet to register? In the end, the survey needs to define a target population on which the survey is conducted.

A researcher who selects a sample actually selects sampling units. *Sampling units* are a non-overlapping collection of elements. The collection of sampling units covers the entire population. The researcher selects sampling units from a list called a *frame*. However, measurements are taken on elements, not on sampling units, although in many applications, sampling units are the same as elements. For example, in sampling employees of a business, it is quite likely that a list of individual employees can be constructed from which a sample can be selected. However, in many different applications, sampling units may contain more than one element. For example, in a survey of purchasing habits of personal hygiene items, a list of individuals may not be available. Instead, addresses of households are used to select individuals. Households are, thus, the sampling units, and individual household members are the elements.

The *sampling frame* is a list of sampling units. The sample is drawn from the sampling frame, not from the population. Thus, sampling frame and population are not the same. Population is defined in terms of elements, whereas sampling frame is defined in terms of sampling units. A sampling frame is like a picture of the population that makes the population accessible, but not the population itself. Population and frame should be identical but rarely are. Sampling units not included in the frame have no chance of being included in the sample. Further reasons for the discrepancies between population and frame are incomplete frames, multiple incompatible frames, overlapping frames, and frames that do not match time periods for which the survey is designed.

In pre-election polling, the difference between the population and the sampling frame is quite striking, because the population of election day voters is not known when most pre-election polls are conducted. Polling organizations filter most likely voters as a way to come up with an operational frame. In telephone surveys of pre-election voting attitudes, the telephone number identifying the household is the sampling unit. Using telephone numbers as the frame requires arrangements for unlisted numbers and households with cell-phone only connections. Figure 2.6 shows that the population and the sampling frame are not the same concept. A large overlap between the population and the frame is, of course, desirable.

Figure 2.6 Population versus Sampling Frame

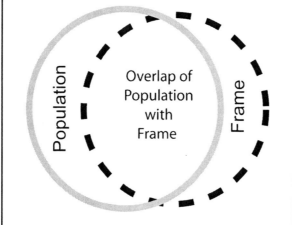

A *sample* is a subset of the population. Samples that are selected by probability methods have a better chance of yielding good samples than those chosen by most other methods. A sample looks at less than 100%. Sampling is usually preferred to a census, but especially in applications in which the cost of sampling is very high or the time required to obtain the proper information on an element is long.

A sample is a collection of elements assembled from sampling units that were drawn from one or more frames. The quality of the sample frame and the selection method influence the quality of the sample. A sample is *representative* if it is an undistorted subset of the population.

Samples drawn from an electronic file using a computer have advantages over manually drawn samples from a physical list. Computers can handle large frames and improve the use of surveyors' time more efficiently. Manual sample selection is often subject to selection errors. Sometimes there is no choice but to draw samples manually, especially when the frame does not list sampling units (for example, the frame consists of records in a file drawer).

Example: Customer Satisfaction after On-Board Program in a Bank

Suppose you want to evaluate an on-board program for new bank customers. On-board programs are intended to welcome new bank customers and to explain the opportunities the bank offers. On-board programs are designed to retain customers and to create customer loyalty. The survey task is to determine the satisfaction of new bank customers who established business relations with a bank during the last three months and were exposed to the on-board program. The survey intent defines the population of interest—namely, all new customers who signed up for an account with the bank during the most recent three-month period. For sampling purposes, you need a list on which these customers appear. The available list might contain the names of customers who opened and terminated the same account during that time. It could contain the names of customers who signed up for a particular new-to-them account but who had other business relationships established with the bank. Thus, the concept of population is the ideal and the list of names (the frame) is the reality. The terms defined earlier can be applied directly to the on-board program example:

Element: An element is a person authorized to access the new account, a new account holder.

Population: The population consists of all new account holders who were exposed to the on-board program within a three-month period.

Sampling Units: A sampling unit is an individual account at a bank. Accounts do not respond to customer satisfaction surveys, but persons authorized to use the accounts do.

In a survey, you would select the account first and then interview persons authorized to use the account.

Sampling Frame: The sampling frame is the list (or lists) of sampling units consisting of new bank accounts where at least one account holder had been exposed to the on-board program.

Sample: A sample is a selection of sampling units from the sampling frame.

2.3.2 Probability Samples

In a *probability sample*, each sampling unit has a known probability of being selected into the sample. In simple random samples, all sampling units have an equal probability of being selected. Probability samples are also called *random* or *statistical* samples. They are random samples because their selection involves random numbers or other random devices. They are called statistical samples because the probability selection process allows us to calculate sampling errors and perform statistical analyses.

Selection of sampling units is carried out by random numbers generated by a random number generator on a computer or, now a bit old-fashioned, by using a table of random numbers. In the next section, the most common probability sample will be outlined. These are simple random samples (SRS), stratified random samples, and cluster samples. Other probability samples are also used, but they are beyond the scope of this book.

Sampling Errors

Probability samples can produce erroneous results and are not entirely risk-free. The sampling error is a measure associated with the process of selecting a probability sample. All other errors are non-sampling errors.

The *sampling error* is the difference between the estimated value of the population characteristic (for example, the sample mean or sample median) and the true value of the population characteristic (for example, the population mean or population median). The *true value* is the value of the population characteristic that would have been obtained in a perfect 100% census. The sample, being a subset of the population, does not contain all sampling units. Sampling errors occur because population characteristics are estimated from sample outcomes. Statistical considerations can be used to manage the magnitude of the sampling error. While a 100% census does not have a sampling error associated with estimates of population characteristics, it often incurs non-sampling errors.

Sampling errors of estimates from probability samples can also be estimated. In non-probability samples, the sampling error usually cannot be quantified. Statistical sampling errors have the following characteristics:

- They decrease with increasing sample size.

- They depend on the variability of the characteristics being measured.
- They can be reduced by using an improved design for the selection of sampling units.

For smaller populations, they depend on the relative sample size (sample size relative to population size). For large populations (for example, over 10,000 elements), the absolute size of the sample tends to be more important in determining the size of the sampling error.

Non-Sampling Errors

Non-sampling errors are all other errors that arise during a survey process but which are not part of the sampling error. Non-sampling errors can be reduced through adequate planning and supervision, but they are usually difficult to measure and occur both in samples and in censuses.

The most common non-sampling errors occur because:

- The population and the frame are inadequately matched.
- The survey asks the wrong questions.
- The questionnaire is poorly designed.
- The survey uses inappropriate measurement scales.
- The survey fails to record the correct responses or is subject to interviewer bias.
- The survey fails to draw a proper random sample.
- The response rate is inadequate because either certain questions are omitted from the questionnaire or those included are returned incomplete.
- Processing errors occur during compilation of the data.
- The survey uses improper statistical estimation methods.
- The results are improperly interpreted.

Non-sampling errors represent the more difficult part of sample surveying because they are often hidden and not obvious to those conducting the survey.

2.3.3 Probability versus Non-Probability Samples

Probability sampling is an objective approach to selecting unbiased samples that are on average representative. Probability samples are subject to sampling error, but the risk of sampling errors can be quantified. Probability samples are considered objective because an external (random) selection algorithm is used in identifying sampling units.

Contrast that with the most common non-probability samples:

- Expert or judgment samples make the selection based on expert understanding of what constitutes a representative sample and experts often disagree.

- Convenience samples select sampling units based on convenience, such as asking passers-by for their opinions.

- Quota samples select certain proportions of subpopulations without ever specifying how they should be selected.

- Volunteer samples are put together with respondents who volunteer to participate in the survey.

Non-probability samples are common in popularity voting for TV shows such as *American Idol*. The selection of respondents is neither from a specific sampling frame nor under the control of those performing the survey. Multiple voting by respondents is a distinct possibility. In the end, there is no way to ensure that the sample of respondents is representative.

2.3.4 Common Types of Probability Sampling

In probability sampling, each sampling unit has a known probability of being selected into the sample. Here we select individual sampling units with the JMP random number generator. Actual selection can be carried out in two ways:

1. Sampling without replacement: when items are selected from the population and not returned until all sampling steps are completed.

2. Sampling with replacement: when items are selected one at a time, measured, and returned to the population with the possibility of being selected again at a future drawing.

For large populations ($N > 10,000$), the distinction is of little consequence. For small populations, sampling without replacement requires adjustments in the estimation formulas. In this book, we treat all samples as coming from sampling with replacement. Next, we outline four types of probability samples.

Simple Random Sampling

A simple random sample (SRS) of size n from a large population of size N is drawn so that every possible sample of size n has the same probability of being selected. A somewhat simpler way of defining a simple random sample is that each element has the same probability of being selected into a sample. JMP selects SRS with the **Subset** function from the **Table** menu to generate a random sample in a new data table.

Figure 2.7 From Frame Containing All IDs to Random Sample with Response to Question 1

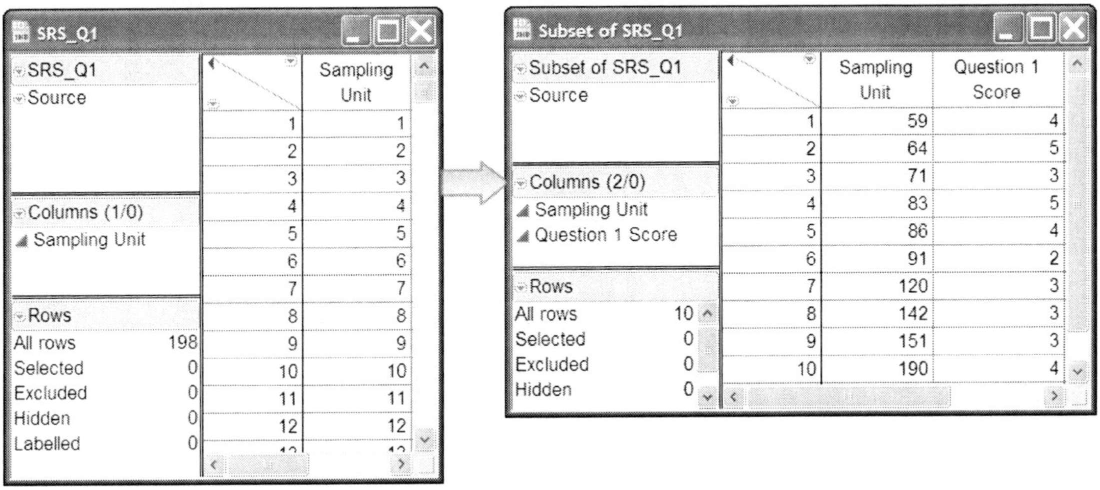

Example: A simple random sample of *n*=10 sampling units will be drawn from a frame containing *N*=198 new customer IDs. The 10 selected customers are asked to respond to Question 1 on a Likert scale from 1 to 7. The frame is stored in the JMP data file *SRS_Q1.jmp* (left in Figure 2.7). The column Sampling Unit contains the IDs for the 198 customers.

Figure 2.8 shows how to draw a random sample of size *n*=10 from the original frame and save it in a new JMP data table called *Subset of SRS_Q1.jmp* using the Subset action from the Table menu. Specify the size of the random sample as *n*=10. Name the output table to which the sample IDs are to be stored.

Figure 2.8 Creating a Random Sample to a New Data File

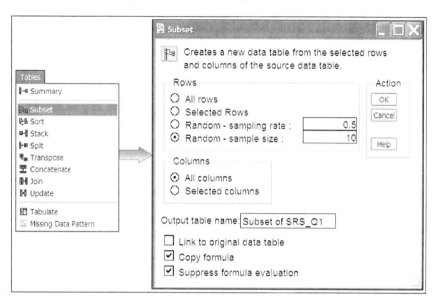

The column Sampling Unit of *Subset of SRS_Q1.jmp* shows the 10 observations selected for the sample. The Likert scale responses of this sample are in the Question 1 Score column (Figure 2.7). Obviously, a sample of size 10 is small and used here to demonstrate the selection tool. In practical surveys, the sample size and the number of questions might be larger.

Systematic Sampling

A *systematic sample* is obtained by randomly selecting one element from the first k-elements in the frame and every k-th element thereafter. Such a systematic sample is called a *1-in-k systematic sample* with random start. Systematic sampling is also called *interval sampling* because sampling units are chosen at intervals of length k. The number k of items skipped is the skip interval.

Systematic samples might be easier to select than simple random samples, if items are selected manually rather than by computer. It is used in a process where each k-th item is selected to ensure the integrity of the production. It is also very convenient when sampling units are arranged in a file drawer.

Systematic sampling requires that the population be in random order with respect to the criterion being investigated. In a file cabinet containing invoices in chronological order, this is not necessarily a problem, as long as the transactions recorded on them occur in a more or less random fashion. If, on the other hand, chronological order reflects seasonal fluctuations, then a systematic sample might lead to biased results.

An example of selecting a systematic sample of size $n=10$ from a frame with $N=198$ sampling units requires a few simple steps. Before generating the sample, recognize that a sample of size $n=10$ from a frame with $N=198$ records requires a skip interval $k = \text{Round } [N/n] = \text{Round}[198/10] = 20$. The skip interval tells us to look at every 20th sampling unit. First, generate a random start that is a number between 1 and 20. In JMP, create a new data table from the **File** menu, as shown in Figure 2.9. Alternatively, click on the New Data Table icon if it is shown on the menu bar.

Figure 2.9 Creating a New Data Table

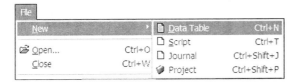

When the empty data file is created, it has a default column but no rows. A single row is needed for the random number that will represent the random start. From the **Rows** menu, select **Add Rows** as shown in the menu on the left of Figure 2.10. This step opens the Add Rows window. Enter the number of rows that are needed. Here add 1 as shown in the middle panel. The resulting data table on the right has a single row and column with the value field marked by a period. It is useful, but not mandatory, to change the heading of the column. Here the column heading is called Random Start.

Figure 2.10 Creating a Data Table with One Row

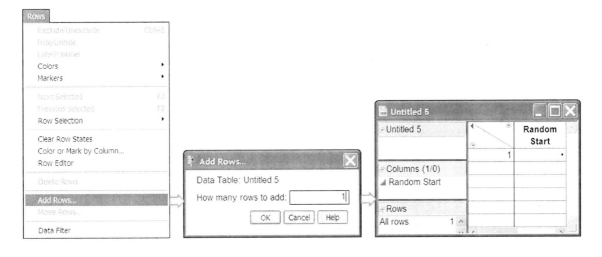

In order to create the random start for the systematic sample, create a formula. Add the Formula property to the Random Start column by right-clicking on its name and selecting **Formula** from the menu. The Formula window for the column appears. From the Function window, scroll down to Random and select **Random Integer**. The resulting pull-down menu is shown on the left of Figure 2.11. Having selected **Random Integer**, you need to specify the range of values from which the integer will be selected. This integer is the skip interval. Because the population size is $N=198$ and a sample of size $n=10$ is desired, every 20th item needs to be selected. This is a so-called 1-in-20 sample.

Figure 2.11 Random Integer from Formula Menu to Generate Random Start

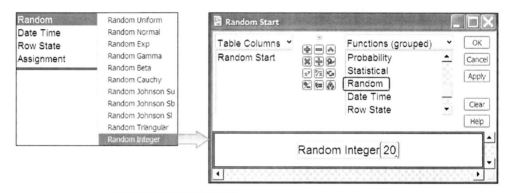

A formula named Random Integer n1 appears in the window. Enter the skip interval value (here $k=20$) into the space between the brackets as Random Integer 20 . The resulting Formula window is shown in the right panel of Figure 2.11.

Click **Apply** and return to the data window. A single number, the Random Start, appears because the data file created has only one column and one row.

The desired data table is shown in Figure 2.12.

Figure 2.12 Single Random Number as Starting Point of Systematic Sample

⊟Untitled 2		Random Start
	1	15
⊟Columns (1/0)		
⊿Random Start ⊕		

The random start number is 15. This means that the first respondent is number 15; the second is 15+20=35. The third is 35+20=55, and so on. The systematic sample of size 10 consists of ID numbers 15, 35, 55, 75, 95, 115, 135, 155, 175, and 195.

In this example, the selection presents no problem as long as the random integer is not larger than 18. However, starting values of 19 (or 20) will yield a sample of only nine units because the last ID number is 199 (or 200) and does not exist. A simple solution to this problem is to work with the smaller sample size or to continue the count from the beginning of the file.

Systematic sampling needs to be used with care:

1. Be aware of any non-random ordering of the items with respect to the criterion in question.

2. In order to avoid the risk of subtle non-random ordering of the population, sometimes you might want to choose more than one starting point.

For a population of N=2,000 records, a systematic sample of n=50 is desired. This can be accomplished by either taking a single 1-in-40 sample or taking two 1-in-80 samples. A 1-in-40 sample chooses a random start in the range from 1 to 40 and then takes every 40th item. Two 1-in-80 samples require two random starts in the range from 1 to 80 and then start selecting each 80th item from each of the two random starts.

Stratified Sampling

Populations with a high variability often can be separated into several, more homogeneous subgroups called *strata*. By sampling and estimating each subgroup (stratum), the overall precision can be enhanced. The precision of the overall estimate by stratification (relative to SRS) is increased when strata are fairly homogenous within (the elements within a stratum differ little from each other with respect to the characteristic of interest), but strata are fairly heterogeneous between (different strata differ with respect to the characteristic of interest).

A stratified random sample is obtained by taking samples from each of L separate, non-overlapping strata of the population. The separation of the population into non-overlapping strata is done with a stratification variable that can be either continuous, such as Account Balance, or qualitative (grouping), such as Region of Country. In a survey of new bank customers, it is often assumed that customers with similar account balances resemble each other in other ways. If new customers are stratified by account balances, then this is the stratification variable. With such a continuous stratification variable, there is flexibility in choosing the stratum boundaries. There are methods to choose boundaries, but they are outside the scope of this book. Stratification can also be based on qualitative grouping variables, such as with stratification by customer type (for example, retail or wholesale).

The samples taken from the strata are typically either simple random samples or systematic samples. In some applications, one of the strata, especially small but high-value strata, might be examined by 100% census. The criteria for proper stratification include the following: every element belongs to only one stratum, the strata must be clearly differentiable from each other, and the size of each stratum N_i must be known. Here are a few reasons for using stratified sampling:

- to obtain a more precise estimate of the population total or parameter with the same sample size relative to simple random sampling
- to estimate subpopulation characteristics with fixed precision
- to combine the results in which samples were drawn from different sampling frames

Before beginning to collect a stratified random sample, keep in mind a few points.

Type of sample allocation: Each stratum is sampled separately. Given an overall sample size, the stratum sample size has to be decided. The most common methods are

- Equal allocation of samples to each stratum: From each stratum, a sample of equal size is drawn.
- Sampling proportional to stratum size: The larger the number of items in a stratum, the larger the sample size allocated to that stratum.
- Sampling proportional to variability: The more variable a stratum relative to the other strata, the higher the allocation of sample items.
- Sampling proportional to cost and variability: In addition to variability consideration, the cost of sampling is also considered. The cheaper to sample, the more a stratum will be sampled.

Stratified samples are essentially different independent samples from each stratum. It is not necessary to apply the same sample selection to each stratum.

Cluster Sampling

A *cluster sample* is a probability sample in which the sampling units are non-overlapping collections (clusters) of elements. The difference between a cluster and a stratum is that a cluster is a sampling unit, whereas a stratum contains sampling units. In other words, in cluster sampling not all clusters are selected for review, whereas in stratified sampling each stratum is sampled.

Cluster sampling might not always be as efficient as other sampling methods. However, there are several types of applications in which it is very useful:

1. Clusters are heterogeneous within but homogeneous between: Each cluster can be viewed as a collection that is nearly representative of the population. For example, you could view each branch of a bank corporation as a small bank. A cross-section of

customers uses the services at the bank. A bank branch is nearly representative of many other bank branches.

2. Clusters are more convenient to sample and, therefore, reduce the cost of sampling: For example, in compliance sampling, records can be kept in different locations. It might be more convenient to sample several, but not all, locations and perform the audit only on those instead of random sampling all possible locations.

The way to conduct a cluster sample is to:

- Identify all clusters so that they represent the entire population.

- Determine the number of clusters to sample.

- Take a random sample of clusters.

- Either take a 100% census of all items in the sample, or sample each cluster by treating each cluster as a subpopulation (two-stage sampling).

2.3.5 Other Sampling Methods

Various applications have spawned a considerable variety of sampling designs. A certain degree of statistical experience and knowledge is required to apply these tools correctly. The more common ones in survey research are those mentioned here.

Sampling with probability proportional to size is different from simple random sampling in that the probability of selection into the sample depends on a size variable. In auditing financial performance of a portfolio, the selection probability of an individual investment could be made proportional to the relative size of an investment. Larger investments have a higher probability of selection than smaller investments. This method is advised whenever the size variable (size of investment) and the variable of interest (for example, the return) are related.

Multi-stage sampling is an extension of the cluster method. Instead of examining all elements within a cluster, a sample of the elements is drawn. This requires two stages of selection. First, the clusters are randomly selected. Then, the elements in each of the selected clusters are sampled. The example of cluster sampling for customer satisfaction in a bank is a perfect candidate for such a two-stage cluster. First, select a sample of bank branches. Then, randomly select customers who do business at these branches. It is possible that an application requires selection at more than two stages. Multi-stage sampling has the advantages of cost reduction and administrative simplicity. For example, it does not require a sampling frame for any of the branches that were not selected into the first-stage cluster sample.

Multi-phase sampling is slightly different. In a multi-phase sample, only very basic data are collected from a large sample. Another smaller sample is taken in which more detailed information is collected. Unlike multi-stage sampling, multi-phase sampling uses the same frame repeatedly. As an example, consider a financial services company that wants to do a survey of lawyers who perform estate planning as a majority of their service. No frame of such lawyers exists. The larger sample would ask a set of questions about the type of legal services that law firms in the sample provide. A second phase sample would then ask for more detailed information of those law firms that qualify by virtue of the larger sample.

Conclusion: Sampling is an important tool in gathering business information. Probability samples are available to match many business applications. Sampling provides objective information efficiently and without bias.

2.4 Summary

- The survey process is complex, requiring several steps.

- Questionnaires are used to collect data in surveys. They relate to a hypothesis or problem statement and consist of words, questions, and the questionnaire format.

- Questionnaires may be administered in direct personal interviews, telephone interviews, and mailed or online questionnaires. Sometimes direct observations of specific characteristics are also possible.

- Questions need careful wording. Double-barreled, leading, or loaded questions should be avoided.

- Response to questions can be open-ended or closed. Open-ended questions offer a wider choice to respondents but often lead to non-response. They may also be more difficult to tabulate. Closed responses offer a fixed set of possible responses.

- Common types of questions are checklists, questions with quantity or intensity scale responses, semantic differentials, ranking questions, and others.

- Sampling is the preferred method to collect survey data.

- Population, sampling frame, population element, and sampling unit are important concepts in sampling.

- Probability samples use random numbers to select respondents from a frame. Non-probability samples use other criteria such as expert knowledge or quotas.

- Probability sampling is also known as random sampling or statistical sampling.

- In probability sampling, both sampling errors and non-sampling errors occur.
- Sampling errors are caused by the random selection of respondents. In probability sampling, they can be quantified.
- Non-sampling errors are caused by inadequate frames, poor questions, non-responses, improper analysis, and other factors. They are difficult to quantify.
- Common types of probability sampling are simple random sampling, systematic sampling, stratified sampling, and cluster sampling. Other types exist. Some are combinations of the common types.
- In a simple random sample, each sampling unit of the frame has an equal chance of being selected into the sample.
- In a 1-in-k systematic sample, a first sampling unit is randomly selected from the first k sampling units in the frame. Thereafter, each k-th sampling unit is selected from the frame.
- In a stratified sample, the population is divided into non-overlapping strata. Then samples are drawn from each stratum and the results are combined into overall estimates.
- In a cluster sample, individual clusters are the sampling units chosen by random selection.
- Simple random sampling is usually easy to design and execute.
- Systematic sampling can even be used when no physical frame exists, as long as the sampling units are in some physical order. Its analysis is similar to simple random samples, although in theory it resembles a cluster sample.
- Stratified sampling works well when differences between elements in individual strata are very small but differences between elements in different strata are large.
- Cluster sampling works well when clusters are representative of the subpopulation they are intended to estimate.

2.5 Problems

1. A cereal manufacturer conducts focus groups to help determine how to position new products in the market place. A new breakfast cereal, Honey Bear Crunch, will be tested at the next focus group. The cereal is high in fiber and vitamins, sweetened primarily with honey, shaped like bears, and available in two flavors—either vanilla or cinnamon. It is anticipated that the cereal will be viewed as a healthy choice that would be satisfying to all family members. Focus group participants will be able to

taste test the two flavors and give their impressions of the cereal. To prepare for the focus group, provide answers to the following questions and include your reasoning:

    a.   What is an appropriate population for this product?

    b.   How do you propose selecting members for the focus group?

    c.   Prepare a list of five questions to use during the session.

2. Write a customer satisfaction questionnaire for a survey of existing bank customers that addresses the following topics:

    a.   Write the introduction to the customer satisfaction questionnaire.

    b.   Establish which bank services the customer actually uses.

    c.   Include some questions concerning the satisfaction of a customer who uses multiple banking services (checking/savings/money market).

    d.   Include some questions concerning the satisfaction of a customer who uses multiple loan services (home equity/installment loan/mortgage).

Make sure to branch the appropriate questions efficiently to avoid tiring the respondent.

3. A human resources department is planning to conduct a sample survey of employee views on workplace diversity issues. This survey will help determine if diversity training programs are needed.

    a.   What type of sampling would you recommend? Specifically discuss the advantages of probability sampling (simple random sampling, systematic, stratified random sampling, cluster sampling) versus non-probability sampling (judgmental, convenience, quota, or volunteer).

    b.   Explain how you arrived at your selection and why the other methods were not chosen.

4. A taxing authority regularly selects taxpayer returns for audits. The selection for this type of audit is random (rather than based on reports or denunciations). Explain why a good sampling frame is important to both the auditor and the taxpayer.

5. A company wants to survey its customers about their satisfaction with technical support for their brand of laptop computer. Consider the following survey questions and statements:

    i.   What is your preferred method for contacting technical support?

    ii.   The technical support representative seemed to act in a very professional manner when I described my computer problem.

    iii.   The technical support representative was good.

iv.   Did you call technical support because you were too lazy to figure out your computer problem yourself?

v.   The technical support representative was not the best I had ever had.

vi.   The technical support representative was very knowledgeable and explained the problem in a way I could understand.

vii. I had to wait too long for my call to be answered.

viii. I used the telephone support line because my baud rate is too slow.

ix.   Should the technical support representatives speak better English?

x.   Did you receive an e-mail confirming that your problem was resolved within 24 hours?

a. For each question or statement, identify deficiencies and which, if any, of the following question-writing principles were violated: use simplicity of language, clarify confusing terms, ask questions that relate directly to anticipated or potential action, avoid compound questions, and avoid leading questions.

b. Rewrite the questions in a way that meets a high question-writing standard.

6. A county composed of three school districts is considering a new method for assessing property values. The new assessment method must be approved by a referendum at the next election. The county legislature would like to conduct a sample survey to assess voter preference for alternative assessment methods. The following table provides information on each of the three school districts:

District	Voters	Students	Area (sq miles)	Total property assessment (million $)
Northwood	4067	5007	12	210
Southside	1251	906	55	270
Eastwind	3275	3689	34	340

a. Which sampling method would you recommend: simple random, stratified, or cluster sampling? Explain your reasoning.

b. Discuss the advantages and disadvantages of each sampling method.

c. If a stratified sample is appropriate, which variable would you select as the stratification variable? Why?

7. A hotel plans to leave a customer satisfaction survey on the nightstand in each room. Review the questionnaire and make a list of five items that need improvement. Suggest how to change these items.

Please Tell Us What You Think!   **Holiday Suites**

Did you recently stay at the Holiday Suites, Oldtown, OH? If so, please let us know how well we served you.

During this stay how satisfied were you with:

	Very dissatisfied	Dissatisfied	Neither dissatisfied nor satisfied	Satisfied	Very satisfied	Did not use
Outside appearance of hotel	☐	☐	☐	☐	☐	☐
Lobby and front desk service	☐	☐	☐	☐	☐	☐
Room cleanliness	☐	☐	☐	☐	☐	☐
Quality and responsiveness of room service	☐	☐	☐	☐	☐	☐
Recreational Facilities	☐	☐	☐	☐	☐	☐
Business Center	☐	☐	☐	☐	☐	☐

1. Was your room ready when you arrived?

Very dissatisfied	Dissatisfied	Neither dissatisfied nor satisfied	Satisfied	Very satisfied
☐	☐	☐	☐	☐

2. The hotel was not unsafe.

Yes	No
☐	☐

3. The temperature of the room was too cold.

Yes	No
☐	☐

4. Please rate your overall satisfaction with this hotel.

1	2	3	4	5
☐	☐	☐	☐	☐

2.6 Case Study: Green's Gym—Part 2

In the Green's Gym—Part 1 case study, you identified the important factors that contribute to customer satisfaction. In Part 2, you will construct a questionnaire to elicit the needed information from Green's members. Refer to Part 1 for the details of the business operation.

Business Goals

Green's Gym is interested in maintaining customer loyalty in the face of new competition.

Strategy

Green's Gym would like to assess customer satisfaction using a questionnaire. The information collected will enable Green's Gym to adjust their services to better meet their customers' needs, thereby retaining members.

Tasks

Perform the following tasks:

1. Determine how Green's Gym will deliver the questionnaire to its members so that the information can be obtained in a timely fashion. Discuss the advantages and disadvantages of the delivery options you identify.

2. Would you recommend a census or a sample survey in this case? Justify your selection.

3. In addition to customer satisfaction information, what demographic information should be collected from the members? Since asking individuals to reveal personal information may lead to non-responses or inaccurate responses, briefly justify why each piece of information is needed for Green's Gym to assess customer satisfaction.

4. Draft a one-page questionnaire that will provide Green's Gym with sufficient information to assess their customer's satisfaction.

5. Propose ways in which Green's Gym can assess member satisfaction on an ongoing basis.

2.7 References

Labaw, Patricia J. *Advanced Questionnaire Design*. 1980. Cambridge, MA: Abt Books.

Payne, S. L. *The Art of Asking Questions*. 1951. Princeton, NJ: Princeton University Press.

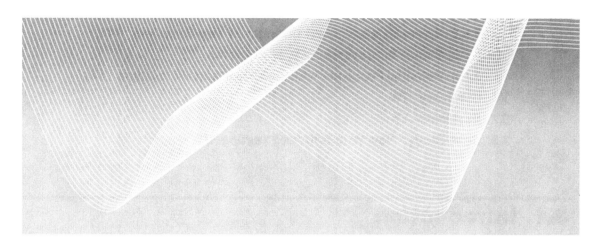

Chapter 3

Describing Data from a Single Variable

3.1   Introduction   54
      3.1.1 Graphs for Continuous Data   54
      3.1.2 Numerical Summaries for Continuous Data   55
      3.1.3 Data Descriptions for Nominal and Ordinal Data   56
3.2   Example: Order Processing in an Herbal Tea Mail Order Business   57
3.3   Descriptive Statistics with the JMP Distribution Platform   58
      3.3.1 Describing a Continuous Variable   58
      3.3.2 Describing a Nominal Variable   62
3.4   Interpretation of Descriptive Statistics   64
      3.4.1 Histogram   64
      3.4.2 Box Plot   65
      3.4.3 CDF Plot   67
      3.4.4 Quantiles   69
      3.4.5 Moments   70
3.5   Practical Advice and Potential Problems   71
      3.5.1 Features of Good Graphs   71

3.5.2 Example: Izod Impact Strength of Two Suppliers   71

3.5.3 Simple Improvements to Histograms   72

3.5.4 Potential Problems with Box Plots and Histograms   74

3.6 Summary   76

3.7 Problems   77

3.8 Case Study: New Web Software Testing   79

3.1 Introduction

Data description is the first step toward getting a feel for the data. To describe data is to represent their essence in an informative and communicative way. Graphs and data summaries need to be matched to the audience's ability to understand. Data description depends on a variety of factors: the information contained in the data, the way they were collected, whether observations are continuous, ordinal, or nominal, and so on. Data description might reveal data defects, such as too many missing values, inappropriate values due to recording errors, or defects arising from data collection. It might also show obvious conclusions that make further analysis redundant.

Graphs are powerful tools of data analysis that lend themselves to quick insights, because they explore data for patterns and deviations from patterns. Graphs are useful for evaluating and modifying models as well as communicating facts in reports and presentations.

Numerical summaries succinctly reveal simple characteristics of data. They are used directly in further inference. Numerical results are an important backup of graphs and can be easily remembered and communicated orally. JMP provides numerical summaries and graphs with great ease.

3.1.1 Graphs for Continuous Data

All of the graphs in this section, except the time plot, reveal static aspects of the data distribution. The time plot looks at data dynamically over time. Each plot highlights some characteristics and neglects others. Which one should be used? JMP generates all of these plots quickly. It is easy to examine as many plots as make sense and to select those plots that best reveal interesting aspects of the data. The following graphs are explored next.

Histogram: Histograms describe the general shape of a data distribution. A histogram is a chart with vertical bars representing counts or relative frequencies. The numeric horizontal axis is divided into bars of equal widths, each measuring the relative frequency on the interval covered by the bar. The widths of the bars influence the vertical appearance. For a given data variable, wider bars tend to result in a flatter appearance of the histogram and vice versa.

Box Plot: A box plot consists of three graphical parts: box, whiskers, and outliers. The box contains the central 50% of the data. The length of the box is the interquartile range. The median is shown by a vertical line in the box. The whiskers are up to 1.5 times the length of the box and show the range of data outside of the box but close to the center. Outliers far from the center are shown as dots or stars.

Box plots show the symmetry or skewness of the box with respect to the median, and the whiskers with respect to the box. Box plots are also useful for identifying outliers, especially when the underlying distribution is symmetric and approximates the bell-shaped curve. Box plots can give a misleading impression for very small data sets.

Time Plot: Time plots show the variable of interest on the vertical scale. Date- or time-related values are shown on the horizontal axis. Time plots are useful for identifying simple time-dependent patterns of data.

CDF Plot: CDF stands for cumulative distribution function, a term discussed in Chapters 5 and 6. The CDF plot shows the scale of the variable of interest on the horizontal axis and a non-decreasing step function of frequencies on the vertical axis. Steps occur at each data value. The height of each step is 1/n. For values where more than one observation occurs, the step size is a multiple of 1/n. CDF plots are smoother than histograms.

Normal Quantile Plot: This plot is a tool to verify graphically whether the data can be assumed to follow a normal distribution. The normal quantile plot is discussed in Chapter 8.

3.1.2 Numerical Summaries for Continuous Data

Measures of location are data summaries representing the position of the data along the dimension of the variable. Several are commonly used:

Mean: The sample mean is a simple average of data values and is the most important measure of location. By multiplying means by the population size, you can estimate population totals.

Quantile: The q% quantile is the value on the Y-scale at which at least q% of data values are below and at least $(1 - q)$% are above. For data distributions, quantiles describe the percentage of observations above or below a value on the Y-scale. In JMP, quantiles are similar to percentiles.

Median: The median is the 50% quantile. In a symmetric distribution, the median and the mean are the same. In skewed distributions, the mean and the median differ.

Quartiles: The lower quartile is the 25% quantile. At least 25% of the Y observations are at or below the lower quartile. At least 75% of the Y observations are at or above the lower quartile. The upper quartile is the 75% quantile. At least 75% of the Y observations are at or below the upper quartile. At least 25% of the Y observations are at or above the upper quartile.

Maximum and Minimum Values: Extreme values are useful for sanity checks of data, but also to get an idea of their range.

Measures of spread are data summaries representing the variability of the data. Here are some common examples:

Variance: The sample variance represents the average of squared deviations of observations around the sample mean. It is rarely used directly in data analysis, but it is important in calculating the standard deviation and derived quantities.

Standard Deviation: The standard deviation (Std Dev) is the square root of the variance. It is a measure of spread of the observations around the mean. Standard deviations are used in calculating confidence intervals and in testing hypotheses. In more advanced applications, the standard deviation is often referred to as Root Mean Square Error (RMSE).

Range: The range is the difference between the largest (maximum) and smallest (minimum) data value, i.e., Range = Maximum − Minimum. The range is a good sanity check of the data. Certain applications such as Control Charting use the range to estimate the standard deviation.

3.1.3 Data Descriptions for Nominal and Ordinal Data

Nominal data of a single variable are summarized by the frequencies of categories. Absolute frequencies measure the number of outcomes in each category, while relative frequencies are proportions in the range from 0 to 1, or percentages from 0 to 100%.

Frequency Chart and Bar Chart: Frequency charts are similar to histograms except that the horizontal axis consists of non-overlapping categories. The heights of bars measure the relative frequency in the outcome categories.

Mosaic Plot: Mosaic plots show relative frequencies on a stacked horizontal scale. Mosaic plots are used like frequency charts, but are preferred to show the proportional content of each nominal category as a part of the total.

Pie Charts: Pie charts are similar to the mosaic plot. They show the proportional content of each nominal category as slices of a total circular shaped pie.

3.2 Example: Order Processing in an Herbal Tea Mail Order Business

This example shows the use of graphs and numerical summaries in JMP.

Situation: A mail order firm distributes herbal teas to retail customers. The firm wishes to expedite mail orders for same-day shipment. The number of orders per day varies. A simple way to expedite orders on any given day is to have sufficient staff working to assemble the requested herbs in boxes and get them ready for shipment. However, demand for tea varies daily and is not known in advance. More workers mean more capacity, but too many workers lead to idleness and increased cost on days with less demand. Too little capacity means that customers might have to wait longer for the herbs. Long waiting times are one of the strongest factors for loss of customer loyalty in the herbal tea business and should be avoided.

Problem: Management needs to know the magnitude and patterns of variation in the number of daily orders. Management would like to find a simple rule that will let it set staff levels so that orders can be properly expedited on a majority of days.

Data Requirements: The data file *Herbal Tea Mail Order.jmp* records the daily number of orders processed and the total daily sales volume in dollars for all orders for the period from Oct. 1 to Dec. 12. The owner of the business also made an annotation defining what she considered a "low," "medium," and "high" volume day. Figure 3.1 shows an excerpt of the data file.

Herbal Tea Mail Order.jmp consists of four variables and 52 historical observations. Each observation is recorded in a separate row. The four columns represent the four variables.

- Column 1, Date, identifies the days. It is a continuous variable in the date format (see Column Info).
- Column 2, Number of Orders, counts the number of orders on each day.
- Column 3, Daily Value $, measures the sales in dollars on each day. Both Columns 2 and 3 are continuous.
- Column 4 is a nominal variable called Sales Volume. It has three levels: Low, Medium, and High. This variable was created by management to help in determining desired staffing levels.

Figure 3.1 Data Table Excerpt from *Herbal Tea Mail Order.jmp*

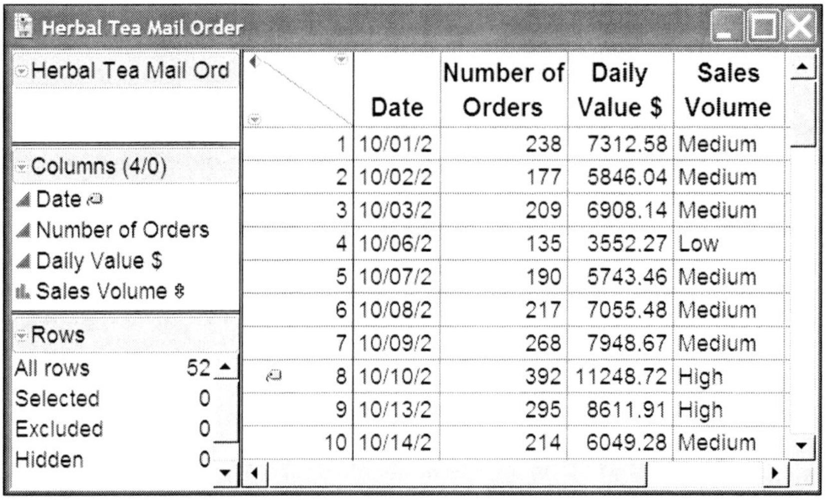

	Date	Number of Orders	Daily Value $	Sales Volume
1	10/01/2	238	7312.58	Medium
2	10/02/2	177	5846.04	Medium
3	10/03/2	209	6908.14	Medium
4	10/06/2	135	3552.27	Low
5	10/07/2	190	5743.46	Medium
6	10/08/2	217	7055.48	Medium
7	10/09/2	268	7948.67	Medium
8	10/10/2	392	11248.72	High
9	10/13/2	295	8611.91	High
10	10/14/2	214	6049.28	Medium

Anticipated Results: The analysis should point to a level of staffing that will allow expedited shipping of orders for a large percentage of days. For this, it is necessary to get a sense of the percentage of days with the number of orders small enough to be accommodated by minimal staffing levels. Of course, to extrapolate these percentages to the future requires the assumption that sales volume remains the same.

3.3 Descriptive Statistics with the JMP Distribution Platform

Descriptive results are often more suggestive than conclusive. Interpretations of descriptive statistics need to be firmed up by inferential statistics discussed in later chapters. The discussion in this section covers one variable at a time. For analyzing a single variable, JMP uses the **Distribution** platform on the **Analyze** menu to present both descriptive statistics and inferential statistics discussed in later chapters.

3.3.1 Describing a Continuous Variable

This section focuses on generating a variety of descriptive statistics and graphs. The interpretation is covered in the next section. The following is a systematic outline for obtaining results for the continuous variable Number of Orders.

Figure 3.2 Analyze Menu for Distribution Platform

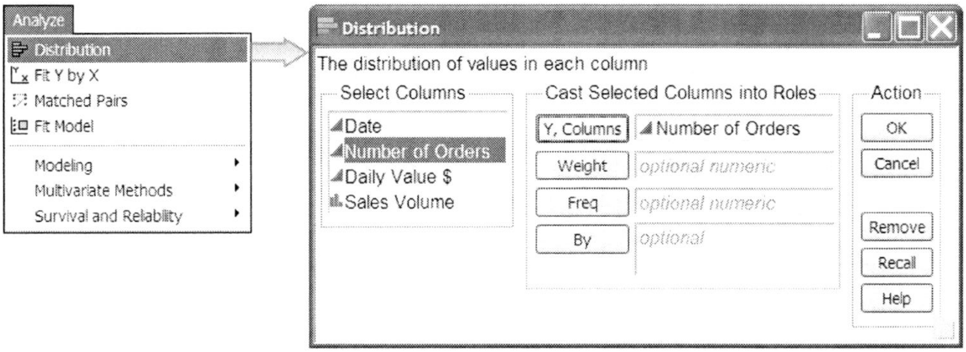

Step 1: Select **Distribution** from the **Analyze** menu, as shown on the left of Figure 3.2.

Step 2: Select the variable Number of Orders by highlighting it in the **Select Columns** list and clicking the **Y, Columns** button shown on the right in Figure 3.2. Click **OK** to obtain basic results. JMP provides general results and allows a great variety of additional results to be selected through options by clicking the red triangle of the variable bar. These options will be shown later.

Step 3: The results from basic JMP output are shown in Figure 3.3. Examine the box plot, the histogram, the quantiles, and the moments, e.g., mean and standard deviation (Std Dev). Select only those parts of the output that help you understand the data. In the **Moments** list, ignore for now the Std Err Mean for the standard error of the mean, and upper and lower 95% Mean for upper and lower 95% confidence limits. These will be discussed in later chapters.

Figure 3.3 Basic Results of Distribution for Number of Orders

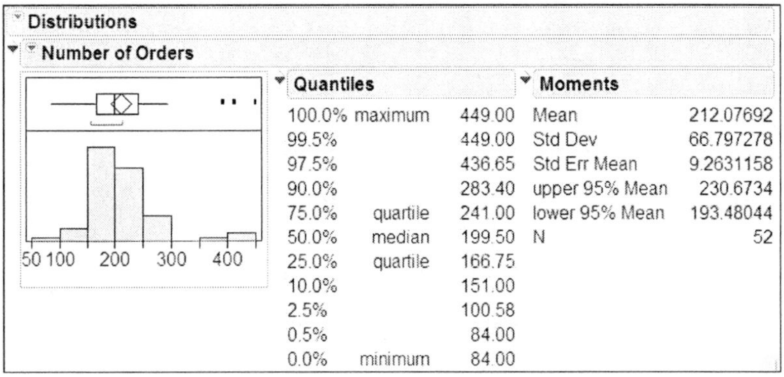

Quantiles

100.0%	maximum	449.00
99.5%		449.00
97.5%		436.65
90.0%		283.40
75.0%	quartile	241.00
50.0%	median	199.50
25.0%	quartile	166.75
10.0%		151.00
2.5%		100.58
0.5%		84.00
0.0%	minimum	84.00

Moments

Mean	212.07692
Std Dev	66.797278
Std Err Mean	9.2631158
upper 95% Mean	230.6734
lower 95% Mean	193.48044
N	52

Time Plot: If the time order in which the data were collected is available, obtain a time plot. The horizontal axis represents the order in which the data were collected. The vertical axis represents the data values.

Figure 3.4 Graph Menu for Overlay Plot to Construct Time Plot

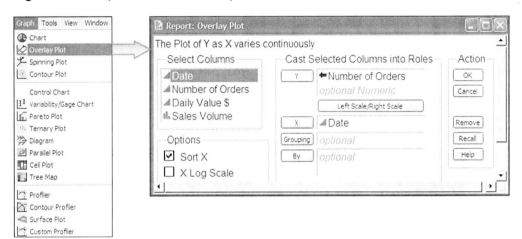

For a simple time plot, use the **Overlay Plot** platform on the **Graph** menu with a date or time order variable as the X-axis (Figure 3.4). Among the options, select **Sort X** whenever the data file is not in a special time order, but dates or time order are stored in a different column. JMP will automatically perform the ordering when **Sort X** is checked. The default appearance of the overlay plot shows only data points. A time plot is greatly improved by connecting consecutive observations with lines. To connect points, click the red triangle of the **Overlay Plot** submenu and select **Connect Thru Missing**, as shown in Figure 3.5.

Figure 3.5 Overlay Time Plot of Number of Orders

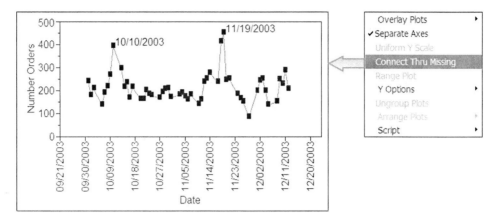

Figure 3.5 labels two individual observations by date using the Label/Unlabel feature. In JMP, there are two Label/Unlabel functions. On the **Row** menu, **Label/Unlabel** operates on individual highlighted rows. The default label is the row number. On the **Column** menu, **Label/Unlabel** allows selecting variables as labels. The menus for two of these Label/Unlabel functions are shown in Figure 3.6. Checked columns and rows are identified by tag marks in the variable list and the row properties list, as shown for Date and row 8 in Figure 3.1.

Figure 3.6 Label/Unlabel for Rows and Columns

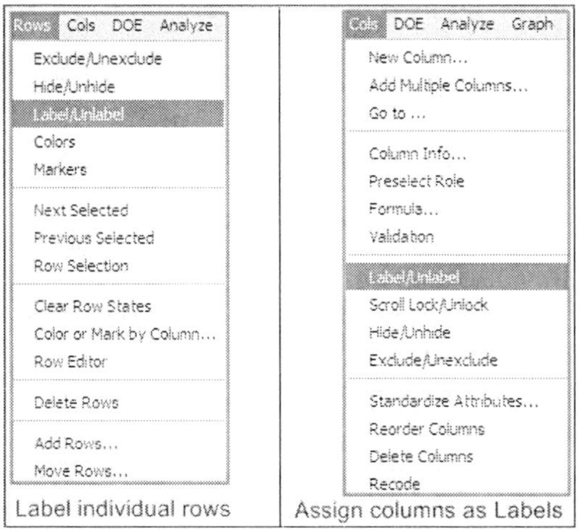

3.3.2 Describing a Nominal Variable

Sales Volume is a nominal variable with three categorical outcomes: Low, Medium, and High. Nominal variables require different output from continuous variables, because certain numerical summaries make no sense. Figure 3.7 shows the variable selection of Sales Volume.

Figure 3.7 Variable Selection Choosing Nominal Sales Volume

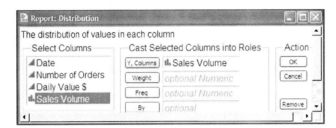

The histogram is now a frequency chart, because the horizontal axis lists nominal categories rather than values grouped from a continuous variable. The basic output can be augmented with several options. For example, it is very instructive to add frequencies of each category to the default histogram by choosing **Show Counts** on the **Histogram Options** submenu, as shown in Figure 3.8. One could add percentages of observations contained in each category.

Figure 3.8 Histogram with Options for Counts on Histogram Bars

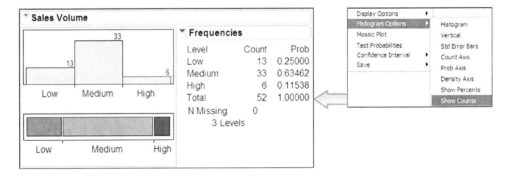

Frequency Histogram: The left of Figure 3.8 shows the frequency chart for Sales Volume. The bar height of each Sales Volume category measures the relative frequency of each category. Each bar is labeled by the count of observations in each category.

Mosaic Plot: The mosaic plot is a separate option from the same menu. It uses a proportional area to represent relative frequencies of observations in each category. The most frequent Sales Volume category is Medium, because it has the highest bar in the histogram and the largest area in the mosaic plot.

The Frequencies table shows numerical counts (i.e., absolute frequencies) and relative frequencies, labeled Prob, for each category. This feature shows that 33 out of 52 were Medium Sales Volume days, representing a proportion of 0.6346 (or 63.46%) of the total. The estimated probability (or relative frequency) of a Medium Sales Volume day is Prob = 0.63, rounded to two decimal places. Notice that the total estimated probabilities add up to 1. Other output for describing nominal variables involves confidence intervals and tests concerning the probabilities of each category.

Pie Charts: Pie charts are widely used in business presentations. They are useful in presenting relative frequencies of nominal categories. The construction of pie charts in JMP requires careful attention. Pie charts are obtained from the **Chart** menu, as shown in Figure 3.9.

Figure 3.9 Menu to Create a Pie Chart

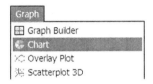

Selecting **Chart** opens the Chart Report window for variable selection. Three steps are necessary to produce a pie chart for the nominal variable Sales Volume. On the left in Figure 3.10, the steps are numbered from 1 to 3.

Step 1: Select **Sales Volume** in the **Select Columns** list and click the **Categories, X, Levels** button (located behind the Statistics menu in Figure 3.10).

Step 2: Keep Sales Volume highlighted and select **N** from the **Statistics** menu. This statistic will count the occurrences of each category.

Step 3: Select **Pie Chart** from the menu in the **Options** panel. (The default is Bar Chart.)

Click **OK** to obtain the desired pie chart on the right in Figure 3.10. The annotations within the pie pieces are added manually. JMP provides the legend below the chart.

Figure 3.10 Chart Report Window and Pie Chart

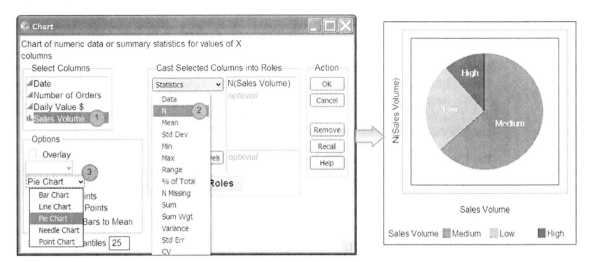

This pie chart shows that more than half of all days are Medium volume days. Only about 11% of the days during that period were High volume days, requiring the highest level of staffing for optimum service. A staffing policy that meets Medium day needs is estimated to be adequate on 89% of all days during that time period. Regularly occurring fluctuations of sales volume might change the observed percentages of High, Medium, or Low volume days.

3.4 Interpretation of Descriptive Statistics

Using the Herbal Tea example, this section contains comments on some of the most useful descriptive statistics.

3.4.1 Histogram

The histogram of Figure 3.11 shows a frequency distribution of the Number of Orders variable. This histogram gives a good impression of the spread of the data. Most observations are concentrated around the 150 to 300 bars. There are a few large values above 350 and a few below 150. It is difficult to say more about the data using the histogram alone.

Figure 3.11 Histogram of Number of Orders

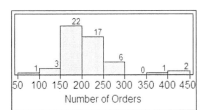

3.4.2 Box Plot

The original idea of the box plot derived from a data summary based on five different characteristics. In JMP, these are the two extremes (minimum, maximum), lower and upper quartile, and the median. The extremes might be outliers, endpoints of the whiskers, or the box itself. The mean is the vertical diagonal (dashed line) of the diamond and is a useful addition to this five-value data summary. The components of a box plot are shown in Figure 3.12.

Figure 3.12 Box Plot Components in JMP

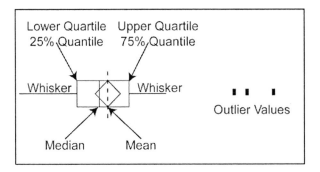

The box incorporates three vertical lines. From left to right, they are (1) the lower quartile (25% quantile); (2) the median (or 50% quantile), represented by the vertical line in the middle; and (3) the upper quartile (75% quantile). The median is usually inside the box, but might on occasion coincide with one of the quartiles. There are a few more additional components:

- Inside the box is a diamond. The upper and lower corners (vertical diagonal of the diamond) represent the mean (as distinct from the median). The left and right corners represent a 95% confidence interval for the mean, discussed in a later chapter. In highly asymmetric data, it could happen that the mean diamond is outside the box.

Figure 3.13 Outlier Box Plot of Herbal Tea Number of Orders Variable

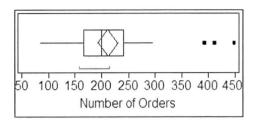

- The horizontal lines to the left and the right of the box are "whiskers." They represent all data values within their range. The maximum length of a whisker is w = 1.5 * (upper – lower quartile). The actual length of the whiskers is determined by the largest (or smallest) value outside the box but within [lower quartile – w] and [upper quartile + w]. The values [lower quartile – w] and [upper quartile + w] represent the inner fence.
- Individually marked outlier values represent data outside the inner fence.
- The horizontal square bracket (not shown in Figure 3.12 but shown in Figure 3.13) indicates the most concentrated (shortest range of) 50% of the data.

The box plot can be used to examine a few characteristics of the data:

Symmetry of the data distribution: In symmetric distributions, (1) the mean and median are nearly the same, and (2) the whiskers are of about equal length. As applied to sample data, these rules hold approximately within the expected sampling variation.

Normality of data: The Normal and other similar unimodal symmetric distributions are characterized by the fact that the central 50% is also the most concentrated 50%. Furthermore, for normally distributed data, the percentage of outliers should approximately be less than 5%. For other distributions, different percentages are appropriate. There are, however, better methods to determine normality, such as the normal quantile plot (also available in JMP Distribution).

Smallest and largest data values: The extreme points on the left and right of the box plot represent the smallest and largest values. In the preceding box plot, the smallest value is still on the left whisker, while the largest is the right-most of the outliers. The box plot in Figure 3.13 identifies three large values (above 350) as outliers with respect to the assumption that the data are symmetrically distributed. The mean and the median differ slightly. These three outliers are partly responsible for this difference.

3.4.3 CDF Plot

The letters CDF stand for cumulative distribution function. The CDF plot is a step function of the cumulative proportion of observations less than or equal to the value plotted on the horizontal axis. This plot is often referred to as the empirical distribution function. The CDF plot of the Number of Orders variable from the Herbal Tea Mail Order data is shown in Figure 3.14. The values of the horizontal axis are those of the Number of Orders variable. The vertical axis is the cumulative proportion, here labeled Cum Prob for cumulative probability, in anticipation of the fact that these proportions are used to estimate the probability of obtaining a value less than or equal to a specific Y-value.

The CDF plot is available as an option from within the Distribution platform. To obtain the options menu on the left of Figure 3.14, click the red triangle next to the variable name and select **CDF Plot** as shown. The CDF plot for the Number of Orders variable is on the right.

Figure 3.14 CDF Plot for Number of Orders

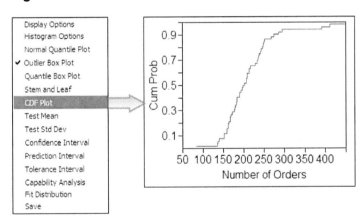

This CDF plot can be used to estimate distribution quantiles graphically. Assume you want to estimate the value of the daily Number of Orders at which 80% are below. Proceed as shown in Figure 3.15.

① Mark the value 0.80 on the vertical axis.

② Using the crosshair tool, draw a horizontal line through the CDF step function.

③ Where the horizontal line meets the CDF, draw a vertical line to the horizontal axis.

④ Read the estimated value of Number of Orders.

Figure 3.15 Estimating Quantiles from CDF Plot

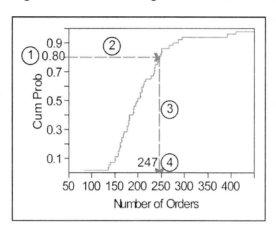

In JMP, the values can be obtained and the lines can be drawn with the crosshair tool. The crosshair tool can be selected from the **Tools** menu or the toolbar as shown in Figure 3.16. In the Mail Order example, the quantile corresponding to an estimated probability is approximately 247 (slightly less than 250). For some Cum Prob values, because of the step function, the resulting Number of Orders value might not be unique.

Figure 3.16 Crosshair Tool from Tool Palette

With the crosshair tool, the value and the lines disappear as soon as you deselect the tool. Reference lines do not disappear until removed and can be added to a plot by double-clicking the pointer tool over the axis for which the line is desired. The Axis Specification window appears. Figure 3.17 shows the Axis Specification window for the X-axis.

Figure 3.17 Add Line to CDF in Axis Specification Window

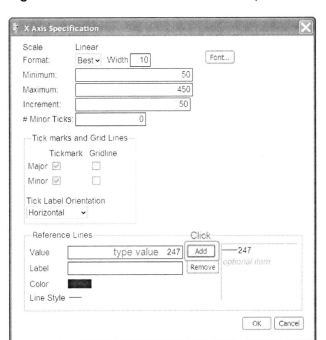

This window enables specification of minimum and maximum axis values and the increments of the axis value label. In this case, using the X-axis, a vertical line is added to the plot at 247. The value 247 needs to be typed in the **Value** field on the left. Then click **Add** and the vertical line appears. More than one vertical line can be added in any graph.

The process can be repeated to add horizontal reference lines by double-clicking the pointer on the vertical axis and entering the values as desired. In the CDF plot, you would enter the value 0.8 to obtain a plot that is close to the one shown in Figure 3.15. You could use the reverse direction, fixing a value on the horizontal axis (Number of Orders), and finding the associated probability value on the vertical axis (Cum Prob).

3.4.4 Quantiles

A q% quantile (or q% percentile) is a data value at which at least q% of the data are at or below and (100 – q%) are at or above that value. In JMP, certain quantile values are interpolated between relevant adjoining data points. Minimum and maximum values are always data points. To understand the quantile table in JMP, start with the data of the Herbal Tea Mail Order example. Order them from smallest (minimum at 84) to largest (maximum at 449).

Minimum 84, 135, 137, 140, 151, 151, 158, 158, 161, 162, 163, 166, 166, (interpolated 25% quantile = 166.75), 169, 171, 177, 178, 180, 181, 183, 183, 190, 190, 193, 196, 198, (interpolated median = 199.5) 201, 206, 206, 209, 209, (mean = 212.077) 213, 214, 217, 229, 234, 235, 236, 238, (interpolated 75% quantile = 241), 242, 244, 247, 249, 250, 251, 268, 275, 287, 295, 392, 411, maximum 449.

Altogether, there are 52 observations. Thirteen, or 25%, are below the estimate of the 25% quantile, 26 or 50% are below the estimated median, and 39 or 75 % are below the estimated 75% quantile. Neither of the two interpolated quartiles is part of the data value set. The box plot shows that the mean and the median do not agree. In fact, five observations are between the mean and the median, setting the mean close to the 60% quantile. This discrepancy may indicate a lack of symmetry of the underlying distribution. You would have to employ a statistical test to prove that this difference is or is not statistically significant. This test is discussed in a later chapter.

The actual list of quantiles gives numeric values for the quantiles used in the box plot and adds additional quantile values. This list is useful to identify precise numbers for the values depicted on the box plot. Thus a reasonable numerical comparison of median and mean is possible by using the median value (199.50) from this list.

Using the Quantiles Table (Figure 3.3), you can calculate the range as the difference of the largest and smallest value in the data, i.e., Range = Maximum – Minimum = 449 – 84 = 365. Especially in Control Chart applications, the range is used to estimate the standard deviation of the population. Note, however, that the range is very susceptible to outlier values. In some instances in the interquartile range, the difference between upper and lower quartile is also used. In this example, the interquartile range is 241 – 166.75 = 74.25. The interquartile range is not susceptible to outliers.

3.4.5 Moments

The **Moments** list (see Figure 3.3) is a collection of various numerical statistics. The most important of these are the mean and the standard deviation. The **Moments** list also shows the Standard Error of the Mean, the "upper 95% Mean," and the "lower 95% Mean, three values that will be discussed in later chapters.

"Std Err Mean" stands for Standard Error of the Mean, an estimate of the precision of the mean estimate. In Figure 3.3, the Standard Error of the Mean is the standard deviation divided by the square root of the sample size n (or N). The "upper 95% Mean" and "lower 95% Mean" represent the 95% confidence limits for the population mean.

The mean ($212.08) is larger than the median ($199.5), which is a discrepancy probably caused by the large outliers. The Std Dev is $66.80. (At this point, it is difficult to place a good interpretation on the Std Dev, unless you are willing to assume that the data follow the Normal distribution.)

You can ask for additional moments by selecting **More Moments** from the **Display Options** submenu. By far the most important statistic on the **More Moments** list is the Variance = (Std Dev)^2 = (66.797278)^2 = 4461.8763. The variance is a measure of spread needed in certain calculations. The Std Dev is the square root of the variance.

3.5 Practical Advice and Potential Problems

3.5.1 Features of Good Graphs

Statistical software has become quite sophisticated in giving users high-quality graphs with ease. Still, a few customization steps can improve a graphical representation. The goal of customization is to produce more effective graphs. What are some useful guidelines on constructing a graph? A good graph is self-contained. This means that everything on the graph should be explained:

- All the axes should be labeled in understandable terms.
- Marker symbols should be clearly identified.
- Smoother lines and connecting lines should be recognizable throughout the range of the graph.
- Marker symbols identifying subgroups of data should clearly differentiate between subgroups.
- Axes scales and tick spacing should facilitate comparisons with other graphs.
- Annotations on the graph should be helpful for interpretation. Unhelpful annotations, sometimes referred to as chart junk, should be avoided.

We demonstrate some of these guidelines with a simple supplier qualification example.

3.5.2 Example: Izod Impact Strength of Two Suppliers

DeGusto and Formago are two established suppliers of automotive plastics. The plastics of this example are molded into car bumpers. A car manufacturer needs to test the impact resistance of each supplier's material. The manufacturer prefers plastics with higher impact resistance.

In the U.S., a common test method for impact resistance follows the standard method of the American Society for Testing of Materials ASTM D256. The method uses a notched plastic rod and measures energy lost per unit of specimen thickness (such as lb/in or J/cm) at a notch. Notch size is specified. Each manufacturer was asked to supply at least 30 notched specimens, each from a different batch. DeGusto submitted 34 and Formago submitted 30 specimens. These were tested by an independent lab. The data are in *Izod Impact Strength.jmp*.

3.5.3 Simple Improvements to Histograms

A few simple modifications to histograms can improve their potential to communicate the proper insights. Here we show that you can make the information from the plots easier to grasp by providing common axes of two histograms that need to be compared and highlighting groups through dynamic linking of windows.

The first step in comparing the two suppliers is to construct histograms for the results by supplier. These are shown in Figure 3.18.

Figure 3.18 Default Histograms of Izod Strength by Supplier

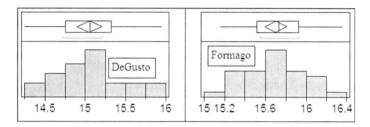

The two histograms do not bring out the differences between the two suppliers. The main reason for this defect is that the horizontal axes have different ranges. The DeGusto axis ranges from 14.0 to 16, while the Formago axis ranges from 15 to 16.4. Also, note that the bars have different widths. DeGusto's bar width is 0.5, while Formago's is 0.2. The horizontal axes are not labeled and no units are given. By adding axis labels and by changing the axes to the same range and the same increments, you can obtain a much clearer picture of the actual performance of each supplier on the notched Izod test. The changes are shown in Figure 3.19. Even the box plots of the two graphs show that the two data sets are predominantly in different ranges of impact resistance.

Figure 3.19 Histograms with Common Axes to Facilitate Comparison

Axes are easily adjusted manually in JMP. Simply double-click the axis that needs adjusting to bring up the Axis Specification window already shown in Figure 3.17. Set the minimum and maximum values as well as the increment to the desired values and click **OK**. In this example, both axes were set to range from 14 to 16.5 with increments of 0.5. This method is most useful if the axis of one graph is changed. A more efficient way to obtain common axes of two or more graphs is to use the uniform scaling option from the red triangle of the Distribution bar.

Adding labels to histograms or to box plots can be done by right-clicking while holding the pointer over the area immediately below the axis. This brings up a menu from which you can choose **Add Axis Label**, as shown in Figure 3.20. These axis labels can be edited further by simply clicking on the label and adding the modification as was done here when the units were added (lb/in) at a later time.

Figure 3.20 Right-Click to Add Axis Label

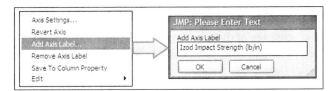

JMP also offers another way to compare two groups of data in the same histogram. Obtain a histogram for the combined data. Next, highlight one of the two groups in the data table; for example, highlight all 30 observations of Formago. This will identify the members of that group in the histogram with a darkened fill. The result is shown in Figure 3.21.

Figure 3.21 Combined Histogram with Formago Highlighted

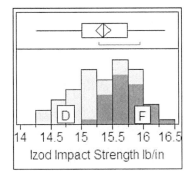

The letters "D" and "F" have been added to the plot using the JMP annotation tool. The annotation tool is the tool marked with a script letter "A" on the **Tools** menu. Click to select the tool and click again in the area of the plot where you want the annotation. The annotation can be resized, moved, and deleted (by right-clicking the annotation and choosing **Delete**).

This histogram vividly shows that Formago plastic has higher impact resistance, since most of their observations are on the right side. One must use some caution to draw definite conclusions from such data. Quite a few things can contribute to a wrong picture, such as different preparation of the Notched Izod Bar specimens by the two suppliers, or a recent spike or dip in performance quality at one of the suppliers. The order of testing the specimens at the lab can also influence the results. Suppose all DeGusto specimens were tested two weeks prior to testing the Formago specimen. The observed difference could be entirely due to measurement or set-up errors.

3.5.4 Potential Problems with Box Plots and Histograms

Box plots are useful summaries of data. However, box plots might not give the right impression of the data in certain situations, such as when data sets are small, data are rounded to a degree that only a few distinct values are recorded, or the distribution is highly asymmetric.

Figure 3.22 Small Data Set for Box Plot

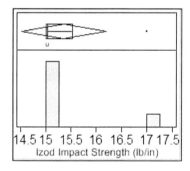

Data Sets Are Small

Observe a box plot from a small data set with six values of Izod Impact Resistance as follows: 15, 15, 15.025, 15.05, 15.05, and 17. Figure 3.22 shows that this box plot is not very informative. JMP uses the 25% and 75% interpolated quantiles for the construction of the box, with 25%q = 15.0 and 75%q = 15.54. This makes the central 50% of the data appear larger than they are, because they contain only the values 15.025, 15.05, 15.05. The median (15.0375) is the average of 15.025 and 15.05, the two central values. The mean (15.354) differs from the median mostly because of the outlier on the right. However, not knowing how the data were collected makes it difficult to ascertain from six observations whether this value really is an outlier. Assuming that the data were collected randomly, you could suggest that this maximum might be an outlier.

Data Are Rounded or Have Very Few Distinct Values

A common problem encountered in compliance sampling is that frequently few values have discrepancies from the norm. In many such applications, the value 0 predominates for discrepancies. Only a few may show positive or negative values.

The following is an example of a sales tax audit of 2112 purchases by a corporation. The actual data are in *Tax Audit.jmp*. The columns include the Merchandize Amount, representing the taxable value of the item purchased; Tax Paid, representing tax payments made for that item; Audited Tax, representing the final value assessed by the tax auditors; and Tax Discrepancy, representing the difference between Audited Tax and Tax Paid. The last column is a derived nominal variable called Error. The value in this column is "error" if there is a discrepancy, and "no error" if Audited Tax and Paid Tax agree.

Figure 3.23 contains both a histogram and a box plot. Because only 23 of the 2112 values are different from 0, the box of the box plot is a single line concentrated at 0. The 23 non-zero values are shown as outliers. There are no whiskers. Similarly, the histogram shows a single bar near zero. The other bars are relatively too small to be visible.

This box plot shows that there are errors, but it is deceiving in the sense that the 23 individual outlier points have more graphic weight than the 2089 instances in which no discrepancy was found. The histogram is also not very informative, showing a single bar at 0, but none of the discrepancies.

Figure 3.23 Tax Discrepancy Histogram and Box Plot   **Figure 3.24** Box Plot of Asymmetric Distribution

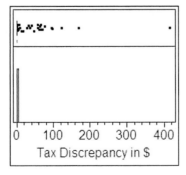

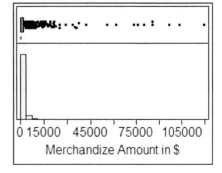

The Distribution of Data Is Highly Asymmetric

In *Tax Audit.jmp,* this is the case with Merchandize Amounts, the purchase values of the 2112 transactions subject to sales tax. Most values are very small, but a few are quite large. The general impression of the box plot or the histogram is that they tell a part of the story, but are not as useful as in other applications. The box plot in Figure 3.24 overemphasizes large values. The box itself appears tiny, although it represents the bulk of the data. The histogram corrects this distorted view somewhat, but also does not properly represent this variable from the large data set.

3.6 Summary

- Data can be described graphically and numerically.
- Continuous and nominal data require different graphs and numerical summaries.
- JMP uses appropriate tools depending on whether a column is either continuous or nominal.
- Continuous data can be graphed by histograms, box plots, time plots, and CDF plots.
- Continuous data histograms show the distribution of values by tabulating the frequency of occurrence per class (or bar) width.
- Box plots summarize the median, the lower and upper quartile, as well as minimum and maximum values.
- Box plots of few data values can distort the data distribution.
- Box plots of data where a large portion of values is concentrated in a narrow band may give undue focus on values outside the concentrated data range.
- Time plots graph data over time. They can be used to discover temporal patterns.
- CDF plots show the cumulative frequency of occurrence.
- Continuous data can be summarized by measures of location (mean, median, and quantiles) and measures of spread (standard deviation, variance, and range).
- The mean is a simple average of the data values.
- The median is the value with at least 50% as large as and at least 50% as small as it.
- The upper and lower quartile split the data into the lower 25%, the middle 50% and the upper 25% respectively.
- JMP uses interpolated quantile values. The median, quartiles, or other quantiles, except the maximum or minimum, need not be data values.
- Nominal data can be graphed in frequency histograms, mosaic plots, and pie charts.

- Nominal data can be numerically summarized by frequency counts and relative frequencies per nominal category. In a nominal frequency histogram, the horizontal axis represents outcome categories.
- Relative frequencies stress the proportional occurrence of each nominal category.
- Relative frequency can be graphed in a mosaic plot or a pie chart.
- A good graph has all axes labeled and does not contain unnecessary information.
- A good graph has easily distinguishable marker symbols.
- A good graph has properly spaced axes.

3.7 Problems

1. The owners of a small business are concerned about the balance they maintain in their checking account. They wish to avoid bank fees when their balance falls below the account minimum set by the bank. In addition, maintaining too much cash in the account prevents them from obtaining a higher rate of return in a different investment. The bookkeeper has analyzed the last 100 weekly account balances with the intention of including the following information in a presentation for a staff meeting. Review the information, making whatever changes you feel appropriate to improve the effectiveness of the data presentation.

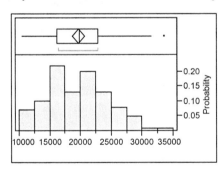

Observations	Mean	Median	Variance	Minimum	Maximum
100	19762.876	19864	25093282	10480	33708

2. Use the Web site of the U.S. Department of Commerce's Bureau of Economic Analysis to download the Total Gross State Product for the state in which you reside. Obtain the annual data back as far in history as available. Review any Web site explanations or cautionary notes that are given for this data series. Prepare a

statistical summary of the data and write an explanatory paragraph including any important definitions and cautions.

3. The numbers of threats removed from a laptop computer by an anti-virus program during its daily scheduled scan are as follows: 17, 25, 24, 14, 4, 29, 16, 15, 11, 66, 15, 13, 5, 0, 7, 13, 21, 12, 7, 9, 0, 17. These data are contained in *VirusThreats.jmp*.

   a. Obtain basic descriptive results.

   b. Comment on the histogram (e.g., symmetry) and compare it to the box plot.

   c. Compare the mean and the median and comment on what this comparison suggests (symmetric, asymmetric).

   d. Are there any outlier observations, and how would you identify them?

4. A California insurance company is required by state law to provide a quarterly summary of claims paid on homeowners' insurance policies. The information technology department has extracted the pertinent information from the insurance company's internal database for the first quarter of the year. The data are given in *CalifClaims.jmp*.

   Prepare a statistical summary, not to exceed one page, of the insurance claim data. Include those graphs and numerical statistics that are effective in summarizing the data set. Include a discussion of unusual observations and patterns or trends in the data.

5. A community is concerned about the traffic near an elementary school. The speed limit is currently posted at 25 miles per hour. A traffic study was conducted, and the data will be used to determine if additional traffic controls or speed limit enforcement are required. The file *TrafficSpeed_2.xls* contains the recorded speeds of vehicles passing near the school between the hours of 8:00 and 10:00 a.m.

   Prepare three presentation slides that summarize the speed data. The first slide should contain the problem statement, the second slide should contain appropriate numerical and graphical summaries of the speed data, and the third slide should contain your interpretation of the data and conclusions.

6. A new method has been developed to measure energy absorption of molded plastic discs. A total of 49 discs were prepared as identically as possible and tested at a rate of one disc/day. The data are contained in the file *MoldedPlasticDiscs.jmp*. Energy absorption is measured in joules.

   a. Assess the variability of the energy absorption using both graphics and numerical statistics.

   b. Find the mean and the median. Compare these two descriptors.

   c. Using JMP, make a time plot of the data by selecting **Graph → Overlay**. Connect the data points using **Y Options → Connect Points**. Are there any discernable patterns?

    d. Plot the data again, this time removing the Connect Points option. Are there any discernable patterns?

    e. Discuss any differences in the appearance of patterns depending on whether or not the data points were connected.

7. The data file *Russian100.jmp* gives the rank, name, net worth (in millions of U.S. dollars), age, and company for the 100 wealthiest Russians as of April 2004. The data were obtained from Forbes.com. Write a paragraph that provides a statistical summary of the Russian 100 using graphs or numerical statistics that effectively describe the data set. Include net worth and age. Also, summarize the five companies that have the largest share of net worth.

8. A help desk for a large software company receives problem reports over the telephone. Problems can be resolved by the help desk staff, systems analysts, or systems programmers. The file *IT_ResolutionTimes.jmp* contains data on the problems reported during the last quarter, the type of analyst that resolved the problem, and the elapsed time in days for resolution. Prepare a table that summarizes the resolution times by analyst type.

9. A regional library system provides summary statistics on its operations for an annual report. The file *LibraryOperations.jmp* contains data from 34 branches. Create a table that contains the mean, median, standard deviation, minimum, and maximum for each of the variables.

3.8 Case Study: New Web Software Testing

A travel agency has contracted with a software company to rewrite the Web interface used by travel agents for booking travel arrangements. Many of the agents complained that the current Web interface was too slow when refreshing the main screen. This was causing delays that were unacceptable to both the customers and to the agents. As part of the acceptance testing, the software contractor has decided to do a side-by-side test of the new Web interface in parallel with the existing software. A small group of agents have been trained and participated in the initial testing of the new interface. The subcontractor has written software scripts that will capture the time required to refresh the main screen. During a two-hour test period, both interfaces will be run simultaneously, with the test group using the new interface and all other agents using the existing interface. The test data are contained in the file *ScreenDelay.jmp*.

Prepare a statistical summary of the screen delay times for each of the interfaces. The summary (including your comments) should fit onto one page. Therefore, select only those numerical statistics and graphs that most effectively describe these two data sets. Comment on whether this is a fair comparison.

C h a p t e r 4

Statistical Models

4.1 Introduction 82

4.1.1 Building Statistical Models 83

4.1.2 Examining Examples of Statistical Models 87

4.1.3 Evaluating Statistical Models 88

4.2 Classification of Statistical Models 90

4.2.1 Models with a Single Y-Variable and No X-Variables 90

4.2.2 Models with a Single Y-Variable and a Single X-Variable 91

4.2.3 Models with a Single Y-Variable and Multiple X-Variables 94

4.2.4 Models with Multiple Y-Variables and No X-Variables 94

4.2.5 Models with Continuous Multiple Y-Variables and Some
       X-Variables 96

4.2.6 Other Approaches to Models 96

4.2.7 Other Models 96

4.3 Model Validation 97

4.3.1 Pitfalls of Model Building 97

4.3.2 Model Validation 98

4.3.3 Residuals in Model Validation   98

4.4   Summary   100

4.5   Problems   101

4.6   Case Study: Models of Advertising Effectiveness   102

4.1 Introduction

Models help us understand the real world as it exists or as it might exist at some time in the future. Architects use models to visualize and communicate their ideas about a building project. Automotive engineers use models to study the aerodynamics of a car. Architectural models are scaled-down versions using light construction materials or computer models. Engineers might create full-scaled models using molded material, or they might create computer models.

Business models are representations of some aspects of the real business world for the purpose of solving business problems. They are used to predict future events such as sales or stock price. They are employed in controlling operations and manufacturing processes.

Statistical models are empirical in nature and assume random outcomes. They establish empirical relationships between variables using data as supporting evidence and quantifying random error. They are more than mere data summaries, because the nature of the relationship is influenced by the expert knowledge of the model builder.

The problem is that the model and the supporting data need to be in some alignment. Not all data are suitable to support specific models, just as not all models are suited to explain specific problems. How do you know which models to use? For specific problems, you can often narrow the choice of models to a particular class. In statistics, regression models or time series are widely used. Given the right data, they can be used to address an astonishing variety of problems. But which of the many regression models should you choose? Even after the type of model has been chosen, you need to determine the variables that should be included and how they should be expressed in the model. Statistical models come in a great variety, from the simple to the very complex. This chapter covers some of the simpler statistical models, those that are written as equations. They can be quite powerful in helping solve business problems.

The famous philosopher and mathematician W. Leibniz recognized an important aspect of modeling that is important in statistical modeling. He distinguished between facts that can be described by some law and those that cannot. Leibniz suggested that in situations where a simpler model explains the facts equally well as a more complex one, the simple model is always better. In statistical terms, you should try to solve a problem with the simplest possible statistical model. Without a statistical model, you must resort to a listing, or a description of the data, a process that does not solve a problem and therefore is not modeling. Nevertheless, in order to grasp the quality of the data available for modeling, most work on data will start with a description of the data.

Whenever a particular data set can be used with several competing statistical models, the choice of a good model follows the principle of parsimony. Under this principle, you always choose the simplest model that will explain the problem adequately. In statistics, simplicity is often defined in terms of the number of variables that are used to predict an outcome. Fewer variables are the mark of a simpler model. Other indicators of simplicity (or complexity) are the degree of the polynomial used in the model, or whether some variables need to be transformed by functions such as logarithms or inverses to achieve the desired explanatory results. Linear terms and no transformations indicate simpler models.

Not all business models are statistical in nature. Deterministic models include mathematical relationships between outcome variables and input variables. In business, deterministic models such as the standard net present value model are especially useful to develop benchmark comparisons for investments. Future returns are discounted to present value using fixed discount rates. These models are correct as long as the discount rates are correct. Deterministic models are formulated using expert knowledge. They are often based on simplifying assumptions such as a single discount rate over the expected life of an investment. A second very useful type of model is the Monte Carlo simulation model. Monte Carlo simulation models tend to be built like deterministic models, but include random outcome generators. They are especially useful to study the random behavior of processes.

4.1.1 Building Statistical Models

Models always need to serve the purpose defined by the problem statement. Business problems are always stated in the language of finance, marketing, operations, information technology, law, or accounting. Such problem statements need to be translated into statistical terminology. Likewise, the results obtained from a statistical model are written in statistical terminology and these have to be interpreted into terms that can be understood by the business world. Figure 4.1 shows a schematic of this two-way translation process between the business problem and the statistical model. This process is by no means unique. As different teams work on translating business problems into statistical language, it is quite possible that they come up with different statistical models. Likewise, the results of statistical models on occasion might be subject to different interpretation, especially when the underlying assumptions are not clearly understood or articulated.

Figure 4.1 Relationship between Business Problem and Statistical Model

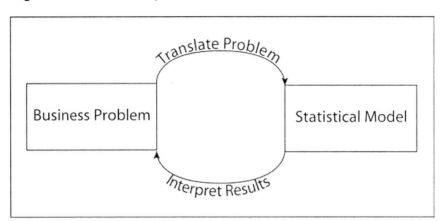

Statistical modeling is a process that starts with problem definition. This is followed by specifying the model and collecting the data needed to fit the model, as shown in Figure 4.2. The four process steps are intertwined. The problem definition and the model specification will suggest the type of data required. In turn, the available or obtainable data will influence model specification. Likewise, statistical methods used in model fitting are largely determined by available data. The interpretation of the model requires an understanding of the problem, the model assumptions, and the process by which the data were collected.

Problem definition is a most important step. The key specification of a problem definition is the performance variable. It is the measure that is compared to the criterion by which the outcome of a process or a product is judged. The problem definition should include a statement about the magnitude of the problem, especially in terms of its impact on business performance. The problem scope refers to the business area where the problem occurs, what the nature of the deviation from the desired value is, and how extensive the problem is. It is also good to indicate, if possible, how long the problem has existed. A problem statement should not yet suggest possible solutions. It is important to keep an open mind about possible solutions.

Figure 4.2 Modeling as a Process

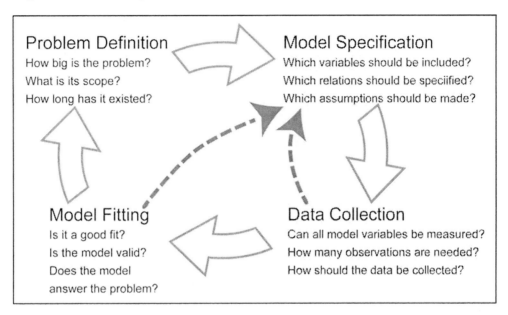

After the problem has been defined, you can think of solution approaches. A good way to start the model-building process is to list the variables that potentially influence or might be related to the performance variable. This is done with the expectation that not all these variables will be needed in the final model. Some variables of interest might be difficult or expensive to measure. The two dashed arrows in Figure 4.2 indicate that revisions of the initial model might be needed. It might be that the data required by the current model cannot be obtained or that the current model is inadequate to answer the questions posed by the problem definition.

Expert knowledge is essential at the model-specification step, because it supplies the theoretical underpinnings of a model. Model building requires that relationships between variables are specified based on current expertise. Expert knowledge in the area of application, e.g., finance, is needed to

- provide insight from similar problems
- suggest relationships between variables
- judge whether model results are consistent with expectations

Data collection consumes resources. One needs to be aware of the quality, quantity, and cost of obtaining data required for problem solving. Too few observations can lead to non-significant results, i.e., the statistical model does not provide results with sufficient precision to distinguish between important solution alternatives. An example of this situation is reflected in the debate about global warming. In climate models, the balance between expert knowledge and empirical evidence has not quite reached a steady state. This might in part explain the discrepancy in the explanations of the causes of global warming. As data become available at a rapid rate, the predictions will become more precise and the debate will be resolved.

Data also need to include all the important variables. Data on some variables may be impossible or too expensive to obtain. In business, speedy solutions are often desirable. This can lead to using available data, rather than data that are needed to solve the problem. However, inadequate data affect the usefulness of model results.

The model-fitting step includes the actual estimation of the model parameters from data. This step also should include an evaluation of the goodness of fit, the validation of the model in view of independent information, and the interpretation of the model in terms of the problem definition. If this cannot be accomplished, you have to go back and change the model, or obtain better data, or review the problem definition for its operational content.

Statistical modeling requires statistical skills that are not to be confused with the computational effort performed by statistical software. JMP provides a broad variety of statistical tools, which were unavailable only a few decades ago. The skill set required for building simple statistical models needs to include understanding the assumptions of the statistical modeling tools, knowing how to interpret computational results in a business context, and knowing how results are influenced by unusual data such as outliers.

In summary, successful statistical models require several ingredients as shown in Figure 4.3. The model requires a specific problem statement that defines the magnitude and scope of the problem and includes criteria for evaluating whether a problem solution has been found. It requires expert knowledge to provide context and interpretation. Data are essential for an empirical model as they provide the factual basis for the model. Statistical skills are required to cope with the necessary assumptions and to develop a proper methodological perspective.

Figure 4.3 Ingredients of Statistical Models

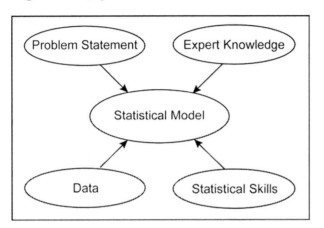

Statistical modeling should be an iterative process by which statistical models evolve over time. As a model is improved, several issues need to be continuously monitored:

- The model might be incomplete by not having all the necessary variables specified or by expressing variables inadequately.

- The model might contain variables that are not needed to solve the problem. This is a case of over-fitting variables and violates the principle of parsimony. It can lead to inadequate results.

- The relationships between variables might change over time, necessitating adjustments in the choice of variables or in model coefficients.

- The available data might be inadequate and lead to wrong results. One common reason for this is that historical data can lead to biased answers.

- Assuming that experts agree on a model, they might disagree as to the extent to which results can be extrapolated to future situations. They might disagree that the data or the model are representative of similar or future situations.

4.1.2 Examining Examples of Statistical Models

What are some problems addressed using statistical models in this chapter? These few examples show some of the ingredients that go into building statistical models. These ingredients include a way to model a random distribution, a method to explain the statistical relationships between the variables, and expert knowledge to facilitate the interpretation of the model in answering business questions.

Package Delivery Times: Is it possible to estimate the proportion of packages delivered late by a package service based on a limited number of observations? The answer is yes, as long as you are willing to make certain assumptions, e.g., the random distribution of the delivery times. Is it symmetric or skewed? One needs to try out different distributions to see which fit the data.

Profitability of Bank Branches: Can we determine differences in average profitability between bank branches of different regions? In this example, you assume that the bank branches in different regions are comparable in some way, but that there are differences between the regions that should be explored to explain branch profitability.

Process Yield of a Chemical Reactor: Can you predict the yield of a polymer produced in a chemical reactor based on the agitation applied to the reactor? Again the answer is yes, if you are willing to make assumptions about the nature of the relationship between process yield and amount of agitation in addition to the random distribution that is present in such processes.

Experience Curve: Can you predict the unit cost of a new product six month from now? Once again, the answer is yes. The relationship between unit cost and amount of experience in building a new product is important in such a model.

4.1.3 Evaluating Statistical Models

Expert knowledge is important in building and applying models. Many statistical models require some understanding of the underlying problem structure. For example, this understanding is useful in determining which variables are important in solving a particular problem. Expert knowledge often postulates certain types of relationships, such as linear increasing or decreasing. It is useful in evaluating whether a statistical model makes sense.

Statistical models serve a purpose. Experts need to decide whether or not a statistical model will meet its designated purpose. They also need to be concerned with how well a model will fit the data or how much of the variability in the data the model is apt to explain. Measures are available to assess how good a model is in statistical terms. For regression models, such measures include the R-Square and the standard deviation of observations around the model, as well as other measures such as variance inflation factors.

These criteria do not always give unambiguous answers. Some interpretive skill is usually needed to make sense of the results. Figure 4.4 shows an example in which four pairs of (X, Y) observations are available. Three models are fitted to the data. The simplest one is the linear equation. In this model, Y is represented as a straight-line function of the variable X. The second model is a quadratic equation, with Y being a function of both X and X^2. The most complex is the cubic equation, with Y being a function of X, X^2, and X^3. R-Square (R^2) and the standard deviation (s_y) around the model equation are two criteria to judge these models. A higher R-Square often, but not always, indicates a better fit, whereas a lower standard deviation around the model is an indicator of a better fitting equation than one with a higher standard deviation.

Figure 4.4 Comparison of Three Polynomial Equations as Models

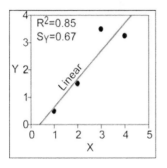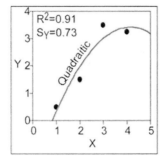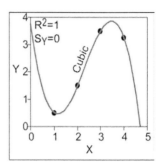

The lowest possible standard deviation ($s_Y = 0$) and highest possible R-Square ($R^2 = 1$) are attained by the cubic equation. It is a perfect fit through the four data points. However, the curve is too complex to interpret for many practical problems. In this case, a perfect fit does not explain anything, because the cubic equation with four parameters does not simplify the four data observations. The cubic equation is no model in the sense of Leibniz, because it does not reduce the complexity of the data. The quadratic equation has the next highest R-Square ($R^2 = 0.91$), but unfortunately also the highest standard deviation around the equation ($s_Y = 0.73$). The simplest model is the linear equation with a standard deviation of 0.67 and R-Square of 0.85. Frequently, but not always, you would choose the linear equation. It is the simplest model and represents a reasonable fit to the data.

If these recommendations seem tentative, it is because the problem underlying the three models has not been clearly defined. A business model is defined explicitly by the context obtained from the problem description and/or implicitly provided by expert knowledge of the team analyzing the problem. You must be aware of problems arising from a purely empirical model—one not based on any understanding. Without a good understanding of the underlying relationship between the variables, such purely empirical models can lead to nonsense models. However, there are exceptions. Recent advances in computing have established data mining as an important model building tool in business applications. Data mining is used when expert knowledge is minimal, but when factual information is plentiful. Data mining is a tool to discover relationships from data. It is largely empirical, and is able to learn from the objective world of data.

4.2 Classification of Statistical Models

Statistical models can be classified by the type of variables used in their construction. Statistical models have one or more output or performance variables designated by the letter Y. These Y-variables are also called response, predicted, dependent, or explained variables. In addition, statistical models may have no, one, or more input variables designated by the letter X. These X-variables are also called predictor, independent, control, or explanatory variables.

JMP characterizes variables by their modeling type as either continuous, ordinal, or nominal. Y- and X-variables can be in any number of combinations of continuous, nominal, or ordinal. The combination of the modeling type of Y- and X-variables determines applicable models in JMP.

4.2.1 Models with a Single Y-Variable and No X-Variables

One of the most basic models concerns a single Y-variable. Such models are important, because many problems focus on a single performance variable. Before any relationships between that Y-variable and other variables are contemplated, you might look at the variable by itself. The questions that can be addressed with such models are different depending on the modeling type. As with all models, the Y-variable can be continuous or nominal/ordinal. In JMP, such models use the Distribution platform for the analysis of data pertaining to these models. JMP chooses the basic analysis and provides options depending on the modeling type of the Y-variable.

Continuous Y: A variable with a continuous modeling type does not have to be continuous in a mathematical sense. Variables of the continuous modeling type need to be numeric and allow calculation of means, standard deviations, etc. Thus even integer-valued variables, which are not continuous mathematically, can be and are of the continuous modeling type.

The standard model for a continuous Y can be represented as a simple equation:

Value of Y = Mean of Y + Random Error

This equation states that each observation Y can be represented by its mean and by the random deviation (or random error) of Y from its mean. The random error distribution is often assumed to follow the Normal distribution. This is by no means the only error distribution. Examples of other error distributions are the Lognormal, Weibull, or the Smallest Extreme Value distribution. Determining the distribution of the Y-values is an important task in any modeling effort.

Random errors are often assumed to be independent of each other. One could say that the outcome of one random error does not influence the random errors of other outcomes. This assumption is not always satisfied, especially when observations are taken over time. Special statistical methods are used to analyze data to support such models. Not surprisingly, these methods are called time series models. This book contains a single chapter on this topic.

One further assumption concerning random errors is often made. Many statistical models assume the distribution of random errors remains constant throughout the entire range of observations. If this assumption is violated, special techniques need to be considered.

Nominal or Ordinal Y: Nominal or ordinal Y-variables have categories or levels as outcomes. For such variables it would make no sense to use a mean. The standard model for nominal or ordinal Y can be represented by assigning probability values to each outcome. A model with k outcome categories for the variable Y can be written as:

$$Y = \begin{cases} \text{category 1 with probability } p_1 \\ \text{category 2 with probability } p_2 \\ \qquad\qquad \vdots \\ \text{category k with probability } p_k \end{cases}$$

The model assigns a probability to each outcome category. It is assumed that the outcome categories are non-overlapping, that the k categories represent all possible outcome values, and that the probabilities sum to 1.

4.2.2 Models with a Single Y-Variable and a Single X-Variable

A simple statistical model is the relationship between a single Y-variable and a single X-variable. The combination of a single Y- and X-variable opens four classes of models that can be used to analyze an amazing array of business problems. In JMP, such models are analyzed with the Fit Y by X platform or with the Fit Model platform. As the name suggests, the Fit Y by X platform is specifically designed to analyze models with one Y-variable and one X-variable. Fit Model is a very general platform that allows an even wider range of models.

Figure 4.5 shows a classification of models with one Y-variable and one X-variable according to the modeling types from JMP. With both Y and X continuous, a bivariate model such as simple regression analysis is used. When both Y and X are either nominal or ordinal, a contingency table model is appropriate. One-way analysis of variance is used for continuous Y and nominal or ordinal X, simple logistic regression for a nominal or ordinal Y and a continuous X-variable. These four model classes will be briefly discussed next.

Figure 4.5 Model Types by X- and Y-Variable

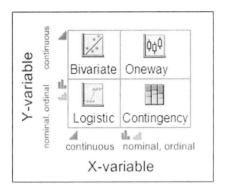

Continuous Y and Continuous X: Continuous Y- and X-variables lead to the simple regression model. The Y-variable is often called the dependent or predicted variable, while the X-variable is called the independent or predictor variable.

The standard simple regression model can be written as a simple equation:

Value of Y = Intercept + Slope * Value of X + Random Error

The slope is the average increase in Y per unit increase in X. The intercept is the average value of Y when X = 0. This value may or may not make sense in the context of the problem. Key assumptions of the simple regression model are (1) the random errors are distributed normally (according to the bell-shaped curve), (2) the errors are independent of each other, and (3) the variation around the model is constant throughout the range of application.

Continuous Y and Nominal or Ordinal X: A single continuous Y-variable and a single nominal or ordinal X-variable lead to the one-way analysis of variance model. Depending on the application, the nominal X-values may be called levels or groups. The one-way analysis of variance model allows us to explain mean differences in the Y-variable as a function of levels of the X-variable. The analysis of variance model fits a separate mean to each X-level. The standard analysis of variance model can be written as follows:

Value of Y = Level Mean + Random Error

Key assumptions of the analysis of variance model are similar to those of the simple regression model. Random errors are distributed normally (according the bell-shaped curve). The errors are independent of each other, and the variation around the model is constant around the level means. If these assumptions are not satisfied, modified approaches are needed to perform a correct analysis.

Nominal or Ordinal Y and Nominal or Ordinal X: When both the Y- and X-variables are nominal or ordinal, a contingency table model is used. The contingency table approach looks for the association between the two variables. Alternatively, it can be viewed as a test of independence (versus dependence). The simplest of such contingency table models is a 2 by 2 contingency table, in which each variable has exactly two outcome levels.

Nominal or Ordinal Y and Continuous X: This simple logistic regression model has a single nominal Y-variable and a continuous X-variable. For example, a bank wishes to predict loan performance with two outcomes: Default and No Default. The bank needs to determine whether the credit rating that is available at the time of loan approval is a good predictor of the outcome. Because there are only two outcome categories, the prediction will be in terms of the probability of Default and of No Default. The simple logistic regression model is nonlinear.

Simple logistic regression models can be used to model simple choice behavior, such as purchase or no purchase, when the purchase decision depends mostly on a single variable. More complex logistic regression models are used to represent situations in which the binary Y-variable is the result of several continuous X-variables. Logistic regression is also used in data mining applications.

4.2.3 Models with a Single Y-Variable and Multiple X-Variables

Statistical models with a single Y-variable and multiple X-variables serve to explain the variation in the Y-variable by using many X-variables simultaneously. A combination of judiciously selected X-variables can be a powerful tool to explain the variation in Y-values. Such a combination might result in a model that explains more than the sum of the single X-variable models. Except in special cases, it is preferable to examine how each X-variable relates to Y in combination with other X-variables, rather than looking at individual effects. For all of these models it is assumed that the relationship between Y and the X-variables is linear, either directly or after transformation. (Nonlinear models are not treated in this book.)

4.2.4 Models with Multiple Y-Variables and No X-Variables

Some models involve multiple Y-variables without any X-variables. These types of models are used to explain the structure of the multiple Y outcomes. Correlation analysis is one tool to understand relationships between many Y-variables. Other tools to perform analyses on multiple Y-variables are principal component analysis and factor analysis. These two are often performed in combination with each other. All are found on the Multivariate platform. They are beyond the scope of this text.

Table 4.1 summarizes the combinations of Y- and X- variables discussed so far. The table also includes the statistical method that is typically used to analyze such situations and the appropriate JMP platform. The next three subsections are given for completeness, but are not treated in this text.

Table 4.1 Y- and X-Modeling Variables with Corresponding Statistical Methods and JMP Platform

Y-Variable	X-Variables	Statistical Method	JMP Platform
1 continuous or 1 nominal or ordinal	0	Descriptive statistics	Distribution
1 continuous	0	Confidence interval (mean and standard deviation)	Distribution
1 nominal or ordinal	0	Confidence interval (proportion)	Distribution
1 continuous	0	Hypothesis test for mean or standard deviation	Distribution
1 nominal or ordinal	0	Hypothesis test for proportion	Distribution
1 continuous	1 nominal or ordinal	Two independent sample t-tests for means and confidence interval for mean difference	Fit Y by X
1 continuous or 2 continuous paired	0	Paired t-test for means & confidence interval for difference	Distribution [Matched Pairs]
1 continuous	1 nominal or ordinal	One-way analysis of variance (multiple comparisons)	Fit Y by X (Fit Model)
1 continuous	2 nominal or ordinal	Two-way analysis of variance (unreplicated, equally & unequally replicated, multiple comparisons)	Fit Model
1 nominal or ordinal	1 nominal or ordinal	χ^2 test for independence	Fit Y by X
1 nominal or ordinal	1 nominal or ordinal	χ^2 test for equality of ≥ 2 proportions	Fit Y by X
1 continuous	1 continuous	Simple regression analysis	Fit Y by X (Fit Model)
At least 2 continuous	0	Simple correlation analysis	Multivariate Methods → Multivariate
1 continuous	≥ 2 continuous, nominal or ordinal	Multiple regression analysis	Fit Model
1 continuous	0	Time series analysis	Modeling → Time Series

4.2.5 Models with Continuous Multiple Y-Variables and Some X-Variables

Some models have multiple Y-variables and might also have one or more predictor X-variables. These X-variables might be continuous, in which case the method is called multivariate regression. The X-variables also might be nominal or ordinal and require multivariate analysis of variance. Such models are beyond the scope of this book.

4.2.6 Other Approaches to Models

Other approaches are models for which the model-specification step is performed incompletely. Potential variables are specified, but functional relationships are sought from the data. Such procedures are useful in data mining. Data mining is used in situations in which only limited expert knowledge is available. It is an approach to data that is used to discover relationships among variables and even among groups of observations. Data mining also establishes the importance of each variable in explaining these relationships.

Partitioning is such a data mining tool. No specification of the functional model is required. Instead, only potential variables are specified. The Y-variable and X-variables can be continuous or nominal. The method generates descriptors of groups of observations after repeated binary splits of the data. Each split is based on the most advantageous X-variable. Each split is performed on an existing group. Partitioning ends with a model consisting of descriptors of groups of observations. The descriptors are in terms of X-variables. There are several other models in this category. They are beyond the scope of this book.

4.2.7 Other Models

Various applications and problem settings require their own models and statistical tools. Many of these are used in business applications. For example, the useful life of a product needs to be estimated for a variety of purposes. This leads to product life models. For most product life models, the Normal distribution is not the first choice. Also, product life data often include censored observations. These are incomplete observations in the sense that a product might continue to work as expected at the time the test is terminated. Product life models require special tools that are beyond the scope of this book.

Many economic and business models are dynamic models with respect to time. For example, orders for a new storage technology evolve over time. Orders in one week are not unrelated to orders in the following week. Such data often violate the assumption of independence. Special models and tools are available to meet the challenges of these applications. Time series analysis is a method that models the dynamic behavior over time. The structure of correlation between observations taken over time is used to identify specific model types. This book covers only exponential smoothing. More sophisticated approaches include Box-Jenkins methods and spectral analysis. Econometric models also find use in many business applications.

There are many more advanced models that go beyond the scope of this book. For example, JMP includes multivariate models like discriminant analysis, partial least squares, and clustering methods. It also offers screening, neural networks, nonlinear, Gaussian, categorical and choice models.

4.3 Model Validation

4.3.1 Pitfalls of Model Building

At almost every step, building statistical models is a process that contains pitfalls that can affect the results and conclusions. Careful attention and a reasonable skill level can avoid many missteps in problem definition, model specification, data collection, and model fitting. A common mistake at the problem-specification stage is that the problem and the associated variables are too broadly defined. This leads to misinterpretation of the original intentions of the modeling process.

Common problems in the model-specification step are (1) a poor understanding of important variables and their underlying relationships, and (2) the omission of important variables or improper functional expressions of variables.

Problems of the data-collection step are discussed elsewhere in this book, but they can be summarized in a few broad categories. Nonrepresentative data might be the result of improperly selected sampling units or of missing responses in surveys. Occasionally a bad random sample can result in poor answers. An inadequate quantity of data is often responsible for inconclusive results. High variability might require a larger sample size than anticipated. For certain prediction problems with high inherent variability, even large data sets are inadequate. For example, the term "asset" in a company's financial statement is often measured to single-dollar precision. However, asset size within a company does not necessarily measure investment. In comparing assets of companies, you would need to know their depreciation schedule and adjust assets accordingly.

In model fitting, the technical challenge is to interpret the results. Model coefficients often are based on the "all others being equal" assumption. In complex models with interrelated variables, such interpretations need to be done with great care.

4.3.2 Model Validation

The model-fitting step requires three different tasks: model estimation, model validation, and model interpretation. Model estimation is largely performed by statistical software. Its issues are very technical and for the most part outside the scope of this book. Model validation and interpretation are important aspects of the modeling process that need to be performed by the model builder or user. Model interpretation will be discussed in conjunction with the various statistical models presented in subsequent chapters. Model validation can be discussed in terms of general principles.

Model validation tests a model's ability to serve its intended purpose. Many times the purpose of a model is to predict future outcomes, such as predicting sales for the next quarter. Other times you need a model to estimate average performance based on a combination of input parameters, such as estimating the average insurance claim per insured automobile for drivers under the age of 25 years. Model validation seeks to compare such model predictions and estimations with information that was not used in model estimation. One way is to use a *hold-back* sample, i.e., to not use a portion of the data in the model-fitting stage and save these data for model validation. This method is sometimes referred to as *data splitting*. Another way is to collect additional data for this purpose.

Model validation should include *checks of the reasonableness* of predictions and estimates. The most important question you can ask is "Does this make sense?" If it does not (for example, because the coefficients have the wrong sign, or because the model estimates and predictions produce values that are outside of expectations), then you need to find out why the model is either inadequate or why the counterintuitive behavior is present. On occasion, unexpected model results might baffle the experts initially. In the end they may lead to fundamentally improved process understanding.

Sensitivity tests provide yet other indications of a valid model. In a sensitivity test, the model parameters or the conditions under which the model has been estimated are perturbed. Reasonable results from a perturbed model are yet one more piece of evidence that the model is valid. Sensitivity analysis can also show model deficiencies of the kind produced by highly correlated predictor variables when minor changes in model inputs produce relatively huge differences in predicted values.

4.3.3 Residuals in Model Validation

Criteria to judge whether or not a model is valid usually involve residuals. Residuals are the difference between the observed values and those predicted by the model. Written as a simple formula:

Residual = Observed value – Predicted value

Residuals are used in graphical displays and also in calculating quantities such as the standard deviation of observed values around the fitted model known as the root mean square error. There are other statistics to measure the validity of a model. These are discussed in connection with specific statistical methods.

Some of the tools to check on model assumptions are graphical. The basic tenet of residuals plots is that a random pattern of residuals shows that the assumption is met. Patterns such as trends or increasing values indicate that the assumptions are not met. Residuals and related quantities are used to test both the appropriateness of the functional aspect of the model and the correctness of the basic model assumptions. The functional aspects of model testing address three main questions:

1. Does the model contain predictor variables that should not be included?

2. Does the model omit predictor variables that should be included?

3. Are the predictor variables in the model properly expressed?

The appropriate chapters discuss scatter plots of residuals or leverage plots as tools to check the need for predictor variables. Most of the models covered in this book satisfy certain simplifying assumptions. The resulting models exhibit simplicity but are very powerful. Usually you can use simpler models in the following cases:

- The data observations are independent.
- There are no noticeable trends over time.
- The variance of the data is constant throughout the region of inputs that is of interest.

For prediction, the assumption of normally distributed residuals is helpful. For models that violate these simple assumptions, there are specialized estimation methods or modeling techniques. However, as a rule they are more difficult to perform and require enhanced statistical expertise. Scatter plots of residuals versus predicted values can be used to find non-constant variance. Residuals against the order in which data were collected might be an indicator of trends over time. The independence of observations can be checked by a variety of other plots, such as the lag plot and the autocorrelation plot. The normal quantile plot of residuals reveals any problems with deviations from the assumption that residuals are normally distributed.

4.4 Summary

- Business models are used to understand the business world.

- Statistical models are empirical models based on data, intended to simplify the information contained in the data. Simplification helps in finding explanations.

- Statistical models are not unique. Several competing models might be available.

- Statistical business models require that the problem statement is translated into statistical language and terminology.

- The results obtained from statistical models need to be re-translated into a language that is understood by the business user.

- Statistical modeling is a process that starts with a problem definition and is followed by model specification, data collection, and model fitting.

- The model-fitting step includes statistical estimation of model parameters, model validation, and model interpretation.

- Simple statistical models can be classified according to the modeling type of the variables involved (see Table 4.1).

- Many statistical models are application specific, such as product life models or time series models.

- Statistical modeling is a process with pitfalls. Models might contain too many or too few variables. Variables might be expressed incorrectly in the model.

- Model validation tests whether a model serves its intended purpose.

- Models can be validated by using independent data.

- Models should be checked for reasonableness.

- Models are tested for the sensitivity of results to changes in the model parameters or data values.

- Residuals are widely used in evaluating the fit of the model.

- Residuals are used in a variety of diagnostic plots.

- Residuals are used to calculate key model characteristics such as R^2 or RMSE.

4.5 Problems

1. For each of the following business performance measures (Y), identify factors (Xs) that might be useful in building a predictive model. Explain why each of the Xs is a potentially useful predictor.

   a. Daily absentee rate for sales associates at a large retailer employing 100 sales associates for each of two shifts (early and late).

   b. Number of new businesses started annually in a county.

   c. Data collected by the biomechanical engineering department of a university on the type and severity of injuries treated at ski resort First Aid stations. The data are used to design safer ski equipment.

2. For each of the following situations, discuss the appropriate boundaries and scope of models.

   a. A producer of packaged food items would like to predict sales of a new fruit drink based on data taken from a focus group of boys and girls aged 6–16 from a suburban community.

   b. An organic strawberry farmer wishes to forecast yield for the next two weeks based on meteorological forecasts. He wants to determine how much temporary help to hire.

   c. A print media publication wants to determine the amount of advertising sold, measured in dollars.

3. Sketch potential models for the uncertainty in the following processes.

   a. Five members of a town council are each equally likely to be appointed to fill a vacancy on a regional planning commission.

   b. An investment firm wants to predict the number of sick days taken by office workers.

4. Moore's Law is an empirical model that states that the transistor count on a computer chip doubles about every two years. Kryder's Law was similarly derived and states that the density of information the can be stored on a hard disk increases by a factor of 1000 every 10.5 years.

   a. Use the Internet to find estimates of the current transistor count on a computer chip and the current storage density for a hard disk.

   b. How long might these laws expected to be valid? What factors might cause these laws to become no longer applicable?

   c. Use Moore's Law and Kryder's Law to predict the transistor count and storage density, respectively, in five years.

   d. Discuss how Moore's Law and Kryder's Law might be of use in budgeting information technology needs.

5. Sketch a model of the distribution of household income for the U.S. Compare the model to data given on the U.S. Census Bureau Web site. How does your model, compare to the facts?

6. A bakery makes handmade French bread. The costs and revenues associated with the number of loaves produced are given in the following table.

Units	Cost	Revenue
20	125	55
50	142.5	108
80	153	175

   a. Plot cost and revenue on the same Y-axis against the number of units produced.
   b. Sketch a model for the cost function.
   c. Sketch a model for the revenue function.
   d. Use the two models to estimate the breakeven point from the graphs.

7. Explore the field of financial engineering. Prepare a written summary including a brief overview of the field and a description of two examples where models are applied in this field. Explain the purpose of these models and how they are applied. (It is not necessary to include mathematical descriptions of the models.)

4.6 Case Study: Models of Advertising Effectiveness

Business Problem
A company must decide how many minutes of television advertising to purchase each day. They have 15- and 30-second commercials available to air.

Tasks
Perform the following tasks:

1. Sketch a possible relationship between the number of seconds of daily advertising and sales (in number of units). Place advertising time on the X-axis and sales on the Y-axis. Write a brief description explaining the relationship.

2. Sketch an alternative relationship between the number of seconds of daily advertising and sales and write a brief description explaining this relationship.

3. The following table gives data from the past five TV advertising campaigns.

Advertising Time	Sales (in 1000 units)
30	9
90	22
120	34
150	26
180	37.5

   a. Plot the data and sketch a model that you think adequately represents the data.
   b. Using your model, quantify the relationship between advertising time and sales.
   c. Use your model to predict the number of units sold for 180 seconds.
   d. What would you expect sales to be for 300 seconds of advertising time? Do you have any reservations about this prediction?

C h a p t e r 5

Discrete Probability Distributions

5.1 Introduction to Distributions   106
    5.1.1 Random Variables   107
    5.1.2 Independent Events   107
    5.1.3 Discrete or Continuous Distributions   109
    5.1.4 Characteristics of Distributions   109
    5.1.5 General Shape of Distributions   110
5.2 Discrete Distributions   111
    5.2.1 Probability Function p(y)   111
    5.2.2 Cumulative Distribution Function F(y)   111
    5.2.3 Mean, Variance, and Standard Deviation   111
5.3 Binomial Distribution   113
    5.3.1 Binomial Distribution Characteristics   114
    5.3.2 Calculating Binomial Probabilities in JMP   115
    5.3.3 Practical Advice   117

5.4 Distributions of Two Discrete Random Variables (Y$_1$, Y$_2$) 118

5.4.1 Return of Two Stocks under Different Economic Conditions 118

5.4.2 Joint and Marginal Probability Functions 119

5.4.3 Covariance Cov[Y$_1$,Y$_2$] and Correlation Coefficient ρ$_{12}$ 121

5.4.4 Calculating E[Y$_1$], E[Y$_2$], E[Y$_1^2$], E[Y$_2^2$] and E[Y$_1$·Y$_2$] 122

5.4.5 Variance and Standard Deviation of a Portfolio of Two Stocks 123

5.5 Summary 124

5.6 Problems 126

5.7 Case Study: Assessing Financial Investments 128

5.1 Introduction to Distributions

Probability distributions model the random nature of many phenomena. Probability models allow businesses to quantify risk in finance, to assess the capability of manufacturing processes, to predict the likelihood and impact of uncertain events on operations, and to incorporate variation in key variables into business decisions.

Probability distributions must not be confused with data distributions. Probability distributions are conceptual models of random processes or phenomena. Data distributions are historically accumulated observations or observations collected in surveys or designed experiments.

Conceptual distributions are theoretical, but based on an understanding of the underlying random process. Given a set of assumptions, the distribution of the resulting random outcomes can be derived by probability methods. For example, the number of sales made in a fixed number of sales calls follows a binomial distribution, if certain assumptions are satisfied.

Sampling distributions are yet another type of distribution derived from test statistics and statistical estimators used in statistical inference. Test statistics and statistical estimators are calculated from sample data and are used to draw conclusions from sample data. Examples of such test statistics include the z-Ratio following the Normal distribution, t-Ratio following the t-distribution, F-Ratio following the F-distribution, and Chi-Square statistic following the Chi-Square distribution.

The coverage in this and the next chapter assumes that you have had a previous exposure to probability. The discrete distributions discussed in this chapter are characterized by finitely many possible outcomes, or infinitely many outcomes that are still countable.

Continuous distributions are covered in the next chapter. We begin by establishing a few terms regarding distributions, including random variables, independent events, discrete versus continuous distributions, and characteristics of distributions.

5.1.1 Random Variables

In business, random variables represent process, economic, or environmental variables. In probability, a random variable Y is a numerically valued function defined on the outcome space, where the outcome space is the collection of all possible outcomes of a random process. The random variable assigns a number to each outcome. Typically, an uppercase letter, such as "Y," denotes the random variable, whereas specific outcomes are denoted by a lowercase "y."

5.1.2 Independent Events

Independence of events is an important probabilistic concept. Two events A and B are statistically independent if the probability of B given that A has occurred (written as P[B given A] = P[B|A]), is equal to P[B], the probability of B unconditional of the occurrence of A (and vice versa). Two events are dependent if they are not independent. A simple example is useful for defining a few probability terms.

Tossing a Fair Coin: Tossing a single coin results in either heads (H) or tails (T). The experiment consists of tossing the coin 10 times and counting the number of heads.

The outcome space S of an experiment or process consists of all possible outcomes. In tossing a coin ten times and observing the binary outcomes H or T, there are altogether $2^{10} = 1024$ possible outcomes. However, in this example, the outcome space can be simplified by counting the number of heads, which cannot exceed the number of tosses. This outcome space S = {0,1,2,3,4,5,6,7,8,9,10} indicates that tossing 10 coins and counting the number of heads might result in an outcome ranging from 0 heads, 1 head, 2 heads, ... , 10 heads. At any one toss, only one outcome occurs.

Events are subsets of the outcome space S and are designated by capital letters such as A, B, etc. The Null-event is the event that none of the outcomes occurs. It is also included in the collection of possible events. Events can be described in several ways. In the example, simple descriptors identify an event: A = {Exactly 2 heads} = {heads = 2} = {2}, or B = {Less than 2 heads} = {heads < 2} = {0,1}, or B^c = {At least 2 heads} = {heads ≥ 2} = {2,3,4,5,6,7,8,9,10}.

B^c is called B complement and is the event of all outcomes not in B. Two events A, B are said to be mutually exclusive if they do not share any outcomes. By definition, B and B^c are mutually exclusive.

An event is said to occur when an outcome in the event occurs. For example, if 10 coin tosses produce two heads, both A and B^c occur, but B does not occur. The union $A \cup B$ of two events A and B occurs if the outcome is either in A or in B. The union of the events A and B above $\{2\} \cup \{0,1\} = \{0,1,2\}$ occurs as long as any one of 0, 1, or 2 occurs. The intersection $A \cap B$ of two events A and B occurs if the outcome is both in A and in B. The intersection of A and B^c is $\{2\} \cap \{2,3,4,5,6,7,8,9,10\} = \{2\}$ and occurs only when the outcome is 2, because 2 is the only outcome contained in both events.

Events have associated probabilities. The probability of event A is P[A] and designates the probability that one of the outcomes in event A occurs. Assuming fair coins such that the probability of heads and tails are the same at 0.5, you can show, using the binomial distribution, that P[A] = .044, P[B] = 0.011, and $P[B^c]$ = 0.989.

Tossing a Fair Coin ctd.: Assume a fair coin such that the probability of heads P[H] and tails P[T] is the same, i.e., P[H] = P[T] = 0.5. Toss the coin once and the probability of heads on the first outcome is $P[H_1]$ = 0.5. Toss the coin for a second time and the probability of heads at the second toss $P[H_2]$ is again 0.5, irrespective of the outcome of the first coin toss. (Subscripts indicate the number of the coin toss). This means that $P[H_2 \text{ given } H_1] = P[H_2 \text{ given } T_1] = P[H_2] = P[H]$ = 0.5. The probability of the outcome of such coin tosses does not change depending on the values of preceding outcomes. Suppose a fair coin is tossed 10 times in a row and all 10 outcomes are heads. The probability of such a sequence is 1 in 1024 or approximately 0.001. However, given that a sequence of 10 heads was just observed, the probability of heads at the eleventh toss is again 0.5. In other words $P[H_{11} \text{ given } H_1, H_2, \ldots, H_{10}]$ = 0.5. The outcome at the eleventh toss is independent of the outcomes at preceding tosses.

Sampling with Replacement: In sampling with replacement, randomly chosen objects are replaced either by replacing the very objects withdrawn or by objects that match them in the characteristics of interest. The observations of such a sample are considered independent.

Sampling without Replacement: In sampling without replacement, drawn objects are not replaced. The sample space changes after each draw, because one object is removed. As a result the outcomes are not independent.

When you sample without replacement from a very large population or from a continuous process, even though the underlying population is finite, the selection probabilities do not change by much. As a result, the observations are treated as independent. Political polls fall into this category of samples.

5.1.3 Discrete or Continuous Distributions

The nature of the random variable and outcome space allow us to classify distributions as either continuous or discrete.

Discrete distributions are associated with random variables defined on outcome spaces with a finite or a countably infinite number of outcomes. Each possible outcome could be assigned to a distinct integer. Typical examples of discrete random variables are a coin toss (1 for heads or 0 for tails), the number of customers waiting to be checked out at a supermarket (0, 1, 2, 3, 4, ...), or the number of defectives per 1 million aluminum cans. Discrete distributions can be characterized by the individual probability function p(y) = P[Y = y], the cumulative distribution function $F(y) = P[Y \leq y]$, and measures of location and spread.

Continuous distributions are for random variables that have real numbers as possible outcomes. There are so many real numbers that it is impossible to count all possible outcomes. As a result, calculus-based techniques are required to calculate probabilities. Fortunately, tables or numerical routines such as those in the Formula Editor of JMP provide the capability to calculate many probability-related tasks. An example of a continuous random variable is the time to failure of a hard drive on a PC. Because the hard drive could fail at any moment, it is easier to use real numbers from zero to infinity for the time-to-failure outcome set. Continuous distributions are characterized by the density function f(y), the cumulative distribution function $F(y) = P[Y \leq y]$, and measures of location and spread. Continuous distributions are discussed in the next chapter.

5.1.4 Characteristics of Distributions

Characteristics of discrete distributions include measures of location (mean, median) and spread (standard deviation or variance).

Measures of Location: Measures of location are values representing the position of the distribution along the dimension of the random variable Y. The most common ones are the mean, the median, and the mode. The mean is the center of gravity of the individual probabilities in discrete distributions or of the density function in continuous distributions. The median is the value that splits the outcomes such that those at or below it have a probability of least 0.5 and those at or above it have a probability of at least 0.5 as well. The mode is the outcome with the highest probability in discrete distributions and of the highest density in continuous distributions. Distributions can have more than one mode (multimodal). Those with one mode are called unimodal.

Measures of Spread: Measures of spread are values representing the variability of the distribution on the scale of the random variable. The standard deviation is the most common measure of spread and is the square root of the variance.

5.1.5 **General Shape of Distributions**

Distributions of a single random variable are classified as either unimodal or multimodal. Unimodal distributions have a single mode and can be classified as either symmetric or asymmetric. Asymmetric unimodal distributions are either left-skewed or right-skewed. Distributions can also be described as bounded or unbounded.

Figure 5.1 shows three discrete distributions with integer outcomes from 0 to 10. The vertical bars indicate individual probabilities P[Y = y] associated with each outcome of Y.

Symmetric Distribution: A discrete distribution is called symmetric if the probability function P[Y = y] is symmetrically distributed around the mean. The center panel of Figure 5.1 shows the probability function of a symmetric distribution. The mean, median, and mode (the outcome value with the highest probability) of a unimodal symmetric distribution are the same.

Distribution Skewed to the Right: A distribution is skewed to the right if the long (thin) end is on the right side. The left panel of Figure 5.1 shows the probability function of a right-skewed distribution. In unimodal right-skewed distributions, the mode is smaller than the median and the median is smaller than the mean.

Figure 5.1 Shapes of Discrete Unimodal Distributions

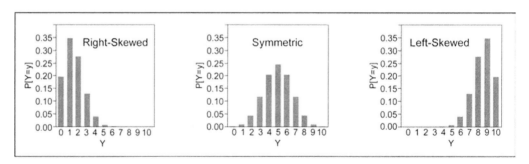

Distribution Skewed to the Left: A distribution is skewed to the left, if the long (thin) end is on the left side. The right panel of Figure 5.1 shows the probability function of a left-skewed distribution. In unimodal left-skewed distributions, the mode is larger than the median and the median is larger than the mean.

Bounded versus Unbounded: Bounded distributions have a limited outcome space, i.e., the random variable Y will be at least a and not more than b, where a and b are two finite numbers with a < b. All three distributions in Figure 5.1 are bounded, because the outcome space is defined by integers from 0 to 10. Unbounded distributions can be unconstrained at the lower or upper ends, or both. The continuous Normal distribution (bell-shaped curve) has an unbounded outcome space from minus infinity to plus infinity $(-\infty < y < +\infty)$.

5.2 Discrete Distributions

Discrete distributions arise from random variables with a finite or countably infinite outcome space. A finite outcome space has a finite number of outcomes. The outcomes can be labeled from 1 to n. The binomial is a distribution with a finite number of outcomes. A countably infinite outcome space has as many outcomes as there are integer numbers. The geometric distribution is an example of a distribution with a countably infinite number of outcomes.

5.2.1 Probability Function p(y)

The values y_i of the discrete random variable Y are associated with the probabilities $p(y_i)$. Together they form the discrete probability function p(y), where $p(y_i) = P[Y = y_i]$ is the probability that the random variable Y takes on the value y_i. The sum of the probabilities of all outcomes is always 1, i.e., $\sum_{\text{all i}} p(y_i) = 1$.

5.2.2 Cumulative Distribution Function *F*(y)

The cumulative distribution function is $F(y) = P[Y \leq y]$ and is useful in calculating the probability of events. The cumulative distribution function is the sum of all probabilities $p(y_i)$ such that $y_i \leq y$, i.e., $F(y) = p(y_1) + p(y_2) + p(y_3) + \ldots + p(y_k)$ such that $y_1 \leq y_2 \leq y_3 \leq \ldots \leq y_k \leq y$ and all other $y_i > y$. The value of the distribution function is always between 0 and 1, i.e., $0 \leq F(y) \leq 1$. $F(y)$ is non-decreasing for y. This means that for two values $y_1 < y_2$, it is true that $F(y_1) \leq F(y_2)$.

5.2.3 Mean, Variance, and Standard Deviation

The Mean[Y] is the weighted average of the y, where the weight of each y_i is $p(y_i)$:

$$\text{Mean}[Y] = \sum_{\text{all obs i}} y_i \cdot p(y_i)$$

The mean is a measure of location representing the probabilistic center of gravity of the random variable.

The variance Var[Y] can be calculated by two equivalent formulas. The first of these two formulas is the direct formula and is the weighted average of the squared deviations of the y_i values from Mean[Y], where each $(y_i - \text{Mean}[Y])^2$ is weighted by $p(y_i)$:

$$\text{Var}[Y] = \sum_{\text{all obs } i}(y_i - \text{Mean}[Y])^2 \cdot p(y_i) = \sigma_Y^2$$

The second formula is the computational formula:

$$\text{Var}[Y] = \sum_{\text{all obs } i} y_i^2 \cdot p(y_i) - (\text{Mean}[Y])^2 = \sigma_Y^2$$

The standard deviation SD[Y] is the square root of the variance $\text{SD}[Y] = \sqrt{\text{Var}(Y)}$.

Why use SD[Y] instead of Var[Y] for interpretation? The Mean[Y] and the standard deviation SD[Y] have the same units. When the random variable is in units of dollars, inches, or tons, the standard deviation will have the same units, whereas the units of the variance are dollars^2, inches^2, or tons^2. However, in order to obtain SD[Y], we first need to calculate Var[Y].

Example—Toss a Fair Coin Once: The outcome space S = {H, T}. A random variable Y associated with this experiment is assigned Y=1 if the outcome is H, and Y=0 if the outcome is T. If a fair coin is used, the probability of either outcome is 0.5, i.e., P[Y = 0] = P[Y = 1] = 0.5. The left panel of Figure 5.2 is a graph of its probability function p(y). In the right panel, the cumulative distribution function $F(y)$ is represented as a step function: $F(y) = 0$ for all values of Y less than zero; $F(y) = 0.5$ for values $0 \le y < 1$; and $F(y) = 1$ for values $y \ge 1$. While the discrete probability function has probability only at two points (Y = 0 and Y = 1), the distribution function has values for y lower, in between, and larger than these two values. The graph shows that $F(.5) = 0.5$ and that the probability is entirely derived from the probability mass at Y = 0.

Figure 5.2 Probability and Cumulative Distribution Functions for
P[Y = 1] = 0.5

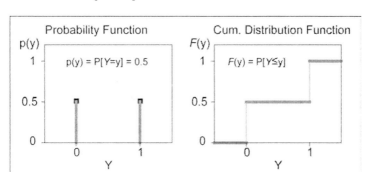

Table 5.1 shows the corresponding numerical representation of this distribution and includes columns that are useful in calculating the mean and the variance. The values in the column for the mean calculation are summed to obtain the distribution mean. Here Mean[Y] = 0 + 0.5 = 0.5. This value is halfway between 0 and 1, because each outcome is weighted by the same probability 0.5. The sum of the last column is used in calculating the variance. The sum (0 + 0.5) = 0.5 is the first term of the second variance formula above. The variance of this random variable is Var[Y] = $0.5 - (0.5)^2 = 0.25$. The standard deviation is the square root of the variance, so SD[Y] = 0.5.

Table 5.1 Table of Coin Toss Distribution

Index	Random Variable	Probability Function	Cum. Dist. Function	Mean Calculation	For Variance Calculation
i	y_i	$p(y_i)$	$F(y_i)$	$y_i \cdot p(y_i)$	$y_i^2 \cdot p(y_i)$
1	0	0.5	0.5	0	0
2	1	0.5	1.0	0.5	0.5

5.3 Binomial Distribution

The binomial distribution is used in many applications. In this book, it is discussed in Chapter 12, Proportions, where methods useful for polling are presented.

5.3.1 Binomial Distribution Characteristics

The binomial distribution is for a discrete random variable having n independent dichotomous outcomes, each with a constant probability of "success" p. A dichotomous outcome allows only two possibilities, e.g., heads or tails, true or false, good or defective, success or failure, etc. Examples of binomial processes include the following:

- A random sample of manufactured items is inspected. Each item is found to be either defective or non-defective. The number or proportion of defectives is recorded.

- Voters are polled prior to a school board election and respond as either in favor or not in favor of the incumbent. The proportion in favor of the incumbent is recorded.

The binomial distribution gives the probability associated with a given number of "successes" in n dichotomous outcomes (for example, the probability that no more than 2 manufactured items were defective in a sample of 10 items). The probability function is

$$P(Y = y) = \binom{n}{y} p^y (1 - p)^{n-y}$$

where p is the probability of success at any trial, n is the predetermined fixed number of trials, and y is the random outcome. The mean of the binomial distribution is Mean[Y] = np and the variance is Var[Y] = $np(1 - p)$. The binomial distribution is symmetric when $p = 0.5$; it is right-skewed when $p < 0.5$ and left-skewed when $p > 0.5$.

Example—Number of Sales in 10 Sales Calls: A sales manager tracks successes of sales calls to customers. The company expects a salesperson to make 10 calls per week. The outcome of any sales call is either a sale or no sale. Long-term experience has shown that on average 20% ($p = 0.2$) of calls result in sales. The sales manager wants to know the probabilities of the number of successful sales per salesperson during a particular week.

The random variable Y measures the number of successful sales in $n = 10$ sales calls. There are 11 possible outcomes {0, 1, 2, . . . 10}. The anticipated success probability for any sales call is $p = 0.2$. The successes at individual sales calls are independent. The probability function P[Y = y] and the cumulative distribution function P[Y ≤ y] are shown in Figure 5.3. The probability function is right-skewed, since p < 0.5. The graph shows that for $p = 0.2$, individual probabilities P[Y = y] for y larger than 6 are nearly 0. As a result, the cumulative distribution function for these values is near 1, because all probabilities of some magnitude are associated with outcomes up to 6. The mean is $np = 10 \cdot 0.2 = 2$ and the variance is $np(1 - p) = 10 \cdot 0.2 \cdot 0.8 = 1.6$. Taking the square root gives the standard deviation SD[Y] = 1.265.

Figure 5.3 Binomial Individual Probability and Cumulative Distribution
Function for n = 10 and p = 0.2

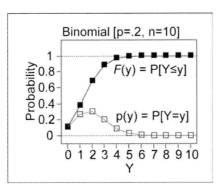

5.3.2 Calculating Binomial Probabilities in JMP

We illustrate the calculation of binomial probabilities with the example of the number of
sales in 10 sales calls. JMP's Formula Editor enables computation of values from both
the individual binomial probabilities and the cumulative distribution function. In order to
calculate binomial probabilities as shown in Figure 5.7, proceed as follows:

1. Create a new data table with three columns. Column 1 shows the possible
   number y of successful sales from 0 to 10. The second column shows
   probabilities of individual outcomes P[Y = y]. The third column is for P[Y ≤ y],
   the probability that there are no more than y successful sales calls.

2. Add 11 rows to the data table to accommodate the outcomes 0, 1, 2, ... , 10 by
   selecting **Add Rows** from the **Row** menu. Enter the value 11 into the **How
   many rows to add** field.

3. Enter the values of y. In row 1 enter 0, in row 2 enter 1, etc.

4. Calculate the probabilities for P[Y = y] using the Formula Editor. Open the
   Formula Editor by right-clicking the variable name and selecting **Formula**
   from the pop-up menu as shown on the left in Figure 5.4.

Figure 5.4 Creating a Formula for a Column P[Y = y]

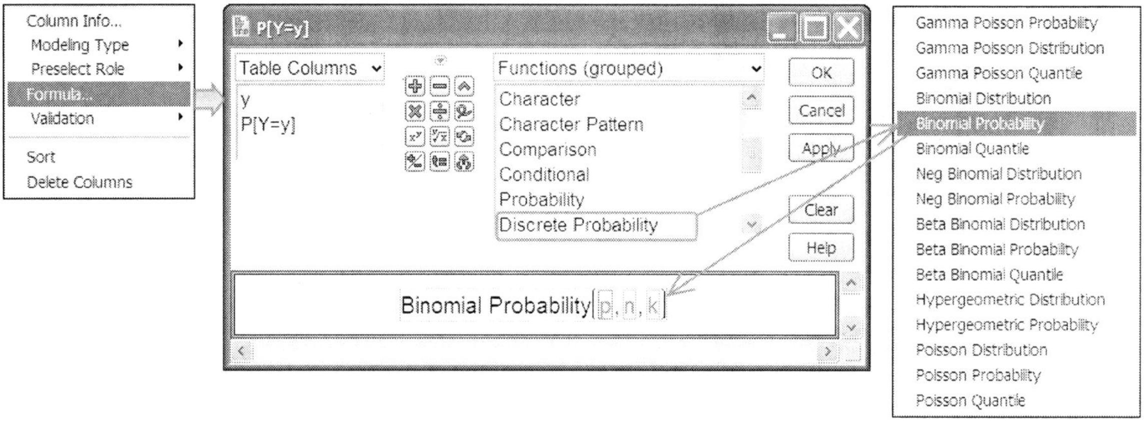

The formula window for the column P[Y = y] appears. From the **Functions** menu, select **Discrete Probability** to obtain the menu on the right of Figure 5.4. This presents a long list of probability functions. In this example, first select **Binomial Probability** for P[Y = y]. (For Column 3, select **Binomial Distribution** for P[Y ≤ y]). The function Binomial Probability appears in the Formula field. The formula shows three open fields labeled with gray letters p, n, and k.

For the sales call example, enter 0.2 for p and 10 for n. For k, click the variable name Y in the **Table Column** list at top left, because it contains the outcome values for which probabilities are desired. The resulting function is shown in Figure 5.5. Click the **Apply** button to calculate the probabilities.

Figure 5.5 Function for P[Y = y] of Binomial with n = 10 and p = 0.2

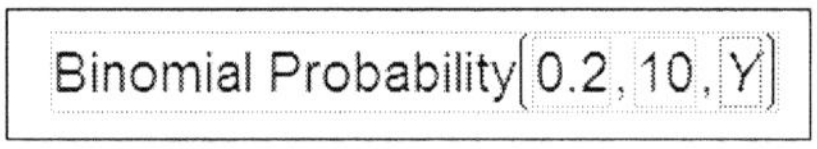

5. To calculate the probabilities for P[Y <= y], repeat step 4 with the following differences. Right-click P[Y <= y] to open the formula window. Select **Binomial Distribution** (instead of Binomial Probability) from the Discrete Probability functions menu and enter the same values for p, n and k. The resulting formula is shown in Figure 5.6. In the formula window, click **Apply** to calculate the probabilities.

Figure 5.6 Function for P[Y <= y] of Binomial with n = 10 and p = 0.2

Binomial Distribution$\left[0.2, 10, Y\right]$

JMP calculates the probabilities shown in Figure 5.7. These probabilities are formatted in the Column Info window to four fixed decimal places.

Figure 5.7 Data Table of Binomial Probabilities with n = 10 and p = 0.2

	y	P[Y=y]	P[Y<=y]
1	0	0.1074	0.1074
2	1	0.2684	0.3758
3	2	0.3020	0.6778
4	3	0.2013	0.8791
5	4	0.0881	0.9672
6	5	0.0264	0.9936
7	6	0.0055	0.9991
8	7	0.0008	0.9999
9	8	0.0001	1.0000
10	9	0.0000	1.0000
11	10	0.0000	1.0000

The column P[Y = y] contains individual probabilities. For example, the probability that in a week with 10 sales calls exactly 3 are successful is 0.2013. The column P[Y <= y] is the cumulative distribution function and represents the sum of all individual probabilities P[Y = y] from Y = 0 to y. For example, the probability that in a week with 10 sales calls no more than 3 (i.e., 0, 1, 2, or 3) are successful is 0.8791. The term *distribution function* always refers to a function that evaluates probabilities for events of the type $\{Y \leq y\}$. This function is also convenient for probabilities for events such as $A^C =$ {3,4,5,6,7,8,9,10} using the fact that the probability that either of two complementary events occurs is 1. The probability of the complementary event in this example is calculated as follows:

$P[Y \leq 2] = 0.6778$ and $P\{3,4,5,6,7,8,9,10\} = P[Y \leq 2]^C = 1 - P[Y \leq 2] = 1 - 0.6778 = 0.3222$.

5.3.3 **Practical Advice**

The terms "less than," "less than or equal," "not more than," "at most," "at least," and "more than" have very precise meanings in determining the set of outcomes included in a particular event. In the sales call example, the correct probability statement for the probability of less than 3 sales in 10 calls is $P[Y < 3]$. This probability is the same as "less than or equal to 2," i.e., $P[Y < 3] = P[Y \leq 2]$. The probability of at most 3 sales in 10 calls would be evaluated with the probability statement $P[Y \leq 3]$.

The binomial is a widely used discrete distribution. However, there are many other discrete distributions with applications in business. Two of these not treated here are the Poisson and the hypergeometric distributions.

5.4 Distributions of Two Discrete Random Variables (Y_1, Y_2)

This section reviews discrete bivariate distributions that are important in calculating the risk of a portfolio of two stocks. An outcome consists of a pair of numbers (y_1, y_2). The pairs of random variables (Y_1, Y_2) have a joint probability distribution $p(y_1, y_2)$. In addition to the mean, variance, and standard deviation for each y_i, you can calculate the covariance, the correlation coefficient, and other expected values from the joint probability distribution. This is demonstrated with a numerical example.

5.4.1 Return of Two Stocks under Different Economic Conditions

This example presents the returns of two stocks A and B and their respective probabilities of achieving returns. In the following discussion, Stock A is assigned the subscript 1 and Stock B the subscript 2. The returns depend on three economic conditions: stagnant economy, slow growth, and fast expansion. In this example, assume that the three economic conditions might be different for the two stocks. Slow growth for Stock A might be paired with fast expansion of Stock B, etc. For both stocks, the returns are -10 under stagnant economy, 0 under slow growth and +10 under fast expansion. For each stock, the probabilities with which each of these economic conditions occurs are given by the marginal probabilities $P[y_1]$ and $P[y_2]$. Table 5.2 shows joint and marginal probabilities of the returns on stocks A and B in a cross-tabulation format.

Table 5.2 Joint and Marginal Probability Distributions of Stocks A and B

Stock B Returns	Stock A Returns			P[y₂]
	−10	0	+10	$P[y_2]$
−10	0.05	0.10	0.15	0.3
0	0.05	0.20	.15	0.4
+10	0.25	0	0.05	0.3
$P[y_1]$	0.35	.30	.35	1

5.4.2 Joint and Marginal Probability Functions

The distribution of two random variables is captured by the joint probability function. It measures the probability that the two outcomes of Y_1 and Y_2 occur together, or in probability notation, $P[y_1, y_2] = P[Y_1 = y_1, Y_2 = y_2]$.

Examples of the joint probabilities contained in Table 5.2 are as follows:

- The probability that both stocks have negative return is $P[y_1 = -10, y_2 = -10] = 0.05$, which is the joint probability that both stocks have a return of −10 dollars.
- The probability that the return on Stock A is −10 dollars and the return on Stock B is +10 dollars is the joint probability $P[y_1 = -10, y_2 = 10] = 0.25$.

The joint probabilities of Table 5.2 can be represented in a list format that is friendlier for calculating means, standard deviations, and covariances. The list format is shown as the first three columns of Table 5.3 These columns are labeled Stock A = y_1, Stock B = y_2, and $P[y_1, y_2]$.

Table 5.2 shows the *marginal probability distributions* $P[y_1]$ and $P[y_2]$ of Y_1 and Y_2. These are the column and row sums of the joint probability distribution respectively. The marginal distribution of Y_1 ignores the outcome values of Y_2. It is the probability distribution of the returns on Stock A without consideration of returns on Stock B. The row sums of the joint probability distribution make up the marginal probability distribution of Y_2, the returns on Stock B. The marginal distribution of Stock B returns, using the random variable Y_2, ignores the returns on Stock A. The marginal probabilities have a different interpretation than the joint probabilities:

- $P[y_1 = -10] = 0.35$ is the probability that the return on Stock A is −10 dollars. This probability is the sum over all the y_2 of all the probabilities with $y_1 = -10$, i.e.,
  $P[y_1 = -10] = P[-10, -10] + P[-10, 0] + P[-10, +10] = 0.05 + 0.05 + 0.25 = 0.35$.

- $P[y_2 = -10] = 0.30$ is the probability that the return on Stock B is -10 dollars. This probability is the sum over all the y_1 of all the probabilities with $y_2 = -10$, i.e., $P[y_2 = -10] = P[-10,-10] + P[0,-10] + P[+10,-10] = 0.05 + 0.10 + 0.15 = 0.30$.

Marginal distributions are used to calculate the mean or expected returns Mean[Y_i] and the standard deviations SD[Y_i] of the individual stock returns. However, these quantities can also be calculated from the joint distributions as shown in Table 5.3. In finance, you might be interested in the expected return of a stock and the standard deviation of returns of a stock as a measure of risk. This requires (1) the expected or average return E[Y] = Mean[Y] of stock, and (2) the standard deviation SD[Y] as a measure of risk.

To avoid confusion, subscripts identify different stocks. Assuming Stocks A and B have the same expected return, a risk-averse investor would prefer Stock A to Stock B if the standard deviation of Stock A returns is smaller than the standard deviation of Stock B returns. With the same mean return, Stock A is less risky than Stock B.

The mean of the marginal distribution of Stock A is its expected return. It is calculated using the usual formula for the mean of a discrete distribution:

$E[Y] = 0.35 \cdot (-10) + 0.30 \cdot (0) + 0.35 \cdot (10) = 0$ dollars.

The expected value of Y^2 is needed for the standard deviation calculation and is $E[Y^2] = 0.35 \cdot (-10)^2 + 0.30 \cdot (0)^2 + 0.35 \cdot (10)^2 = 70$. Using E[Y] and $E[Y^2]$, the standard deviation of the marginal distribution of the return on Stock A is

$$SD[Y] = \sqrt{E[Y^2] - E[Y]^2} = \sqrt{70 - 0^2} = 8.37 \text{ dollars.}$$

Similarly, for Stock B, $E[Y] = 0.30 \cdot (-10) + 0.40 \cdot (0) + 0.30 \cdot (10) = 0$ dollars. The expected value of Y^2 is $E[Y^2] = 0.30 \cdot (-10)^2 + 0.40 \cdot (0)^2 + 0.30 \cdot (10)^2 = 60$. E[Y] and $E[Y^2]$ are used to calculate the standard deviation of the marginal distribution of the return on Stock B as

$$SD[Y] = \sqrt{E[Y^2] - E[Y]^2} = \sqrt{60 - 0^2} = 7.75 \text{ dollars. This shows that both stocks A and}$$

B have the same expected return = 0 dollars, but that Stock B is less risky because its standard deviation is lower than that of Stock A.

When the joint probability function $P[y_1, y_2]$ is the product of the row and column values of the marginal probabilities, i.e., $P[Y_1 = y_1, Y_2 = y_2] = P[Y_1 = y_1] \cdot P[Y_2 = y_2]$, the two random variables Y_1 and Y_2 are probabilistically independent. In the example of the return of two stocks, this is obviously not the case. $P[Y_1 = -10, Y_2 = +10] = 0.25$, but the product of the two marginal probabilities is $P[Y_1 = -10] \cdot P[Y_2 = +10] = (0.35) \cdot (0.30) = 0.11$. This means that the returns on these two stocks are not independent and that the covariance and correlation coefficient will most likely be different from 0.

Another important question is whether owning a portfolio of both Stock A and Stock B might be less risky than owning only one of the two stocks. For this, you need the covariance in addition to the standard deviations of the returns.

5.4.3 Covariance Cov[Y₁,Y₂] and Correlation Coefficient ρ_{12}

Joint probability functions are required to calculate the covariance and the correlation coefficient. The covariance of Y_1 and Y_2, denoted by $Cov[Y_1, Y_2]$, is a measure of joint variation of the two random returns. Unlike the variance, a covariance might be positive or negative depending on the relationship between Y_1 and Y_2. The easiest way to calculate the covariance of two random variables Y_1 and Y_2 is by using $Cov[Y_1,Y_2] = E[Y_1 \cdot Y_2] - E[Y_1] \cdot E[Y_2]$, where $E[Y_1 \cdot Y_2] = \sum_{\text{all } y_1} \sum_{\text{all } y_2} y_1 y_2 P[y_1, y_2]$ is called the crossed expectation between Y_1 and Y_2. $E[Y_1]$ and $E[Y_2]$ are the means of the Y_1 and Y_2 distributions respectively. A second formula, used mostly for insight and less often for computations, is $Cov[Y_1,Y_2] = E[(Y_1-\mu_1) \cdot (Y_2-\mu_2)]$. This equivalent formula shows that the covariance measures the joint variation of (Y_1, Y_2) around the two respective means $\mu_1 = E[Y_1]$ and $\mu_2 = E[Y_2]$. The unit of the covariance is the product of the units of (Y_1, Y_2). In the stocks example, the units of both stock returns are in dollars, so the units of the covariance are dollars squared. Crossed units make covariances difficult to interpret across applications, because the (Y_1, Y_2) might have different ranges and measurement units. To resolve this problem, you calculate the correlation coefficient. It can be thought of as a standardized covariance. It measures linear association between two random variables.

The correlation coefficient between Y_1 and Y_2 is the *covariance between* (Y_1, Y_2) *divided by their two standard deviations:*

$$Corr[Y_1, Y_2] = \rho_{12} = \frac{Cov[Y_1, Y_2]}{SD[Y_1] \cdot SD[Y_2]} \text{ with } -1 \le \rho_{12} \le +1.$$

The Greek symbol ρ is called rho. If $\rho_{12} = 0$, then there is no linear association between Y_1 and Y_2. When $\rho_{12} = -1$ or $\rho_{12} = +1$, the linear association is perfect. Division by the standard deviations $SD[Y_1]$ and $SD[Y_2]$ ensures that the correlation coefficient is a unitless quantity. This characteristic allows comparison of correlation coefficients from a variety of applications with originally altogether different measurement units.

Figure 5.8 Concentration Ellipses for Positive and Negative Correlation

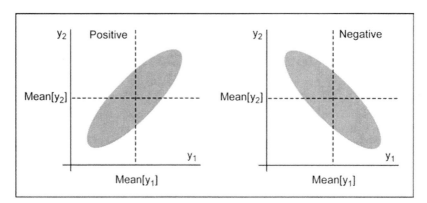

The concentration of the joint probability in the (Y_1, Y_2) plane determines whether the covariance or correlation are positive or negative, as shown in Figure 5.8. The two panels are more appropriate for correlations between continuous random variables, but suggest the nature of correlation in the discrete case as well. The left graph shows a positive correlation. As the value of one variable increases, the other variable tends to increase as well. Many common stocks are positively correlated. The right graph shows a negative correlation. As the value of one variable increases, the other variable tends to decrease.

5.4.4 Calculating E[Y₁], E[Y₂], E[Y₁²], E[Y₂²] and E[Y₁·Y₂]

For manual calculations, use a spreadsheet approach to calculate $E[Y_1]$, $E[Y_2]$, $E[Y_1^2]$, $E[Y_2^2]$ and $E[Y_1 \cdot Y_2]$ from the joint probability distribution. In this example, for the calculations of means and standard deviations, expand the 3-by-3 table of joint probabilities into a table of 9 rows. Two columns identify the combination of returns of each stock; the third column lists the corresponding joint probabilities. To these three columns add 5 more calculated from the first three. The headings indicate the kind of row operation: $y_1 P[y_1,y_2]$ multiplies y_1 by $P[y_1,y_2]$, $y_1^2 P[y_1,y_2]$ multiplies y_1^2 by $P[y_1,y_2]$, and $y_1 \cdot y_2 P[y_1,y_2]$ multiplies y_1 by y_2 by $P[y_1,y_2]$. The column sums are given in the last row of Table 5.3. They are the sums of the values of each function multiplied with $P[y_1,y_2]$ and represent the expected values of these functions.

$E[Y_1] = \text{Sum}\{y_1 \cdot P[y_1,y_2]\}$, $E[Y_1^2] = \text{Sum}\{y_1^2 \cdot P[y_1,y_2]\}$, and $E[Y_1 \cdot Y_2] = \text{Sum}\{y_1 \cdot y_2 \cdot P[y_1,y_2]\}$.

The sums involving Y_2 are similar.

Table 5.3 Detail Calculations for Expected Values

Stock A=y_1	Stock B=y_2	$P[y_1,y_2]$	$y_1 P[y_1,y_2]$	$y_2 P[y_1,y_2]$	$y_1^2 P[y_1,y_2]$	$y_2^2 P[y_1,y_2]$	$y_1 \cdot y_2 P[y_1,y_2]$
−10	−10	0.05	−0.5	−0.5	5	5	5
0	−10	0.1	0	−1	0	10	0
10	−10	0.15	1.5	−1.5	15	15	−15
−10	0	0.05	−0.5	0	5	0	0
0	0	0.2	0	0	0	0	0
10	0	0.15	1.5	0	15	0	0
−10	10	0.25	−2.5	2.5	25	25	−25
0	10	0	0	0	0	0	0
10	10	0.05	0.5	0.5	5	5	5
Column Sums		1	0	0	70	60	−30

The Column Sums of Table 5.3 can be summarized as follows:

- The average return on both stocks is 0 dollars, i.e., $E[Y_1] = 0$, $E[Y_2] = 0$.
- The expectations of the squared returns are $E[Y_1^2] = 70$ and $E[Y_2^2] = 60$ dollars squared.
- The crossed expectation between Stock A and Stock B is $E[Y_1 \cdot Y_2] = -30$ dollars squared.
- From these quantities you can calculate the variances $V[Y_1] = 70 - 0 = 70$ and $V[Y_2] = 60 - 0 = 60$ dollars squared and the respective standard deviations $SD[Y_1] = 8.3666$ and $SD[Y_2] = 7.7460$ dollars.
- The covariance $Cov[Y_1, Y_2] = -30 - (0)\cdot(0) = -30$ dollars squared, based on $E[Y_1 \cdot Y_2] - E[Y_1] E[Y_2]$.
- The unitless correlation coefficient is $Corr[Y_1, Y_2] = \rho_{12} = -30/[8.3666 \cdot 7.7460] = -0.4629$, a negative correlation.

5.4.5 Variance and Standard Deviation of a Portfolio of Two Stocks

This section shows how to calculate the variance or the standard deviation of the return on a portfolio of two stocks. The standard deviation is a good measure of relative risk of the portfolio. A portfolio with a smaller standard deviation of returns has less risk than a portfolio with a relatively higher standard deviation. The calculations require the means, standard deviations, and covariance (or correlation coefficient) of the returns of the two stocks.

As an example, consider an investor with a fixed amount of money to invest, say 10,000 dollars, who needs to make a decision as to how much of each stock to buy. The investor could buy only Stock A or only Stock B or a combination of both stocks. Assume that the investor decides to use a proportion, a, of the 10,000 dollars to buy Stock A and the complementary proportion $(1 - a)$ to buy Stock B. What is the standard deviation of the return on this portfolio of two stocks?

Designate the actual returns of Stock A and B by Y_1 and Y_2 respectively. In order to find the standard deviation of a portfolio of two stocks, you need to be able to find the expected returns of each stock $E[Y_1]$ and $E[Y_2]$, the standard deviations of the returns $SD[Y_1]$ and $SD[Y_2]$, and the covariance $Cov[Y_1, Y_2]$.

The expected return of a portfolio of two stocks is

$$E[aY_1 + (1-a)Y_2] = aE[Y_1] + (1-a)E[Y_2]$$

The variance of the return of the two stocks is as follows:

$$V[aY_1 + (1-a)Y_2] = a^2 V[Y_1] + (1-a)^2 V[Y_2] + 2a(1-a)Cov[Y_1, Y_2]$$

where in this case $0 \le a \le 1$ is the proportional allocation to Stock A. The standard deviation is the square root of the variance, i.e.,

$$SD[aY_1 + (1-a)Y_2] = \sqrt{V[aY_1 + (1-a)Y_2]}.$$

The preceding example has the expected returns of each stock at 0, so the combined return of the portfolio is also 0. The variances and covariance are $V[Y_1] = 70$, $V[Y_2] = 60$, and $Cov[Y_1, Y_2] = -30$. A portfolio that allocates exactly 50% of the total funds to each stock has a standard deviation

$$SD[0.5Y_1 + 0.5Y_2] = \sqrt{0.25 \cdot 70 + 0.25 \cdot 60 + 2 \cdot 0.5 \cdot 0.5 \cdot (-30)} = \sqrt{17.5} = 4.183.$$

The standard deviation of the portfolio of the two stocks is 4.183 dollars. That is considerably less than the standard deviations of either Stock A $(SD[Y_1] = 8.37)$ or B $(SD[Y_2] = 7.75)$ alone. A smaller standard deviation of returns indicates a smaller risk. A portfolio with equal allocation of funds to stocks A and B has a smaller risk than one that invests in either stock alone.

5.5 Summary

- Conceptual distributions are derived from theoretical considerations using assumptions about the random outcomes.

- Sampling distributions are for statistical measures such as means, standard deviations, and similar statistics.
- Distributions can be continuous or discrete, depending on the type of the outcomes.
- Discrete distributions have integer numbers as outcomes. Continuous distributions use real numbers as outcomes.
- Discrete distributions are characterized by their individual probability functions $p(y) = P[Y = y]$ and cumulative distribution functions $F(y) = P[Y \leq y]$.
- Distributions are distinguished by the shape of their probability functions, their measures of location, and their measures of spread (or variation).
- Distributional shape is symmetric or asymmetric, unimodal or multimodal, with finite or infinite range.
- The mean and the standard deviation have the same units. If the mean is measured in dollars, the standard deviation is also measured in dollars.
- Means and standard deviations can be calculated for the sum and weighted averages of several random distributions.
- The binomial distribution is a discrete distribution useful in quality methods and polling.
- The risk of an investment can be measured by the standard deviation of its returns.
- The expected return of a security is the mean return.
- A marginal probability of the return of a stock is the probability that a stock has a specific return without regard to the return of the other stock.
- The correlation coefficient can be viewed as a standardized covariance. It measures linear association between two outcome variables such as the returns of two stocks.
- Positive correlation means that the two outcomes tend to increase or decrease in tandem.
- Negative correlation means that outcomes of the two variables tend to move in opposite directions.
- The covariance or the correlation coefficient between two securities can be used, together with their standard deviations, to calculate the risk of a portfolio.
- The variance of a portfolio of two stocks is usually smaller than the variance of either of the individual stocks. Thus a portfolio of two stocks is less risky than holding either of the stocks individually.

5.6 Problems

1. For each of the following processes, identify the random variable and the outcome space.

   a. An observer posted at a specified location on a lake counts the number of waterfowl within his field of view during a one-hour period.

   b. A doctor reviews the outcome of a strep throat test.

   c. An operator counts the number of vehicles that pass through a toll booth in an eight-hour period.

   d. A candy company produces individually wrapped mints. The quality control inspector removes and counts the pieces where the wrapping is not completely sealed.

2. The number of small cosmetic defects found in a manufactured product is given by the following probability distribution. Enter this probability distribution into a JMP datasheet. Create the necessary columns and formulas in JMP to calculate the average, variance, and standard deviation of the number of defects.

Y	P(Y=y)
0	0.30
1	0.35
2	0.25
3	0.05
4	0.05

3. Calculate the mean, variance, and standard deviation of the following distribution of financial returns.

Economy	Probability Function p(y)	Return Y in 1000 $	$y*p(y)$	$(y - Mean[Y])^2*p(y)$
Recession	0.30	−15		
Slow growth	0.40	15		
Boom	0.30	30		
SUM	1.0	n.a.	Mean[Y] =	V[Y] = SD[Y] =

4. An office supply retailer is planning the grand opening of a new store. Flyers containing information about the grand opening event and coupons for three complementary items are mailed to households within a 25-mile radius. The regional manager believes there is an equally likely chance that a customer will redeem none, one, two, or three coupons. Consider that the random variable Y = the number of coupons redeemed by a customer.

   a. Sketch the probability function.

   b. Sketch the cumulative distribution function.

   c. What is the probability that a customer will redeem at least one coupon?

5. Periodically, a warehouse club randomly mails out coupons for a free three-month membership. A household has a 20% chance of receiving the coupon. An apartment building has 10 apartments, all of which are occupied.

   a. What is an appropriate discrete distribution for the random variable Y?

   b. Use JMP to find the individual probability distribution function for Y, the number of apartments that receive the coupon in a given mailing.

   c. Use JMP to find the cumulative distribution function.

   d. What is the probability that none of the apartments receive a coupon?

   e. What is the probability that either one or two of the apartments receive the coupon?

   f. What is the probability that three or more apartments receive the coupon?

6. Every hour, five recorded conversations from a financial institution's telephone help line are reviewed to determine if customer concerns were adequately addressed. Historically, 20% of such calls have been judged to provide inadequate information to the customer.

   a. Find the probability that all of the conversations in an hourly sample provided adequate resolution to customer concerns.

   b. Find the probability that at most one conversation in the hourly sample provided inadequate resolution.

7. The joint probability distribution shown below quantifies a financial analyst's belief in the possible outcomes for two stocks—one from the health care sector and the other from the banking industry.

	Returns from health care stock		
Returns from bank stock	**−5**	**0**	**5**
−5	.15	.15	.05
0	.10	.20	0
5	.05	.05	.25

a. Set up a table to compute the means, variances, and covariance, and calculate the quantities needed to determine the portfolio variance.

b. What are the expected returns and risk for each of the individual stocks?

c. Are the returns of these two stocks independent? Explain how you arrived at your answer.

d. Compute the standard deviation for a portfolio with a 50% allocation of each stock. How does it compare to the standard deviations of the two individual stocks?

8. An investor is considering changes to her asset portfolio. Since she is approaching retirement, she would like to minimize risk. Of particular concern are two stocks from the personal products market. The joint probability distribution is given in the following table. Find the expected returns and risk for the two individual stocks and a portfolio with equal allocation of the two stocks. Based on your results, make a recommendation as to how the investor should adjust the portfolio that is consistent with her investment goals. Explain how you arrived at your answer.

	Returns from Stock 1	
Returns from Stock 2	**−5**	**5**
−4	.05	.35
6	.25	.35

5.7 Case Study: Assessing Financial Investments

Background

In the context of investing, probability models can assist with the evaluation of risk and expected return. Investors often consider both the expected return and risk when making investment choices. The mean of the distribution is the expected return, which represents the return of the stock over a long time horizon. The standard deviation (square root of the variance) measures the risk associated with the stock. Lower variability in the returns indicates less risk as compared to a return with higher variability.

Business Problem

Two stocks are available in which to invest. The returns and associated probabilities for possible economic conditions are given in the following tables.

Returns from Woodside Corporation

Economic Condition	Probability Function *f*(y)	Return Y in %
Recession	0.1	−10
No growth	0.2	−1
Slow growth	0.5	5
Boom	0.2	12

Returns from Brookside Corporation

Economic Condition	Probability function *f*(y)	Return Y in %
Recession	0.1	−2
No growth	0.2	0
Slow growth	0.5	3
Boom	0.2	6

Based on these probability distributions, find the expected return and risk for each of the two stocks. Compare their expected returns and risk in light of investor risk preferences (i.e., risk seeking, risk aversion). Summarize your findings in a one-page white paper. White papers are reports for purposes such as educating customers or demonstrating successful applications of a company's products. White papers should be free of jargon, short in length (a page or two), and easily understood by the average consumer.

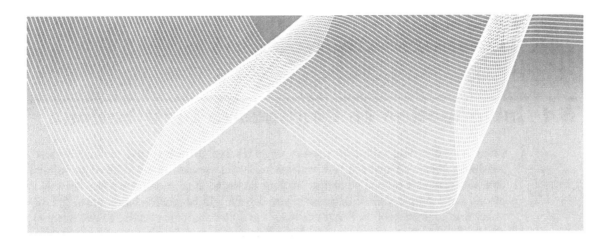

Chapter **6**

Continuous Probability Distributions

6.1 Introduction to Continuous Distributions   132

6.2 Characteristics of Continuous Distributions   132

    6.2.1 Density Function f(y)   132

    6.2.2 Cumulative Distribution Function F(y)   133

    6.2.3 General Shape of Continuous Distributions   133

    6.2.4 Measures of Location and Spread   134

6.3 Uniform Distribution   135

6.4 The Normal Distribution   136

    6.4.1 General Normal Distribution with Mean[Y] = μ and SD[Y] = σ 136

    6.4.2 Standard Normal Distribution   139

6.5 Central Limit Theorem   146

6.6 Sampling Distributions   148

    6.6.1 Student's t-Distribution   149

    6.6.2 Chi-Square Distribution   150

    6.6.3 F-Distribution   151

6.7 Summary   152

6.8 Problems   154

6.9 Case Study: Julie's Lakeside Candy   156

6.1 Introduction to Continuous Distributions

Continuous distributions are for random variables that, in theory, have real numbers as possible outcomes. However, few business variables use the full potential of the infinite precision provided by real numbers. They are recorded to no more than a few decimal places. For example, failure times on a hard drive might be recorded as hours to failure rather than in nanoseconds. Other examples are dollar amounts rounded to two decimal places, physical dimensions measured to a precision of 0.001 inches, length of cracks given in one-tenth of an inch, and strength of a material in pounds per square inch. Even though real numbers are not required to model these variables, in the end they are used because these models are more convenient to use than equivalent discrete ones.

Many sampling distributions, i.e., the distributions of test statistics and statistical estimators, are continuous. Examples of continuous distributions are the uniform, Normal (bell-shaped curve), Student's t, F, and Chi-Square. For continuous distributions, calculus-based techniques are required to calculate probabilities. Fortunately, tables or functions such as those in the Formula Editor of JMP minimize these tasks.

6.2 Characteristics of Continuous Distributions

Continuous distributions arise from random variables with an outcome in some interval of real numbers. They are characterized by a density function f(y) and the cumulative distribution function $F(y) = P[Y \leq y]$. Continuous distributions have measures of location (mean, median) and spread (standard deviation, variance). Their shapes are associated with the density function and can be skewed or symmetric, bounded or unbounded, unimodal or multimodal.

6.2.1 Density Function f(y)

Instead of individual probability functions such as $P[Y = y]$, continuous random variables have density functions f(y). Density functions do not measure probabilities. Instead, the area under the density function of the interval (a, b) measures the probability of an outcome between a and b, i.e., P[a $\leq$ Y $\leq$ b]. This area is the integral of the density from a to b. With a continuous random variable P[Y = y] = 0, because the area under the interval (y, y) is 0. A density function is never negative and the area under the density of the entire outcome space equals 1, ensuring that the total probability is 1. The density function characterizes the shape of a distribution.

- The uniform distribution has a flat density over the entire range of admissible values, i.e., the interval (a, b).

- The Normal distribution has a bell-shaped density ranging from $-\infty$ to $+\infty$.

- The exponential distribution has an exponentially decaying density function from 0 to ∞.

- All continuous distributions in this text have density functions.

6.2.2 Cumulative Distribution Function *F*(y)

The cumulative distribution function $F(y_i) = P[Y \leq y_i]$ is the area under f(y) for all values less than and including y_i. As with discrete random variables, the value of $F(y_i)$ is always between 0 and 1, i.e., $0 \leq F(y) \leq 1$. $F(y)$ is non-decreasing for y, meaning that for two values $y_1 < y_2$, $F(y_1) \leq F(y_2)$. Also, $F(-\infty) = 0$ and $F(\infty) = 1$.

The permissible outcome sets of continuous random variables are intervals of the form (a, b). The area under the density for the interval (a, b) is P[a $\leq$ Y $\leq$ b] = $F(b) - F(a)$, where a < b. This formula is useful in calculating probabilities of outcome sets. Distribution functions are also used to calculate distribution quantiles (percentiles).

6.2.3 General Shape of Continuous Distributions

The shape of a continuous distribution is associated with its density. Depending on the shape of the density f(y), distributions of a single random variable are either unimodal or multimodal. Unimodal distributions have a single mode and are either symmetric or asymmetric. Asymmetric unimodal distributions are either left-skewed or right-skewed.

Bounded distributions have a limited outcome space, i.e., the random variable Y will be at least a and not more than b, where a and b are two finite numbers with a < b. Unbounded distributions are unconstrained at the lower or upper ends, or both. The Normal distribution (bell-shaped curve) has an unbounded outcome space from minus infinity to plus infinity ($-\infty < y < +\infty$).

A continuous distribution is symmetric if the density function f(y) is symmetric. In a symmetric unimodal distribution, the mean, median, and mode are the same. The center panel of Figure 6.1 shows the density f(y) of a (continuous) symmetric distribution. A distribution is skewed to the left (shown in the left panel), if the long (thin) end of the density f(y) is on the left side. In unimodal left-skewed distributions, the mode is greater than the median, and the median is greater than the mean. A distribution is skewed to the right (shown in the right panel), if the long (thin) end of the density f(y) is on the right side. In continuous unimodal right-skewed distributions, the mode is less than the median and the median is less than the mean.

Figure 6.1 Shapes of Unimodal Distributions

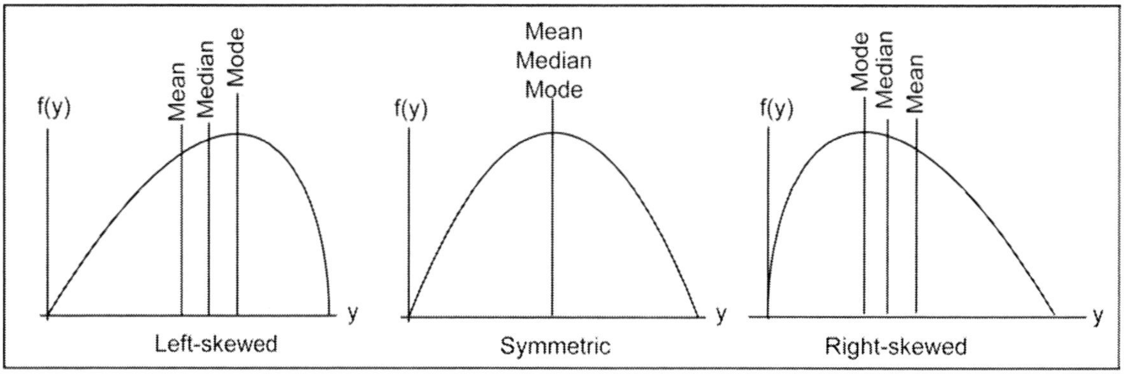

6.2.4 **Measures of Location and Spread**

As with discrete distributions, the common measures of location are the mean and the median. However, given that continuous distributions use density functions, the mean is the center of gravity not of the probabilities, but of the density. The median splits the area under the density into two, with each half having probability 0.5. This split of probabilities might not result in a unique median. The mode is the outcome with the highest density. It is quite possible that distributions have more than one mode. Distributions with a single mode are called unimodal; those with more than one mode are called multimodal.

The most useful measure of spread is the standard deviation. It has the same measurement units as the mean. Its interpretation for the Normal distribution will be discussed later. It can be used in connection with the empirical rule, which holds approximately for unimodal symmetric distributions. It says that approximately 68% of probability is within ± one standard deviation, 95% within ± two standard deviations, and 99% within ± three standard deviations of the mean. The empirical rule must be used with caution when the distribution is not normal.

Mean, Variance, and Standard Deviation: The Mean[Y] is the weighted average of the y_i, where the weight of each y_i is the density $f(y_i)$. In analogy to discrete random variables, the mean of a continuous variable is calculated as the integral over all outcomes of the function $(y_i \cdot f(y_i))$. The variance Var[Y] can be calculated by formulas analogous to those of discrete random variables, except that integral calculus is required. The standard deviation SD[Y] is the square root of the variance $SD[Y] = \sqrt{Var(Y)}$. Closed form expressions are available for many useful distributions.

6.3 Uniform Distribution

The uniform distribution is widely used in computer simulation. It has a sample space bounded by its smallest value a and its largest value b. The standard range of the uniform distribution is the interval from a = 0 to b = 1. In general, a and b are both finite numbers with a < b. Figure 6.2 shows the uniform distribution on the interval from 0 to 1. The left panel shows the density f(y), which is constant at 1 over the entire interval and zero outside. This ensures that the total area under the density is exactly 1. The right panel shows that the distribution function F(y) is always between 0 and 1, because F(y) represents a probability and is never less than 0 or greater than 1.

Figure 6.2 Uniform Distribution on (0,1); Density f(y) and Distribution Function F(y)

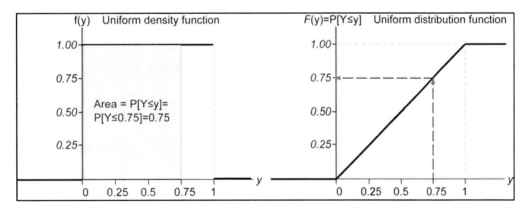

The formula for the distribution function of the uniform distribution is F(7) = y for $0 \le y \le 1$. In order to read a value of F(y) and see its relationship to the area under f(y), do the following: On the horizontal y-axis, mark the value .75. The area under f(y) for 0 < y < 0.75 is 0.75, representing ¾ of the total probability. The distribution function value for y = 0.75 is F(.75)=0.75 as shown by the dashed line in the right panel of Figure 6.2. For the standard uniform with a = 0 and b = 1, the Mean[Y] = 1/2 and standard deviation is $SD[Y] = 1/\sqrt{12} = 0.289$.

6.4 The Normal Distribution

The Normal distribution is the most important probability distribution, because of its use as a sampling distribution and because many random phenomena can be modeled with a Normal distribution. The Normal distribution is applicable to physical measurements such as the weights of adults or blood pressure, as well as to psychological measurements such as intelligence quotients. The Normal distribution is used in many statistical techniques. The reason for its importance can be attributed in large part to the central limit theorem.

6.4.1 General Normal Distribution with Mean[Y] = μ and SD[Y] = σ

The general Normal distribution has two parameters, the mean and the standard deviation. The mean is Mean[Y] = μ and can be positive or negative. The standard deviation SD[Y] = σ is always positive. [Note: μ and σ are Greek letters called mu and sigma.]

Figure 6.3 Comparison of Two Normal Densities with the Same Mean and Different Standard Deviations

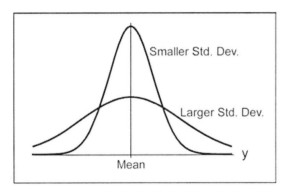

Figure 6.3 shows two superimposed Normal densities with the same mean, but with different standard deviations. The smaller standard deviation results in a tighter distribution. Normal densities are symmetric. Random outcomes have a theoretical range from $-\infty$ to $+\infty$. However, on average all but about 1 in a billion outcomes will occur within a range of the mean plus or minus six standard deviations.

Standardizing Random Variables

Because there are infinitely many combinations of means and standard deviations for the Normal distribution, the general Normal random variable Y is transformed into the standardized random variable Z. One such standardization from Y to Z is depicted in Figure 6.4. The left panel shows the density function of a general Normal distribution with mean = μ and standard deviation = σ. The right panel shows the density function of a standard Normal distribution with mean = 0 and standard deviation = 1. Standardization allows conversion of any general Normal distribution to a standard Normal distribution. Standardizing Y is a two-step process:

1. Subtract the mean from the random variable, i.e., $Y - \mu$. The mean of $(Y - \mu)$ is Mean$[Y - \mu] = 0$. The standard deviation σ has not been affected by subtracting the mean.

2. Divide $(Y - \mu)$ by the standard deviation σ to obtain

$$Z = \frac{Y - \mu}{\sigma}$$

The mean of Z is Mean$[Z] = 0$ and the SD$[Z] = 1$. The result of the completed standardization is shown on the right in Figure 6.4. The inverse $Y = \mu + Z \cdot \sigma$ can be used to express any Y as a deviation from its mean μ in terms of the number of standard deviations.

Figure 6.4 General Normal and Standard Normal Densities

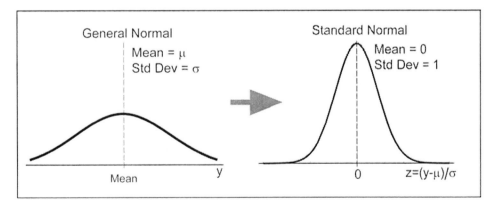

Example: Lifetime of Light Bulbs

The following example illustrates the standardization process, a necessary step in finding Normal distribution probabilities. Suppose the useful life Y of a new light bulb has a Normal distribution with mean $\mu = 10000$ hours and a standard deviation $\sigma = 1500$ hours. Find the standardized Z values for various Y. In Table 6.1, the first row and first column shows a useful life for a light bulb Y = 7000. The second column shows the standardization to the standard Normal Z-scale by subtracting the mean and dividing by the standard deviation. The third column shows the Z value corresponding to the Y of column 1. In row 1, the correspondence is between Y = 7000 hours and Z = –2. The Y-scale is in hours, while the Z-scale has no units. The fourth column of this row expresses Y = 7000 hours as a value that is two standard deviations ($-2 \cdot 1500 = -3000$ hours) below the mean of 10000 hours.

Table 6.1 Relationship between General Normal Y and Standard Normal Z

Original Y	$Z = (Y - \mu)/\sigma$	Standardized Z	$Y = \mu + Z \cdot \sigma$
7000	$-2 = \dfrac{7000 - 10000}{1500}$	–2	$10000 - 2 \cdot 1500$
8000	$-1.33 = \dfrac{8000 - 10000}{1500}$	–1.33	$10000 - 1.33 \cdot 1500$
10000	$0 = \dfrac{10000 - 10000}{1500}$	0	$10000 - 0 \cdot 1500$
11500	$1 = \dfrac{11500 - 10000}{1500}$	1	$10000 + 1 \cdot 1500$
12000	$1.33 = \dfrac{12000 - 10000}{1500}$	.33	$10000 + 1.33 \cdot 1500$
14000	$2.67 = \dfrac{14000 - 10000}{1500}$	2.67	$10000 + 2.67 \cdot 1500$

6.4.2 **Standard Normal Distribution**

The standard Normal distribution has a Mean[Z] = 0 and a standard deviation SD[Z] = 1. The standard Normal distribution is often denoted by N(0,1). The left panel of Table 6.2 shows the area under the Normal density for $Z \leq 1$ representing the probability $P[Z \leq 1]$. However, to calculate the probability, we use the distribution function $F(z)$ with its characteristic S-shape (on the right), available as the Normal Distribution function from the JMP Formula Editor or for selected Z values from Table 6.2. For $P[Z \leq 1]$, use the distribution function $F(z)$ graph, select the value of Z = 1 on the horizontal axis, and read the corresponding value $F(1) = P[Z \leq 1] = 0.84134$ on the vertical axis. Tables and functions work similarly, as will be shown next.

Figure 6.5 Standard Normal Density f(z) and Distribution Function F(z)

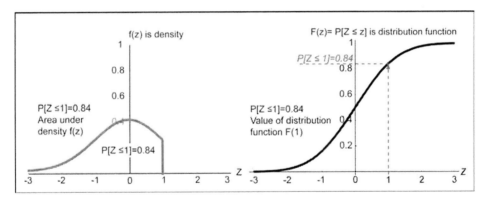

Table 6.2 shows the three frequently used Normal probabilities. In column 2, $P[Z \leq -z_0]$ = $P[Z > z_0] = 1 - P[Z \leq z_0]$ gives the distribution function for negative z_0 and is also the complement of the distribution function. Column 3, $P[Z \leq z_0]$ is the distribution function, and column 4 gives $P[-z_0 < Z < z_0] = 1 - 2 \cdot P[Z > z_0]$, which is the probability of the center area.

Table 6.2 Some Convenient Standard Normal Tables

z_0	$P[Z \leq -z_0] = P[Z > z_0]$	$P[Z \leq z_0]$	$P[-z_0 < Z < z_0]$
3.090	.001	.999	.998
3.000	.00135	.99865	.9973
2.667	.00383	.99617	.99234
2.576	.005	.995	.99
2.326	.01	.99	.98
2	.02275	.97725	.9545
1.960	.025	.975	.95
1.645	.05	.95	.90
1.5	.0668	.9332	.8664
1.333	.0913	.9087	.8175
1.282	.1	.9	.8
1	.15866	.84134	.68269
.8416	.2	.8	.6
.5	.3085	.6915	.3829
0	.5000	.5000	0

Recall that in the previous example, the useful life of a new type of light bulb is normally distributed with a mean $\mu = 10000$ hours and a standard deviation $\sigma = 1500$ hours. First, calculate the standardized Z values. These are used to look up probability values in the Normal distribution table of Table 6.2. Find the following probabilities:

- That a light bulb will fail before 7000 hours
  Answer: $Z = [7000 - 10000]/1500 = -2$ and $P[Y < 7000] = P[Z \leq -2] = .02275$

- That a light bulb will last more than 13000 hours
  Answer: $Z = [13000 - 10000]/1500 = +2$ and $P[Y > 13000] = P[Z > +2] = .02275$

- That a light bulb will last more than 14000 hours
  Answer: $Z = [14000 - 10000]/1500 = +2.667$ and $P[Y > 14000] = P[Z > 2.667] = .00383$

- That a light bulb will fail between 8000 and 12000 hours
  Answer: $P[8000 < Y < 12000] = P[-1.333 < Z < +1.333] = P[Z \leq 1.33] - P[Z \leq -1.33] = 0.9087 - 0.0913 = 0.8174$

Calculating Normal Probabilities in JMP

The Formula Editor in JMP enables computation of both the Normal distribution function and the quantile function. For continuous distributions, only the cumulative distribution function is needed to find probabilities. In order to calculate Normal probabilities, proceed as follows:

1. Create a new data table from the JMP **File** menu with four columns. One column will contain the values of the random variable Y needed for the probability calculation, i.e., 7000, 13000, 14000, 8000, 12000. The second column calculates the corresponding standardized Z values (Figure 6.6). The third column calculates the distribution function $F(z) = P[Z \leq z]$. In case the probability $P[Z > z]$ is required, it is added in a fourth column. The final results are in Figure 6.9.

2. In the second column, perform the standardization from Y to Z using a mean of 10000 hours and a standard deviation of 1500 hours. First, right-click the **Z** column in the **Table Columns** list, and select **Formula** from the resulting menu, as shown in Figure 6.6. Then use the Formula Editor to create the standardization formula.

$$\frac{Y - 10000}{1500}$$

Enter the variable name Y in the formula by clicking **Y** in the **Table Columns** list, and then subtract 10000 and divide by 1500. Click **Apply** to calculate the associated Z values.

Figure 6.6 Formula Window for Z-Variable

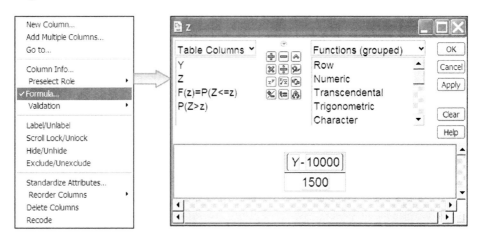

3. Right-click column **F(z) = P[Z <= z]** to bring up another formula window. To calculate the probabilities for P[Z ≤ z], the Normal Distribution function is used. Figure 6.7 shows how to find the function from the **Probability** submenu of the **Function** menu. In JMP, Normal Distribution calculates probabilities P[Z ≤ z] for standard Normal variates z. The function has only one argument. Enter the variable name Z by selecting it from the **Table Columns** list. Click **Apply** to generate the probabilities in the data table.

Figure 6.7 Formula for P[Z ≤ z]

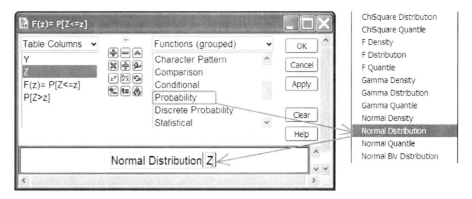

4. Column P[Z > z] requires that F(z) = P[Z ≤ z] be subtracted from 1. The formula window for the column P[Z > z] is shown in Figure 6.8.

Figure 6.8 Formula for P[Z > z]

1 - Normal Distribution(Z)

The completed table with Y, Z, P[Z ≤ z] and P[Z > z] is shown in Figure 6.9.

Figure 6.9 Normal Probabilities for Life of Light Bulbs Example

LightbulbsNormal

	Y	Z	F(z) = P(Z<=z)	P(Z>z)
1	7000	-2	0.022750	0.977250
2	13000	2	0.977250	0.022750
3	14000	2.66666	0.996170	0.003830
4	8000	-1.33333	0.091211	0.908789
5	12000	1.33333	0.908789	0.091211

Columns (4/0): Z, F(z) = P(Z<=z); Rows

Figure 6.9 reveals that when Z = –2, P[Z ≤ –2] = 0.02275 and P[Z > –2] = 0.97725. As required by the rule of complementation, these two probabilities sum to 1. Probabilities calculated in JMP might have minor discrepancies with values calculated from tables due to rounding. The probabilities, calculated previously using tables, are as follows when obtained from JMP:

- That a light bulb will fail before 7000 hours:
  P[Y ≤ 7000] = P[Z ≤ –2] = .022750
- That a light bulb will last more than 13000 hours:
  P[Y > 13000] = P[Z > +2] = .022750
- That a light bulb will last more than 14000 hours:
  P[Y > 14000] = P[Z > 2.66666] = .003830
- That a light bulb will fail between 8000 and 12000 hours:
  P[8000 < Y < 12000] = P[–1.33333 < Z < +1.333] = P[Z ≤ 1.33333] – P[Z ≤ –1.33333] = 0.90879 – 0.09121 = 0.81758

Calculating Normal Distribution Quantiles in JMP

The cumulative distribution function (CDF) gives probabilities as a function of the random variable Y as shown by $F(y) = P[Y \le y]$. The quantile function is the inverse of the CDF, as shown by $y_p = F^{-1}(p)$. By choosing a probability p, you calculate the quantile y_p of the random variable Y such that it satisfies $p = F(y_p)$. The p-quantile y_p is the value of Y such that the probability of being less than or equal to y_p is p.

Several quantiles are important enough to have their own name. The median is the 50% quantile. The lower quartile is the 25% quantile, while the upper quartile is the 75% quantile. The lower decile is the 10% quantile; the upper decile is the 90% quantile. Quantiles are also called percentiles. For example, the lower decile corresponds to the 10th percentile and the upper quartile to the 75th percentile.

In the light bulb life example, quantiles could be used in warranty statements. For example, the probability that a light bulb fails by time $y_{0.01}$, the 1% quantile or first percentile, is 0.01. On average, 1 percent of light bulbs will fail by $y_{0.01}$. With an average life of 10000 hours and the standard deviation of 1500 hours, that value is $y_{0.01} = 6510.48$ hours. The following steps are required to obtain that value in JMP:

1. Start a new data table with two columns. Assign the column headings p to column 1 and p-Quantile to column 2. As shown in Figure 6.10, column 1 contains the probability 0.01, while column 2 will contain the calculated quantile. The final results are in Figure 6.11.

2. Right-click the variable name **p-Quantile** and select **Formula** from the resulting menu. This action opens the JMP Formula Editor. In the **Functions** list, right-click **Probability** and select **Normal Quantile** from the menu. This action enters the function Normal Quantile() into the formula field, but without the argument.

3. The argument is entered by clicking **p** in the **Table Columns** list. As shown in Figure 6.10, this results in the standard Normal quantile. Then multiply the function by the standard deviation (1500) and add the mean (10000). In the formula window, click **Apply**.

Figure 6.10 Formula for Normal p-Quantile with Mean 10000 and Standard Deviation 1500

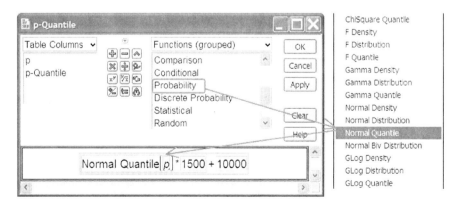

Review the final result in the data table as given in Figure 6.11. The 1% quantile is 6510.478 hours for Normal distribution, with mean = 10000 hours and standard deviation = 1500 hours. For other means and standard deviations, the formula has to be modified accordingly.

Figure 6.11 1%-Quantile for Light Bulb Example

	p	p-Quantile
1	0.01	6510.47819

Distribution of the Mean of Normally Distributed Outcomes

A very common statistical operation is to take averages of random outcomes. In statistics, the sample mean is the simple average of the outcomes of a sample. It is often used to estimate the mean of a distribution. In order to evaluate how far the sample estimate can be from the actual mean, the distribution of the sample mean is needed.

What is the distribution of the mean of n outcomes from a Normal distribution with Mean[Y] = μ and Standard Deviation[Y] = σ? The mean of n independent outcomes Y_1, $Y_2, \ldots, Y_n$ is $\overline{Y} = \dfrac{Y_1 + Y_2 + \cdots + Y_n}{n}$

The random variable $\overline{Y}$ is called the sample mean.

It can be shown that the distribution of $\overline{Y}$ is another Normal distribution with

$$\text{Mean}[\overline{Y}] = \text{Mean}[Y_i] = \mu \text{ and}$$
$$\text{Standard Deviation}[\overline{Y}] = \frac{\text{Standard Deviation}[Y_i]}{\sqrt{n}} = \frac{\sigma}{\sqrt{n}}$$

The standard deviation of the sample mean distribution can be viewed as a measure of precision of the sample mean as an estimate of the actual population mean. It is smaller than the standard deviation of an individual observation by a factor $1/\sqrt{n}$. As the sample size increases, the standard deviation of the sample mean distribution gets smaller depending on the square root of the sample size. As the sample size approaches infinity (gets very large), the standard deviation of the sample mean distribution goes to zero.

In calculating probabilities relating to a distribution of a mean, you have to standardize with $Z = \dfrac{\overline{Y} - \mu}{\sigma/\sqrt{n}}$ and $\overline{Y} = \mu + Z \cdot \dfrac{\sigma}{\sqrt{n}}$

Using the light bulb life example in which individual light bulbs have a mean life of 10000 hours with a standard deviation of 1500 hours, calculate the probability that the average life of four light bulbs is less than 8500 hours. To answer this question, note that $n = 4$, $\mu = 10000$, and $\sigma = 1500$. The probability that the average life of four light bulbs is less than or equal to 8500 hours is

$$P\left[\overline{Y} \leq 8500\right] = P\left[Z \leq \frac{8500 - 10000}{1500/\sqrt{4}}\right] = P\left[Z \leq -2\right] = .02275$$

This probability is smaller than the probability that the life of an individual light bulb is less than or equal to 8500 hours with $Z = [8500 - 10000]/1500 = -1$ and $P[Y \leq 8500] = P[Z \leq -1] = .15866$.

In addition to the uniform and Normal distributions, there are several other continuous distributions that find frequent use in business applications:

Exponential: The exponential distribution is continuous with non-negative outcome values. The exponential distribution is important in the fields of reliability, survival studies, telecommunications, and modeling queuing behavior. It models phenomena with constant failure rates.

Lognormal: The lognormal distribution is continuous with non-negative outcome values. The lognormal distribution serves as a conceptual model in a wide variety of disciplines, including product life, economics, biology, pharmaceuticals, and materials science. It is often appropriate for outcomes that span orders of magnitude (powers of 10).

Triangular: The triangular distribution is continuous with bounded outcome values. The triangular distribution is employed in project management to specify the distribution of task times. This distribution is always triangular in shape, but can be symmetric, left-skewed, or right-skewed.

6.5 Central Limit Theorem

The central limit theorem (CLT) is the theoretical reason for the practical importance of the Normal distribution in probability and statistics. The central limit theorem proves mathematically that, for distributions with means and standard deviations, linear combinations of random variables, such as sums and weighted averages, tend to be normally distributed.

A simple version of the CLT is as follows: Take a sample Y_1, Y_2, ... , Y_n of n independent observations from any distribution with Mean$[Y_i] = \mu$ and Standard

Deviation$[Y_i] = \sigma$. Then as the sample size *n* approaches infinity, the distribution of the standardized sample mean, i.e.,

$$\frac{\overline{Y} - \mu}{\sigma/\sqrt{n}}$$

approaches the standard Normal distribution with mean $\mu = 0$ and standard deviation $\sigma = 1$. Similarly, the distribution of the standardized sample sum, i.e.,

$$\frac{\left(Y_1 + Y_2 + \cdots + Y_n\right) - n \cdot \mu}{\sqrt{n} \cdot \sigma}$$

approaches the standard Normal distribution.

Table 6.3 Convergence of Normal, Uniform, and Exponential Random Variables

Original Distribution	1000 Individual Values	1000 Sums of 12 Standardized Values
Normal The sum of any number of normally distributed outcomes is normally distributed.		
Uniform The sum of 12, properly standardized, is a simple way to get random numbers that are nearly normally distributed.		
Exponential Only the exponential fails to achieve symmetry of the histogram, although it is rapidly approaching it.		

As its name indicates, the CLT is a limit theorem and applies strictly speaking to linear combinations of a number *n* of random variables that approaches infinity. Even though, in theory, the sample size has to increase to infinity, in practice, a much smaller number of observations can suffice to produce an approximate Normal distribution. The speed of convergence depends on the shape of the distribution of the original individual values. Observations from symmetric distributions converge faster than those from asymmetric ones. Table 6.3 shows that the practical effects of the CLT can be seen for linear combinations of an astonishingly small number (n = 12) of random variables.

Table 6.3 shows six different simulated distributions of 1000 values each. The first column gives a brief description of the distributions used in each row of the table. The first row looks at Normal distributions, the second at uniform distributions, and the third row at asymmetric exponential distributions. The second column contains simulations of 1000 individual values of the standard Normal, uniform, and exponential distributions. The third column contains simulations of 1000 standardized sums each of 12 standard Normal, 12 uniform, and 12 exponential random numbers. The sums are standardized so that means and standard deviations of the sums are 0 and 1 in each case. These standardized distributions can be compared directly, especially to the standard Normal distribution. Column 3 of Table 6.3 shows that the histograms of the Normal and uniform sums compare closely to the standard Normal distribution. In theory, the sum of Normal random outcomes is also normally distributed. The uniform distribution is a symmetric distribution, but it is flat. Yet it converges to the Normal distribution with only 12 observations. The individual exponential distribution is skewed. However, even sums of the 12 exponential random variables begin to converge to a more symmetric shape.

6.6 Sampling Distributions

The following sampling distributions will be encountered in the remainder of this book. Values of related test statistics often appear on JMP output. Sampling distributions are probability distributions of derived statistical quantities, such as test statistics, which are useful in statistical inference. They are used for analyzing experimental (and observational) data, in particular in calculating and comparing test statistics and confidence intervals. The following are the most common sampling distributions, with some examples of their use:

- Standard Normal distribution

    □ For analyzing the population or process, average [μ] based on one or more sample means with known process variance σ^2 (see Chapter 8).

- Student's t-distribution

  □ For analyzing the population or process, average [μ] based on one or more sample means when the process variance σ^2 is unknown and is estimated by the sample variance s^2 (see Chapters 8 and 9).

  □ For testing for the significance of a regression coefficient and for many other applications (see Chapters 14–18).

- Chi-Square distribution (χ^2)

  □ For testing hypotheses or finding confidence intervals for a single variance [σ^2] assuming the underlying distribution is Normal (see Chapters 7 and 8).

  □ For testing the association between two qualitative variables or testing the equality of several proportions (see Chapters 12 and 13).

- F-distribution

  □ For testing the overall significance of the model in analysis of variance and regression analysis (see Chapters 10, 11, 16, 17, and 18).

  □ For comparing two population or process variances based on two sample variances (see Chapter 9).

Users of JMP do not need to select test statistics and calculate their values. This task is largely done by the software. Users must learn to interpret these values. In many applications, the interpretation is facilitated by the *p*-values associated with values of the test statistic. JMP will calculate these *p*-values using the appropriate sampling distribution. The interpretation of *p*-values (and the related test statistics) is the subject of the following chapters.

The remarkable thing about using statistical software is that many users can avoid detailed knowledge of sampling distributions. Only the expert statistician needs to be intimately familiar with the distributions, how they are derived, and what potential alternative distributions are. Users of statistics should be familiar with the assumptions underlying the distributions, and whether they are crucial or not.

6.6.1 Student's t-Distribution

The Student's t-distribution is characterized by *degrees of freedom*. With the Student's t-distribution, the number representing degrees of freedom (df) is usually a positive integer, although there are non-integer approximations as well. Degrees of freedom typically depend on the number *n* of observations and the number *p* of parameters to be estimated. For example, in statistical inference from a single sample, the degrees of freedom ($n - 1$) could be thought of as referring to the number of observations effectively available in calculating the sample standard deviation. In simple regression analysis, the error degrees of freedom is $(n-2)$ and refers to the n observations minus the two parameters (slope and intercept) that have to be estimated before the standard deviation observations around the regression line can be estimated.

The density function of the t-distribution looks like the Normal distribution. In fact, as the degrees of freedom approach infinity, it approaches the Normal distribution. A Student's t-distribution with ∞ degrees of freedom is the standard Normal distribution. For all degrees of freedom, it is symmetric around 0 and appears bell-shaped.

To obtain values of the Student's t-distribution, use the JMP formula facility. For example, in order to find P[t(5) ≤ 2.01505], i.e., the probability that a t-distribution value with 5 degrees of freedom is less than 2.01505, create a new data table with three columns: column 1 for the degrees of freedom (here df = 5), column 2 for the t-distribution value (here t(df) = 2.01505), and column 3 for the resulting probability P[Y ≤ t(df)]. The formula for column 3,

	df	t(df)	P[Y<=t(df)]
t Distribution $t(df)$, df	5	2.0150	0.9500001

evaluates the probability statement P[t(5)≤2.015] = 0.95. By the symmetry of the t-distribution density, you can say that P[t(5) ≤−2.015] = 0.05. This probability could be verified by entering the value −2.01505 in column t(df).

The 90% quantile of a t-distribution with 25 df can be found from P[t(25)≤____] = 0.90. The underlined space needs to be filled in by the 90% quantile. In this case, a different formula is used because, instead of specifying t(df), it needs to be calculated from P[Y ≤ t(df)] = 0.90. Proceed by creating a data table with three columns, column 1 containing the degrees of freedom df = 25, column 2 the probability associated with the quantile Q(df), and column 3 the calculated 90% quantile. The formula is slightly different, because the t Quantile function is used instead of the t Distribution as above:

	df	P[Y<=Q(df)]	Q(df)
t Quantile $P[Y<=Q(df)]$, df	25	0.9	1.31634507

The result is Q(df) = Q(25) = 1.31634507 for 25 df or Q(25,0.9) = 1.316 (rounded). By symmetry, P[t(25) ≤ 1.316] = 0.90 implies that P[t(25) ≤−1.316] = 0.10. By the complementation rule, P[t(25) > 1.316] = 0.10.

6.6.2 Chi-Square Distribution

The Chi-Square distribution is asymmetric (skewed to the right) and characterized by a single parameter called degrees of freedom. The term *Chi-Square* is interchangeably used with the Greek symbol χ^2. The Mean$[\chi^2]$ = degrees of freedom, the variance $V[\chi^2]$ =

2·degrees of freedom. In statistical inference for a variance or in finding a confidence interval for the variance, df depend on the number of observations. In the tests for the equality of proportions or in tests for the association between qualitative variables, df depend on the number of categories (or level combinations) involved in the test. These will be discussed in subsequent chapters. In statistical tests, JMP calculates the appropriate df and uses them in finding the *p*-values and confidence intervals.

JMP formulas can be used to obtain Chi-Square probabilities and quantiles, similar to the t-distribution. For example, in order to find the probability P[ChiSq(5) ≤ 1.145476], i.e., the probability that a Chi-Square distribution value with 5 degrees of freedom is less than or equal to 1.145476, use the following JMP formula to calculate the resulting probability P[Y ≤ ChiSq(df)]:

ChiSquare Distribution(ChiSq(df) , df)⟶

df	ChiSq(df)	P[Y<=ChiSq(df)]
5	1.145476	0.04999998

The resulting probability statement is $P[\chi^2(5) \leq 1.145476] = 0.04999998 \approx 0.05$.

For a quantile of the Chi-Square distribution proceed similarly. For example, find the 90% quantile of a Chi-Square distribution with 25 df, as expressed by probability statement $P[\chi^2(25) \leq \underline{\quad}] = 0.90$, use the following ChiSquare Quantile formula:

ChiSquare Quantile(P[Y<=ChiQ(df)] , df)⟶

df	P[Y<=ChiQ(df)]	ChiQ(df)
25	0.9	34.381587

This results in a probability statement $P[\chi^2_{25} \leq 34.381587] = 0.90$. The desired 90% quantile with 25 df is ChiQ(df) = ChiQ(25) = 34.381587 ≈ 34.382.

6.6.3 F-Distribution

The asymmetric F-distribution is often referred to as the variance ratio distribution. The following F-ratio:

$$F_{df1,df2} = \frac{(n_1 - 1)s_1^2 / \sigma_1^2}{(n_2 - 1)s_2^2 / \sigma_2^2}$$

is used to compare two variances for equality. The two variances σ_1^2 and σ_2^2 are associated with the variances of two independent samples from a Normal distribution. The F-ratio is also encountered in analysis of variance and in regression analysis, where it is used to test the overall significance of the proposed model and of individual terms.

The F-ratio depends on numerator df1 and denominator df2 degrees of freedom derived from the numerator and denominator sum of squares. In the test for the equality of two variances, the sum of squares are $(n_1 - 1)s_1^2$ and $(n_2 - 1)s_2^2$ and with degrees of freedom df1 = $(n_1 - 1)$ and df2 = $(n_2 - 1)$.

JMP formulas can be used to obtain F-distribution probabilities and quantiles. For example, in order to find the probability P[F(5, 6) ≤ 9.5], i.e., the probability that the F-distribution value with 5 numerator and 6 denominator degrees of freedom is less than 9.5, create a data table with four columns and one row: column 1 for the numerator degrees of freedom (here df1 = 5), column 2 for the denominator degrees of freedom (here df2 = 6), column 3 for the F-distribution value (here F(5, 6) = 9.5). The probability P[Y ≤ F(df1, df2)], evaluated with the JMP formula, is in the fourth column and completes the table, i.e.,

F Distribution $F(df1, df2), df1, df2$	df1	df2	F(df1, df2)	P[Y<=F(df1, df2)]
	5	6	9.5	0.99188448

This results in the probability statement P[F(5, 6) ≤ 9.5] = 0.99188448.

For F-quantiles, proceed similarly. For example, in order to find F(5, 6)0.95 = 4.38737, the 95% F-distribution quantile with df1 = 5 (numerator degrees of freedom) and df2 = 6 (denominator degrees of freedom), create a data table with four columns, column 1 containing the degrees of freedom df1 = 5, column 2 the degrees of freedom df2, column 3 the probability associated with the quantile FQ(df1,df2) – here 0.95, and column 4 the calculated 95% quantile. The formula for the fourth column is the F Quantile function, i.e.,

F Quantile $P[Y<=FQ(df1, df2)], df1, df2$	df1	df2	P[Y<=FQ(df1, df2)]	FQ(df1, df2)
	5	6	0.95	4.38737419
	6	5	0.95	4.95028807

The resulting probability statement is P[FQ(df1, df2) ≤ 4.3873742] = 0.95. The desired 95% quantile with df1 = 5 and df2 = 6 is FQ(5, 6) = 4.3873742 ≈ 4.387. In the preceding example, a second row has been added with the degrees of freedom reversed to show that reversing the degrees of freedom produces a different F-distribution quantile. Note that FQ(5, 6) = 4.3873742 is not FQ(6, 5) = 4.9502881.

6.7 Summary

- Continuous distributions use real numbers as outcomes. However, even noncontinuous scales can be modeled with continuous distributions.

- Continuous distributions are characterized by their density functions f(y) and cumulative distribution functions $F(y) = P[Y \leq y]$.

- Continuous distributions are distinguished by the shape of their density functions, their measures of location, and their measures of spread (or variation).

- Distributional shape is symmetric or asymmetric, unimodal or multimodal, with finite or infinite range. The Normal distribution (bell-shaped curve) is a symmetric, unimodal distribution with an infinite range.

- Parameters determine the shape, location, and spread of distributions. The parameters of the Normal distribution are the mean and the standard deviation. The parameters of the uniform distribution are the two extreme values, a and b.

- Distributions are used to calculate probabilities.

- The mean and the standard deviation have the same units. If the mean is measured in dollars, the standard deviation is also measured in dollars. The variance has the units of the mean squared (dollars squared).

- The uniform distribution comes in a discrete and a continuous version. Both are useful in computer simulation.

- The Normal distribution is also called the bell-shaped curve, because of the shape of its density.

- Normal distributions with any mean = μ and standard deviation = σ can be standardized (transformed) to the standard Normal with mean = 0 and standard deviation = 1.

- The standard Normal distribution can be found in tables and in JMP formula functions.

- Standardization can be viewed as a two step process. Step 1 subtracts the mean. Step 2 divides by the standard deviation of the original distribution.

- The exponential distribution is continuous and is used to model the time to failure of complex repaired systems and many other phenomena.

- The central limit theorem underscores the importance of the Normal distribution. Under very general conditions, the distribution of sums or of weighted averages approaches a Normal distribution for most underlying non-Normal distributions.

- Sampling distributions are used as distributions of statistics.

- The most common ones (other than the Normal distribution) are the Student's t-distribution, the Chi-Square distribution, and the F-distribution.

- The Student's t-distribution is used for statistical analyses of means and regression coefficients.

- The Chi-Square distribution is used for statistical analysis of single variances, for goodness of fit, and for the association between categorical variables.

- The F-distribution is used for the ratio of two variances and in analysis of variance (ANOVA).

6.8 Problems

1. For each of the following random variables, identify the outcome space and whether the distribution is discrete or continuous.

   a. The price of a stock.

   b. The annual rate of return on a money market fund.

   c. The time taken to complete a marathon when the race will be terminated after six hours.

   d. The number of counts per hour of particulates less than 10 microns recorded by an air sampler.

   e. The change in weight, measured to the nearest pound, one month after the start of a diet.

2. The monthly unpaid balance on a credit card follows a Normal distribution with $\mu = \$1500$ and $\sigma = \$210$. Find the standardized z value that corresponds to each of the following unpaid balances:

   a) $1800 b) $1600 c) $1500 d) $1400 e) $1200

3. Use the standard Normal distribution to find the following probabilities and quantiles:

   a. $P[Z > 1.75]$

   b. $P[0.3 < Z < 2.1]$

   c. $P[Z < 1.26]$

   d. $P[-0.85 < Z < 1.30]$

   e. $P[Z > ?] = 0.92$

   f. $P[Z < -0.55]$

   g. $P[Z > -2.24]$

   h. $P[Z < ?] = 0.25$

   i. $P[Z = 1.0]$

   j. $P[-1.8 < Z < -1.0]$

4. Bank fees to refinance a home mortgage follow a Normal distribution with mean $3050 and standard deviation $200.

   a. What is the probability that the bank fees on a refinance are between $2900 and $3150?

   b. What bank fee will be exceeded with probability 0.05?

    c. The bank refinanced four loans on Friday. What is the probability that the average of the bank fees for the four loans exceeds $3250?

5. The time required to complete a computer-based training program follows a Normal distribution with a mean of 20 hours and a standard deviation of 1.5 hours. Using Normal quantiles, find the time by which the following percentages of the students have completed the training:

    a. 10%

    b. 25%

    c. 50%

    d. 75%

    e. 90%

6. Historically, loans to small businesses held by a local bank follow a Normal distribution with mean $10500 and standard deviation $2000.

    a. What is the probability that the amount of the loan is between $8000 and $12000?

    b. What loan amount will be exceeded only 8% of the time?

    c. The bank granted nine loans this week. What is the probability that the average of the nine loans is less than $9200?

7. A vending machine is designed to fill a 9-ounce capacity cup with 8 ounces of coffee. The filling process follows a Normal distribution with a mean of 8 ounces and a standard deviation of 0.5 ounces.

    a. What is the probability that the cup will overflow?

    b. What is the probability that the machine will deliver between 7.75 and 8.25 ounces of coffee?

    c. What is the probability the cup will be filled with exactly 8 ounces of coffee?

    d. What is the probability that a cup of coffee will contain either less than 7.2 ounces or more than 8.6 ounces?

    e. Find the 90th quantile for this distribution. Write a sentence that gives an interpretation of the 90th quantile.

8. The critical dimension of a machined part is normally distributed with a mean of 150 mm and a standard deviation of 2 mm. The part cannot be used in further assembly if it is greater than 158 mm or less than 146 mm.

    a. What proportion of parts is expected to be too large?

    b. What proportion of parts is expected to be too small?

    c. What proportion of parts will not be usable?

    d. (Optional) One thousand parts are made each day. The cost to discard a part that is not within specification is $42. What is the daily cost to dispose of rejected parts?

9. Find the t value corresponding to the 95th quantile for each of the following degrees of freedom:

   a. 2

   b. 10

   c. 15

   d. 20

   e. 30

   f. 60

   g. 120

10. Find the z value for the 95th quantile of the Normal distribution. How do the t values found above compare to this z value?

11. The time it takes for a telephone sales associate to complete a customer order is referred to as "handle time." The handle times for a catalog company that sells casual clothing have historically followed a Normal distribution with a mean of 6 minutes and a standard deviation of 0.8 minutes. The company has set a goal to complete a telephone order within 7.5 minutes.

    a. Use the Normal distribution model to find the probability that a handle time exceeds the goal.

    b. How much reduction in variation, as measured by the standard deviation, is needed to achieve a probability of 0.015 that handle time exceeds the goal? (Draw a graph of the problem.)

    c. Assuming the original process standard deviation, evaluate the effect on the probability of exceeding the target if the average handle time is reduced by 15 seconds.

6.9 Case Study: Julie's Lakeside Candy

Reliability is important in manufacturing and production settings. It plays a role in establishing equipment maintenance programs, planning production schedules, and designing manufacturing facilities. Basic probability concepts can be employed to assess the reliability of simple systems. More sophisticated techniques, such as simulation, can be applied to evaluate the system reliability of complex configurations of components.

The reliability of a component or machine can be defined as the complement of the probability of failure:

Reliability = 1 – Prob[Component Failure]

A series system is a configuration of components such that the failure of one component causes failure of the entire system. For components connected in series, the system reliability is the product of the component reliabilities:

System Reliability = Reliability$_{\text{Component 1}}$ * Reliability$_{\text{Component 2}}$ *** Reliability$_{\text{Component n}}$

This definition of system reliability assumes that component failures are independent.

Business Problem

Julie's Lakeside Candy operates in a popular resort town during the summer season. Saltwater taffy is the featured product and is made in the storefront where customers can watch the process. Twenty-five different flavors are made, one batch at a time. The candy mixture is cooked to the proper temperature and allowed to cool. It is then placed on a pulling machine that aerates the taffy, causing it to be light and chewy. This is followed by chopping, where the candy is cut into bite-sized pieces. The final step is wrapping the individual pieces in wax paper. Failure of any one of the machines causes the entire process to stop while the failed machine is repaired. Mechanical failures are a particular problem due to the sticky nature of the candy. The three machines operate in series, and their mechanical failures are considered independent. Julie is concerned that the taffy-making system is down for repair too frequently. This affects throughput and causes some of the 25 advertised flavors to be unavailable. The operating machinery is visible from the street and draws customers into the shop.

The failure probabilities of the three machines are Prob[pulling machine failure] = 0.05, Prob[chopping machine failure] = 0.10, and Prob[wrapping machine failure] = 0.25.

1. Find the component reliabilities for each of the machines.

2. Find the system reliability for the taffy-making system.

3. Suppose that the reliability of the wrapping machine could be raised to 0.85. What is the effect on the system reliability?

4. Julie is considering purchasing a new wrapping machine. What wrapping machine reliability would be required to raise the system reliability to 0.80?

5. What is the best system reliability that could be achieved if only the wrapping machine reliability is addressed?

158

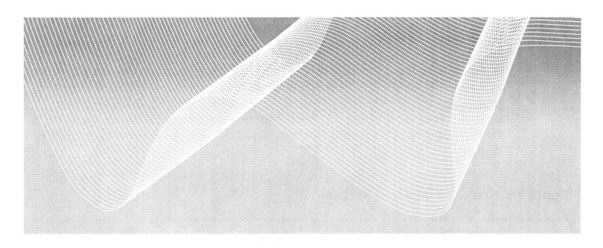

C h a p t e r 7

Confidence Intervals

7.1 Introduction 160

7.2 Point Estimates of Mean and Standard Deviation 160

    7.2.1 What Is a Point Estimate? 161

    7.2.2 Parameters and Estimates 162

    7.2.3 Standard Error of Estimate 164

7.3 Confidence Intervals for Mean and Standard Deviation 165

    7.3.1 What Is a Confidence Interval? 166

    7.3.2 $(1 - \alpha)100\%$ Confidence Level 167

7.4 Detail Example: Package Delivery Times of Herbal Teas 169

7.5 JMP Analysis of Herbal Tea Package Delivery Times 170

    7.5.1 Confidence Intervals for Mean and Standard Deviation 171

7.6 Prediction and Tolerance Intervals 174

    7.6.1 Prediction Intervals 174

    7.6.2 Tolerance Intervals 178

7.7 Summary 180

7.8 Problems 181

7.9 References 183

7.1 Introduction

Statistical inference is a method for drawing conclusions about processes or populations on the basis of sample data. There are three related approaches to statistical inference: point estimation, confidence intervals, and hypothesis testing. Statistical inference requires several steps. Figure 7.1 shows these steps as a straight sequence. However, as the understanding of the problem evolves, steps might have to be repeated to adjust to changes.

Figure 7.1 Statistical Inference Process

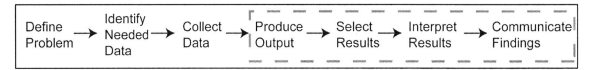

The data used in statistical inference have to satisfy certain minimum standards of quality. The following chapters cover statistical methods based on independent sample observations from random or probability samples. These chapters focus on the steps from Producing Output to Communicate Findings, as marked by the dashed lines in Figure 7.1.

Point estimates are used with both confidence intervals and hypothesis testing. Confidence intervals and hypothesis testing provide different, but mostly equivalent, interpretations of the same sample data. In this chapter, we discuss how to use random samples of a single variable to perform point and confidence interval estimation. Hypothesis testing is covered in the next chapter. JMP presents the results from the three approaches side-by-side.

7.2 Point Estimates of Mean and Standard Deviation

A point estimate of a process or population characteristic is a single value calculated from sample data. Point estimates communicate what the data tell about a characteristic of a population or a process. Different samples from the same process or population usually give different point estimates. However, by giving only a single value per characteristic, point estimates do not express their inherent variability and precision.

7.2.1 What Is a Point Estimate?

For a single variable, a point estimate is a value calculated from the data, a statistic that estimates an *actual* population or process characteristic. Common point estimates are the sample mean or the sample standard deviation that estimate the actual mean and standard deviation respectively. Point estimates exist for other distribution characteristics, such as the median, quantiles, and other parameters such as regression coefficients. There might be several point estimates for a parameter. For example, to estimate the actual median in a symmetric distribution, you could use either the sample median or some average of lower and upper quantiles. You could also construct many other reasonable estimates.

Point estimates of the population mean give a single value. Here are some examples:

- Estimate the mean hours that elapse between order and delivery of a book from Dot.com bookseller to U.S. customers based on a random sample of $n = 400$ shipments.

- Estimate the strength of a material used for car bumpers based on five test specimens.

- Estimate the average percentage of sales tax that is due on purchases by ABC Company based on a sample of 3000 purchases.

- Estimate the average approval rating of a politician based on a poll of 1000 likely voters.

In quality management and other disciplines, the standard deviation is also an important characteristic. The following examples require estimates of the standard deviation:

- Estimate the variability of a measurement device.

- Estimate the batch-to-batch variability of the strength of a plastic material used in automotive applications.

- Estimate the variability in the thickness of plastic catheters.

Some estimates are preferred to others because they are more precise, unbiased, or less sensitive to outliers. To estimate a population mean of a Normal distribution, you could use the average of all values in a sample, or a sample median because the Normal distribution is symmetric, or the average of the central 50% of the data. However, for normally distributed data, the sample mean is more precise than either the sample median or the central 50%. That is the reason for using the sample mean to estimate the population mean. Likewise, the sample standard deviation is the preferred estimate of the actual standard deviation. However, in quality control charting, a standard deviation estimate based on the range of the data is widely used.

Example: Estimating the Mean Number of Orders per Day

Let us show the mean and the standard deviation as two of the most common point estimates using the mail order business selling herbal teas. For staffing purposes, the business needs to estimate the mean number of orders per day. The file *Herbal Tea Mail Order.jmp* contains data for 52 observations.

The sample mean of the 52 observations of Number of Orders is 212.08 with a sample standard deviation of 66.80 (rounded to two decimals). These are the point estimates of the mean and standard deviation of the process that generates these orders each day. Assuming the order process is stable, without trends or extraordinary changes, the sample mean and standard deviation estimate the current process of orders per day and can be used in predicting future orders. Because the data were collected as a random sample, the point estimates are subject to sampling error; another sample of 52 observations might give us a very different sample mean and standard deviation.

Point estimates are subject to variability between samples. The larger the relative variability from sample to sample, the less precise the estimate. The point estimate on its own fails to express how precise it is. The sample mean of 212.08 orders could have been based on four days, 52 days, or 300 days of data. The point estimate itself does not communicate the difference in precision between the different sample sizes. There are at least two ways to express how precise a point estimate is: through standard errors and confidence intervals. These two measures are related. In fact, many confidence intervals require the standard error in their calculation.

7.2.2 Parameters and Estimates

Conventionally, Greek symbols designate actual population or process parameters, while Latin letters designate sample estimates. Table 7.1 shows some population parameters and their estimates based on sample data. In this table as elsewhere, the symbol used for the summation of all y_i is

$$\sum_{i=1}^{n} y_i$$

with the index ranging from 1 to n, where n is the sample size. Summations of other functions of y_i can be interpreted analogously.

Table 7.1 Parameters Point Estimates

Parameter	Symbol of Parameter	Point Estimate
General Parameter	θ = theta	$\hat{\theta}$ = theta hat
Mean	μ = mu	$\bar{y} = \dfrac{1}{n}\displaystyle\sum_{i=1}^{n} y_i$ = ybar
Variance	σ^2 = sigma squared	$s^2 = \dfrac{1}{n-1}\left[\displaystyle\sum_{i=1}^{n} y_i^2 - \left(\displaystyle\sum_{i=1}^{n} y_i\right)^2\right]$
Standard Deviation	σ = sigma	$s = \sqrt{s^2}$

Sample estimates of the mean and the standard deviation can be calculated manually, as long as the sample is not too large. As an example of the manual calculations, use four daily observations of the number of orders for herbal teas: 238, 177, 209, and 135 orders per day. What is the point estimate of the mean based on these four observations? The four observations are shown as bullets in Figure 7.2.

Figure 7.2 Estimating the Mean from a Sample of Four Observations

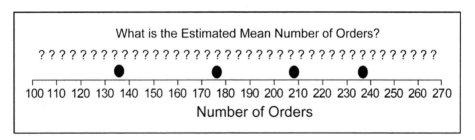

Table 7.2 shows the manual calculations for the sample mean, variance, and standard deviation. The column sums with $n = 4$ are inserted in the formulas of Table 7.1. The sample mean = 759/4 = 189.75 orders, the sample variance $s^2 = (149879 - 759^2/4)/(4 - 1) = 5858.75/3 = 1952.9167$, and the sample standard deviation s = $\sqrt{1952.9167}$ = 44.1918.

Table 7.2 Example of Manual Calculations for the Mean and Standard Deviation ($n = 4$)

Observation	y	y^2
y_1	238	56644
y_2	177	31329
y_3	209	43681
y_4	135	18225
Column Sum	759	149879

The point estimate of the mean Number of Orders is 189.75. It is marked by the arrow in Figure 7.3.

Figure 7.3 Point Estimate of the Mean Is a Single Value (Shown by Tip of Arrow)

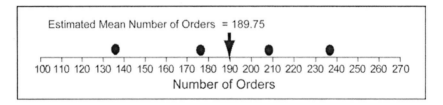

7.2.3 **Standard Error of Estimate**

The variability of a point estimate can be measured by its standard error, representing the standard deviation of the point estimate distribution. (The point estimate distribution could be thought of as the distribution of point estimates from many samples.)
The standard error for the sample mean based on a single sample is as follows:

$$\text{Standard Error of Mean} = \frac{\text{Sample Standard Deviation}}{\sqrt{\text{Sample Size}}} = \frac{s}{\sqrt{n}}$$

In the previous example with $n = 4$ and $s = 44.1918$, the standard error of the mean is as follows:

$$\text{Standard Error of Mean} = \frac{44.1918}{\sqrt{4}} = 22.0959$$

The standard error of the mean is smaller than the standard deviation s, because it is the standard deviation divided by the square root of the sample size n. A smaller standard error implies greater precision. You can estimate the actual mean more precisely with the sample mean than with an individual outcome. Furthermore, with samples taken from the same population, a sample mean based on a large sample size *n* is more precise than one based on a smaller *n*. The reduction in standard error is proportional to the square root of the sample size, and not to the sample size itself. The marginal reduction in the standard error decreases with increasing sample size *n*.

7.3 Confidence Intervals for Mean and Standard Deviation

Confidence intervals, instead of giving a single value as point estimates do, give a range of values determined by the distribution of the point estimate. A confidence interval for the mean expresses the uncertainty inherent in estimating the mean from the sample data with its lower and upper confidence limit. The four observations of the herbal tea mail order example result in a 95% confidence interval for the mean from 119.43 to 260.07 orders.

Instead of the single point estimate of 189.75, the confidence interval gives a range for the mean number of orders from 119.43 to 260.07. Figure 7.4 shows that the confidence interval based on only four observations admits a rather wide range of values. The confidence interval is 189.75 ± 70.32. Thus, the entire confidence interval width is 70.32 + 70.32 = 140.64 orders. The width of the confidence interval for the mean is an indication of how well the point estimate can be trusted to be close to the actual mean.

Figure 7.4 95% Confidence Interval for Mean Is a Range of Values

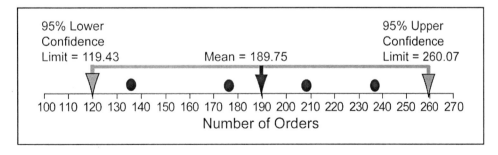

The confidence interval for the mean number of orders based on only four observations is very wide. Taking a larger sample, other things being the same, shortens the confidence interval. Using all 52 observations of *Herbal Tea Mail Order.jmp* results in a point estimate of Mean Number of Orders = 212.08 and a 95% confidence interval from 193.48 to 230.67. Figure 7.5 shows that both the point estimate of the mean number of orders and the confidence interval change considerably. Figure 7.5 indicates that the width of a confidence interval depends on the sample size and that sample estimates are subject to sampling error.

Figure 7.5 Sample Size Effect on Length and Center of 95% Confidence Interval for Mean Number of Orders

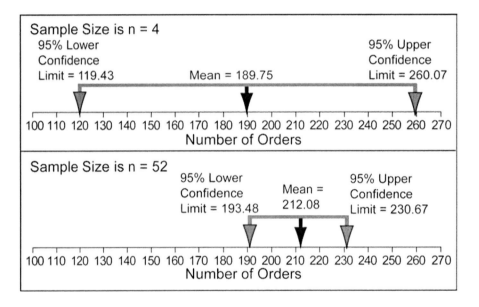

7.3.1 What Is a Confidence Interval?

A confidence interval is a range of values calculated from sample data that contains the true but *unknown* parameter θ with a stated level of confidence $(1 - \alpha)100\%$. Theta = θ is a general parameter and can be the actual mean, standard deviation, or regression coefficient. Confidence intervals for means and other similar statistics are of the following form:

Lower Confidence Limit = LCL < True Parameter θ < UCL = Upper Confidence Limit

In practical applications, the point estimate is within the confidence interval. The length of the confidence interval measures the precision with which the true parameter value is estimated. Two-sided confidence intervals have a lower and upper confidence limit. One-sided confidence intervals do not specify either the lower limit for upper confidence intervals or the upper limit for lower confidence intervals.

7.3.2 (1 – α)100% Confidence Level

The probability that a confidence interval is wrong is α; the probability that it is correct is $(1 - \alpha)$. The confidence level $(1 - \alpha)100\%$ is that latter probability expressed as a percentage. The choice of the value of α depends on the consequences of an incorrect confidence interval.

A confidence interval, being based on sample data, might or might not include the actual parameter value. Many confidence intervals will include the actual parameter, but some will not. When a confidence interval does not contain the parameter, the sample outcome is misleading. The true answer to whether an interval contains the parameter is unknown. However, with statistics the probability of a correct answer can be calculated. The confidence level is thus a property associated with the confidence interval method. Figure 7.6 shows five possible situations. The vertical line indicates the value of an actual population mean. The horizontal intervals show possible confidence interval outcomes.

- The intervals A, B, and C all contain the actual mean. Although A and B are very close to missing the mean line, these confidence intervals are correct.

- The interval D is too high; the interval E is too low. Both miss the actual mean. Confidence intervals D and E are incorrect.

Figure 7.6 Cases of Confidence Interval Containing or Not Containing Mean

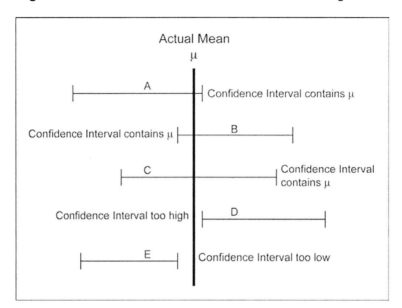

In a practical situation, it is not known which of these cases has occurred. However, the confidence level is related to the probability of the confidence interval being correct. The $(1-\alpha)100\%$ confidence level can be considered the long-run relative frequency that the confidence interval contains the true parameter in repeated computations of such intervals. Everything else being the same, higher confidence levels result in wider confidence intervals.

Confidence intervals are always stated with a confidence level. In the herbal tea mail order example, the confidence level is 95%. A 95% confidence interval implies that there is a 5% chance that the interval is wrong. For example, in the herbal tea mail order example, the sample mean is 212.08 orders per day. The 95% confidence interval for the mean number of orders is from 193.48 to 230.67. This suggests that there is a 5% chance that the actual mean either is lower than 193.48 or higher than 230.67. Other confidence levels are quite common; however, the 95% level is most common. In the Distribution platform, JMP enables the user to select confidence levels.

7.4 Detail Example: Package Delivery Times of Herbal Teas

Situation: A mail order firm selling Herbal Teas is shipping its orders with a package service. The time to delivery is important particularly during the hot summer months. The mail order company tries to get its shipment to customers in as short a time as possible. The current package service takes an average of 75 hours to deliver packages with the cheapest ground service. It has instituted several process simplifications with the intention of reducing the current average time of 75 hours.

Problem: Examine package delivery data from a sample after the process changes have been instituted. Determine the new mean time to delivery.

Data Requirements: From long experience, the delivery time of the old package service is known to average 75 hours. Data are collected only from the simplified process. These data are in *DeliveryTimes.jmp* and are excerpted in Figure 7.7. The file contains only one column of delivery times for 25 randomly chosen packages shipped and delivered by the new service.

Anticipated Results: It is anticipated that the data will show that the mean delivery time is shorter than 75 hours. Estimating the mean delivery time will provide objective evidence that the process simplifications resulted in shorter delivery times.

Figure 7.7 Delivery Times of Herbal Tea Packages in Hours

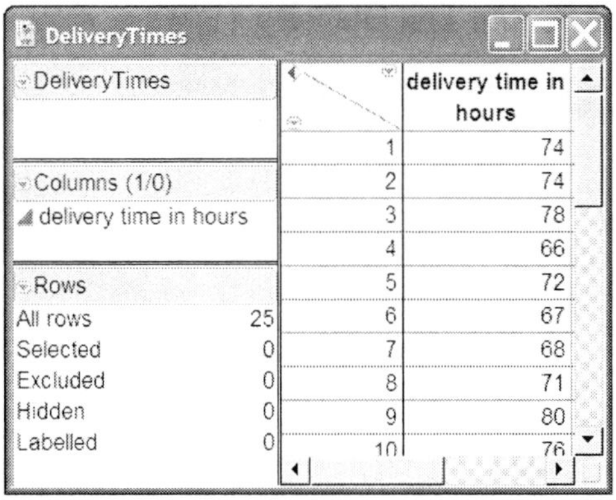

7.5 JMP Analysis of Herbal Tea Package Delivery Times

In JMP, the analysis of a single variable starts by choosing the Distribution platform from the **Analyze** menu. In the resulting Distribution window, select **delivery time in hours** as the Y-variable and click **OK** to obtain standard output (see Figure 7.8).

Figure 7.8 Report Window for Distribution Platform

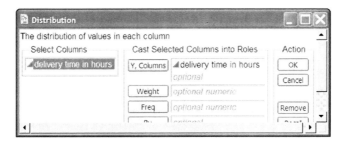

The basic results of the Distribution platform are shown in Figure 7.9. They include a histogram, a box plot, quantile point estimates, and a **Moments** list containing a summary of key point estimates, such as the sample mean and standard deviation, the standard error of the mean based on $n = 25$, and also the upper and lower 95% confidence limits for the mean.

Figure 7.9 Basic Results for Herbal Tea Delivery Times in Hours

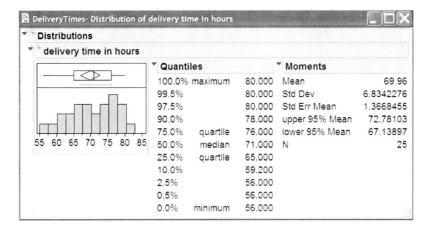

In order to prepare a meaningful report, the numerical output should appear in easily understandable format. It is hardly necessary to present the hours of mean delivery time to more than two decimal places. Even though the mean only shows two decimal places, the standard deviation and confidence intervals show up to seven decimal places. JMP offers a convenient way to reformat numerical output. First, double-click the column for which the format needs to be changed. In the window shown in Figure 7.9, double-click the numerical column of the **Moments** list. The dialog box shown in Figure 7.10 appears, enabling you to change the default format to **Fixed Dec** and **2** decimal places. The field width can be adjusted to accommodate numbers with many or very few digits. No change in field width is needed in this example. Similarly, you can change the number of decimal places for the **Quantiles** list to a different number of decimal places. The resulting **Moments** list is shown on the right of Figure 7.10.

Figure 7.10 Formatting Column Display to Two Fixed Decimal Places

7.5.1 Confidence Intervals for Mean and Standard Deviation

The basic results include a 95% confidence interval for the mean, but do not allow the flexibility to use other confidence levels or to have one-sided confidence intervals. The confidence interval option (see Figure 7.11) provides one-sided and two-sided confidence intervals for the mean and standard deviation. To invoke the option menu, click the red triangle to the left of the variable name (delivery time in hours).

Selecting the **Confidence Interval** option and one of the predetermined confidence levels (**0.90**, **0.95**, or **0.99**) produces two-sided confidence intervals for the mean and the standard deviation. The confidence interval will be based on the sample standard deviation Std Dev = 6.83 shown in the **Moments** list.

The confidence intervals will appear in the results window shown on the right of Figure 7.11 with the number of decimal places adjusted to two. In the Estimate column, two point estimates identical to those of the **Moments** list are shown:

- The sample mean of delivery time = 69.96 hours.

- The sample standard deviation of delivery time = 6.83 hours.

The column Lower CI contains lower confidence limits; the column Upper CI contains upper confidence limits. The last column labeled 1-Alpha shows the confidence level expressed as a proportion rather than a percentage. The resulting confidence intervals are as follows:

- 95% confidence interval for the mean is from 67.14 to 72.78 hours.

- 95% confidence interval for the standard deviation is from 5.34 to 9.51 hours.

Figure 7.11 Distribution Options with Selected Confidence Level and 95% Confidence Intervals

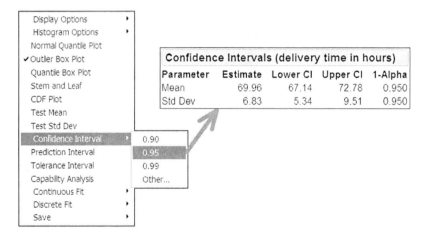

One-sided confidence intervals are useful when unidirectional changes, such as a lowering of the mean delivery time or increasing the material strength, are anticipated. Since the example asks us to determine if the delivery times have been lowered, it makes sense to ask for a one-sided upper confidence limit. One-sided confidence limits are obtained by selecting **Other** from the menu shown in Figure 7.11. The dialog box shown in Figure 7.12 appears, enabling you to select any confidence level by overtyping the default 0.95. You can also select two-sided as well as one-sided lower and upper confidence limits. (If the option **Use known Sigma** is selected, you are asked to enter a value for the standard deviation. This option is unselected by default, so that the sample standard deviation is used.)

Figure 7.12 Selecting a 95% One-Sided Upper Limit

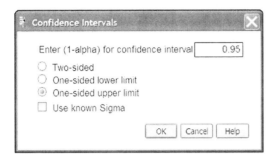

The upper confidence limit gives an upper bound for the mean delivery time. If the largest value of the one-sided confidence interval is smaller than the current value, it can be concluded that process improvements lowered the mean delivery time. For example, in Figure 7.13, the upper 95% one-sided confidence limit is 72.30. It states that with a 95% confidence level, the mean delivery time in hours is below 72.30. The lower limit is not given, because it was not requested. The 95% one-sided upper confidence limit places the total 5% chance of being wrong on the upper side. For that reason, it is different from the 95% upper two-sided limit of 72.78.

Figure 7.13 95% One-Sided Upper Confidence Limit

One-sided Confidence Interval (delivery times in hours)				
Parameter	Estimate	Lower CI	Upper CI	1-Alpha
Mean	69.96	.	72.30	0.950
Std Dev	6.83	.	9.00	

The problem was to determine whether the average delivery time of the new service is lower than that of the old one. The 95% upper confidence limit is lower than the mean (= 75 hours) delivery time of the old service. This is a strong indication that the new service provides on average a shorter delivery time.

7.6 Prediction and Tolerance Intervals

The previous section discussed confidence intervals for the mean and the standard deviation. Confidence interval methods are available for other parameters, such as a distribution quantile or the probability that an outcome is within specified limits. Confidence intervals are intended to contain unknown population parameters or other population characteristics.

Some applications ask questions that require different intervals. In the herbal tea mail order example, an important question for determining staff requirements is to know an interval that contains at least 90% of Number of Orders per day. This interval would give management an idea of the range within which a large proportion of the population of Number of Orders per day will fall. Such intervals are called tolerance intervals. Both confidence intervals and tolerance intervals are descriptive. Other applications require prediction of future outcomes. For example, management might want an interval for a single future observation, i.e., Number of Orders of a future day. Such intervals are called prediction intervals, because they predict a range for future outcomes with a stated confidence level.

This section gives simple examples of prediction intervals and tolerance intervals that are available in JMP. For these methods, *the data need to be normally distributed* in order to achieve valid results. A great variety of other special intervals that answer many different questions are available. For these, see the book *Statistical Intervals: A Guide for Practitioners*, by Gerald J. Hahn and William Q. Meeker (1991).

7.6.1 Prediction Intervals

As the name suggests, prediction intervals predict functions of future outcomes. In this section, we discuss prediction intervals that contain the mean of k future outcomes. We begin with the special case where k = 1, a prediction interval for a single future observation.

Prediction Interval for a Single Future Observation

Suppose a random sample of *n* observations is available from a normally distributed population. How can we use the sample data to predict a single future outcome? As a point estimate, we can use the sample mean. However, as has been discussed, point estimates do not express the inherent uncertainty of this estimate. Confidence intervals for the mean are easily obtained, but they describe the mean and do not predict a future outcome. For this, we use a *prediction interval for a single future observation,* which is an interval that contains a single random future observation with a specified confidence level.

The prediction intervals presented here require that the data are normally distributed. Deviations from this assumption result in prediction intervals with "wrong and unknown" confidence levels that would be difficult to interpret. (Confidence intervals for the mean are much less dependent on the normality assumption.)

Example: Delivery Times

Using the data in *DeliveryTimes.jmp*, find an interval that contains a single future Number of Orders per day. A Normal Quantile plot of the available data shows that the assumption of normality is satisfied. It is OK to proceed with prediction intervals.

To obtain a prediction interval for the Number of Orders on a single day, enter **Number of Orders** into the Distribution platform. Click the red triangle of the **Number of Orders** title bar to reveal the options menu already shown in Figure 7.11. Selecting **Prediction Interval** opens a dialog box to specify the type of prediction interval, the confidence level, and the number of future samples. In Figure 7.14, we choose a two-sided, 95% prediction interval for a future sample with k = 1. Entering **1** in the **Enter number of future samples** field will result in a prediction interval for a single future observation.

Figure 7.14 Specifications for a 95% Prediction Interval for a Single Future Observation

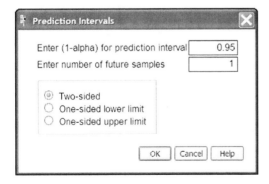

Figure 7.15 shows that the 95% prediction interval for the Number of Orders of a single (individual) future day is from 55.58 to 84.34 hours (rounded). This is a wide range, despite the fact that the sample was based on 52 observations. The 95% prediction interval tells us that with 95% confidence, we can expect that a future observation will be in the interval from 55.58 to 84.34 Number of Orders. Figure 7.15 also shows a prediction interval for the mean of a single future observation, which is the same as for a single future observation. The two intervals (Individual versus Mean) differ for

predictions of two or more future observations. A prediction interval for the standard deviation of the future sample is not shown, since it will be impossible to calculate a standard deviation of a single future observation.

Figure 7.15 95% Prediction Interval for a Single Future Observation

Prediction Interval for single future observation				
Parameter	Future N	Lower PI	Upper PI	1-Alpha
Individual	1	55.57551	84.34449	0.950
Mean	1	55.57551	84.34449	0.950
Std Dev	1	.	.	0.950

The 95% prediction interval is considerably wider than the 95% confidence interval for the mean, which ranges from 67.14 to 78.78 hours. The reason for this is that the standard error of prediction is larger than the standard error of the mean. Recall that the estimated standard error of the mean is

$$\text{se}(\text{Mean}) = s\sqrt{\frac{1}{n}}$$

i.e., divide the standard deviation by the square root of the sample size n. As the sample size increases, the standard error of the mean gets smaller and smaller. The standard error of prediction of a single unit is

$$\text{se}(\text{Pred\_1}) = s\sqrt{\frac{1}{n} + 1}$$

The standard error of prediction not only contains the term $\frac{1}{n}$ that expresses the uncertainty inherent in the sample, but the larger term "1" that expresses the uncertainty of selecting a future observation. The larger term cannot be reduced by increasing the sample size n. As a result, prediction intervals are wider than confidence intervals in similar circumstances.

Because the mail order business is concerned only about packages that take too long, it might be useful to calculate a one-sided upper prediction limit. Figure 7.16 shows that a 95% upper prediction limit on a single future delivery time is 81.88 hours. This is less than the upper limit of the two-sided interval (84.34 hours). The interval predicts with 95% confidence that a single future package will take no more than 81.88 hours.

Figure 7.16 95% One-Sided Upper Prediction Limit for a Single Future Observation

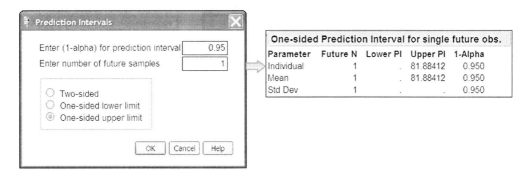

One-sided Prediction Interval for single future obs.				
Parameter	Future N	Lower PI	Upper PI	1-Alpha
Individual	1	.	81.88412	0.950
Mean	1	.	81.88412	0.950
Std Dev	1	.	.	0.950

Prediction Intervals for k Future Observations

These prediction intervals are useful when decisions are based on the future outcomes. For example, a chemical business might be concerned about meeting the permit limit of a particular effluent from the wastewater treatment plant. The protocol for the permit limit specifies that a random sample of five measurements is averaged. If the average is larger than the permit limit, then the business is in violation. In order to predict the outcome of such a test, a prediction interval for the mean of five future observations is required.

In the delivery times example, a prediction interval for all of nine randomly selected parcels might be of interest. The initial steps are the same as for a prediction interval for a single future observation. However, we are asking for prediction intervals for nine future observations (see Figure 7.17). The confidence level is 95% and the requested interval is two-sided.

Figure 7.17 shows that the 95% prediction interval for nine future individual observations is from 48.7274 to 91.1926. This interval is designed to contain every one of nine future observations. The 95% prediction interval for the mean of nine future observations is from 64.4769 to 75.4431 hours. The interval for nine individual observations is wider, because nine individual observations are expected to have more spread than their average. Both intervals are wider than the 95% confidence interval for the mean.

Figure 7.17 also contains a 95% prediction interval of the standard deviation of a future sample of size k. In the delivery times example, the predicted standard deviation of a future sample is the point estimate of the standard deviation s = 6.83 hours. The 95% prediction interval states that the future standard deviation is between 3.44 and 11.39 hours, with 95% confidence.

Figure 7.17 95% Prediction Intervals for k = 9 Future Observations

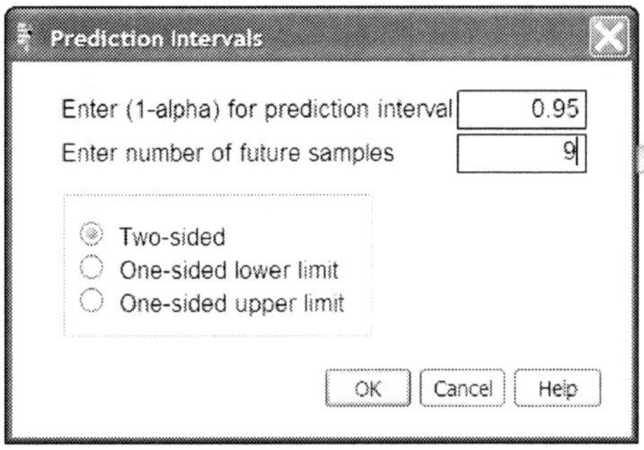

7.6.2 Tolerance Intervals

Tolerance intervals are used to describe intervals that contain at least a proportion p of the population at the $(1 - \alpha)100\%$ confidence level. Tolerance intervals are not concerned with means or standard deviations, but with individual population values. The proportion p refers to population values that are enclosed within the tolerance interval. The $(1 - \alpha)100\%$ confidence level is a property of the procedure and indicates that if $(1 - \alpha)100\%$ tolerance intervals are calculated repeatedly, on average $(1 - \alpha)100\%$ of tolerance intervals will contain at least a proportion p of the population values. The width of the tolerance interval depends on the size of the sample from which the interval is calculated, the proportion p, and the confidence level $(1 - \alpha)100\%$.

In the delivery times example, we want to calculate an interval to contain at least 90% of all delivery times. In the menu shown in Figure 7.11, select **Tolerance Interval** to open the specification dialog box. In this example, we choose a two-sided 95% tolerance level and specify a proportion to cover of at least p = 0.9 (Figure 7.18).

Figure 7.18 Specifications for a 95% Tolerance Interval

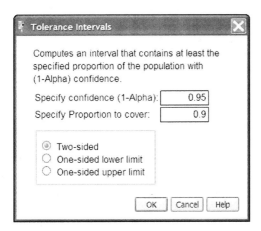

Click **OK** and the results are given as in Figure 7.19. The 95% tolerance interval is from 54.82 to 85.10 hours. Expect that at least 90% of the population values are enclosed with 95% confidence.

Figure 7.19 95% Tolerance Interval to Contain at Least 90% of Population Values

Tolerance Intervals for 90% of Population			
Proportion	Lower TI	Upper TI	1-Alpha
0.900	54.82161	85.09839	0.950

As with confidence and prediction intervals, it is possible to calculate one-sided lower or upper tolerance limits. In the delivery times example, it makes sense to ask for a one-sided upper tolerance limit to contain at least 90% of all values. Figure 7.20 shows that this upper tolerance limit is 82.52 hours. At least 90% of all delivery times are expected to be less than 82.52 hours, again with 95% confidence. This value is lower than the upper limit of the two-sided interval (85.10 hours). The reason for this is that the upper limit specifies a one-sided confidence interval that is not bounded from below, only from above. The mail order firm now has a good idea that no more than 10% of parcels will take longer than 82.5 hours to deliver, assuming that the sample of 25 delivery times remains representative of the population. The tolerance intervals covered here require that population values are normally distributed.

Figure 7.20 95% One-Sided Tolerance Limit to Contain at Least 90% of Population Values

One-sided Tolerance Interval for 90% of Population			
Proportion	Lower TI	Upper TI	1-Alpha
0.900	.	82.522	0.950

Conclusion

The choice of confidence, prediction, or tolerance interval depends on the problem that is being solved or the question that is being asked. Confidence and tolerance intervals are descriptive of the population. They are for parameters, such as means or standard deviations, or for proportions of the population. Prediction intervals, as the name suggests, are for predicting future outcomes. The prediction intervals discussed here are for one or more observations or the average of k future observations. There are many other prediction intervals. We will encounter prediction intervals in connection with regression analysis.

7.7 Summary

- Point estimation, confidence intervals, and hypothesis tests are approaches to statistical inference.

- Point estimates are single values, calculated from the data. They usually estimate population characteristics such as the means, medians, or standard deviations.

- Point estimates do not express the uncertainty associated with them.

- The standard error of the mean is the standard deviation of a sample mean. It is a measure of the uncertainty of the point estimate.

- The standard error of the mean from a simple random sample depends on the sample size. Larger sample sizes result in smaller standard errors.

- Confidence intervals give a range for the estimate of the actual population characteristic. The range is from the lower confidence limit to the upper confidence limit.

- Confidence intervals are always reported with a confidence level. The confidence level is usually reported as a percentage, representing the chance that the confidence interval is correct. A 95% confidence level implies that, on average, only 1 out of 20 confidence intervals does not contain the actual population characteristic.

- Different population characteristics require different formulas. There may be more than one approach to confidence intervals for a given population characteristic. JMP performs most calculations treated in this book.

- A prediction interval for a single future observation is an interval that contains a single random future observation with a specified confidence level.

- A tolerance interval specifies an interval that contains at least a proportion p of the population at the $(1 - \alpha)100\%$ confidence level.

7.8 Problems

1. Consider a problem in your job, career, or personal life where a confidence interval could assist in making a decision. Write a paragraph describing the problem, the point estimate selected, and how the confidence interval is useful in the decision-making process.

2. A fabric used for insulation in winter boots is manufactured continuously and must be monitored for thickness. A random sample of 10 measurements (in mm) was taken. Find a 95% confidence interval for the mean thickness. Also, calculate a 99% confidence interval. In a few sentences, give an interpretation of these confidence intervals. The data can be found in *winter_boots.jmp*.

3. A company with 21 regional offices is preparing an advertising campaign emphasizing their environmental consciousness. Various offices were asked to increase their use of reusable and recyclable office products (e.g., toner cartridges, paper, and batteries). The file *office_recyclables.jmp* contains the reported increases in "green" office supplies as a percent of the office supply budget.

   a. What point estimator for the increase in "green" office supplies would you recommend? Why?
   b. Compute a two-sided confidence interval. What significance level would you recommend? Why?
   c. In a few sentences, give an interpretation of the confidence interval in general terms.

4. A corporate trainer conducts classes in business ethics. Each quarter, she places an order with the printer for student workbooks. To help the corporate trainer determine the number of workbooks to order, find a one-sided upper confidence interval for the mean class size. Use the results of a random sample of last year's class sizes as found in *ethics_class_sizes.jmp*.

5. One production line in a bottling plant fills two-liter bottles with ginger ale. Plant operations personnel believe it is important to monitor and control the variability in the filling process. They are interested in monitoring the standard deviation of the amount of ginger ale in the bottles. The file *ginger_ale_bottles.jmp* contains the volume of liquid from 50 randomly selected bottles.

   Find a 98% confidence interval for the standard deviation of the volume of ginger ale, measured in milliliters, and interpret it.

6. Melissa has decided to obtain a master's degree to help her advance in her profession. She plans to work full-time while studying part-time until she receives her degree. Melissa is concerned about the impact of tuition increases, especially since she expects it will take 5 to 6 years to complete her studies. Melissa has compiled data on recent tuition increases for colleges offering the master's degree in her field. Find an appropriate statistical interval that would be useful to Melissa in her financial planning. Give an interpretation of this interval. Discuss any limitations or assumptions associated with the interval. The data can be found in *tuition_increases.jmp.*

7. A random sample of 45 aluminum shafts with a nominal length of 100.50 millimeters was obtained. Their lengths were precisely measured in a metrology lab by a qualified inspector. The results are given in *shaft_length.jmp*.

   a. Describe the data with a histogram.
   b. What conceptual distribution would you suggest as an appropriate model for shaft lengths? Why?
   c. Find and interpret a 98% confidence interval for the mean aluminum shaft length.

8. A firm is planning a new construction project. One concern is the potential for schedule overrun. The schedule overruns (in number of days) for similar projects undertaken within the past five years are recorded in *project_overrun.jmp*. Find an interval that would help determine a reasonable number of days to schedule for the project overrun. Explain your choice.

9. Polyvinyl chloride (PVC) pipe is used in a variety of applications for building, plumbing, and conduits. PVC pipes are sold in nominal 10-foot lengths. The file *PVC_pipes.jmp* records the actual lengths of 36 randomly selected pipes from a manufacturer. Find confidence intervals for the mean and standard deviation pipe length. Select an appropriate confidence level. Give a brief interpretation of these intervals and justification for your choice of confidence level.

10. A charity holds a 50/50 raffle at their monthly fundraising breakfast. In a 50/50 raffle, numbered tickets are sold and the winning number receives 50% of the proceeds, with the charity retaining the other 50%. The file *raffle50.jmp* contains the total proceeds from 24 previous raffles.

    a. What is the best estimate of the true mean raffle proceeds?

    b. What is the best estimate of a single future raffle?

    c. Find a 95% confidence interval for the mean raffle proceeds.

    d. Is the normality assumption satisfied for this data? Explain how you arrived at your answer.

    e. Find a 95% prediction interval for the proceeds from a single future raffle.

    f. Which interval would be of interest to a person playing the raffle? Why?

    g. Which interval would be of interest to the charity sponsoring the raffle? Why?

    h. Find a 95% prediction interval for the average proceeds from two future raffles.

    i. Compare the three intervals and discuss their differences.

11. A bank is running a promotion on low-interest credit cards. The file *credit_card_application.jmp* contains the time to fill out a credit card application for 30 randomly selected customers. Time is recorded in minutes.

    a. Find a one-sided upper tolerance interval to contain 90% of the application times with 95% confidence.

    b. Find a one-sided upper tolerance interval to contain 95% of the application times with 95% confidence.

    c. Find a one-sided upper tolerance interval to contain 99% of the application times with 95% confidence.

    d. How do these three intervals compare?

    e. How would a one-sided upper tolerance interval on application time be useful to the bank?

7.9 References

Hahn, Gerald J., and William Q. Meeker. 1991. *Statistical Intervals: A Guide for Practitioners*. New York: Wiley-Interscience.

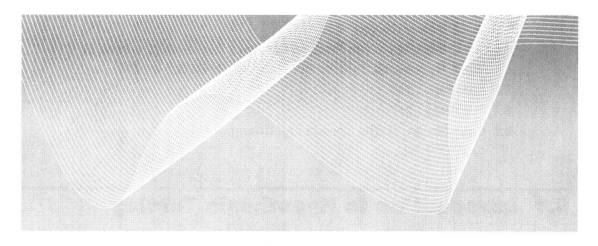

Chapter 8

Hypothesis Tests for a Single Variable Y

8.1 Introduction to Hypothesis Testing   186

    8.1.1 Accept or Reject Decisions for the Mean: H_0 versus H_A 186

    8.1.2 Significance Level α 190

    8.1.3 Test Statistic   191

    8.1.4 *p*-Value   191

    8.1.5 Decision Rule to Accept or Reject H_0 192

    8.1.6 Example: Order Processing Times   192

    8.1.7 Test Statistic, Significance Level, Critical Value, and *p*-Value   198

    8.1.8 Example: Order Processing with JMP   200

    8.1.9 Confidence Intervals and Two-Sided Hypothesis Testing   202

8.2 Sample Size Needed to Test H_0: Mean = $Mean_0$
versus H_A: Mean = $Mean_A$ 203

    8.2.1 Introduction   203

    8.2.2 Example: Sample Size Calculations for Mama Mia's
Pizza Parlor   205

    8.2.3 Sample Size in JMP   206

    8.2.4 Approximate Formulas for Sample Size   208

8.2.5 Power Curves 209

8.2.6 Sample versus Process Standard Deviation (s versus σ) 211

8.2.7 Hypothesis Test for the Standard Deviation σ 215

8.3 Summary 216

8.4 Problems 217

8.5 Case Study: Traffic Speed Limit Change 221

8.1 Introduction to Hypothesis Testing

Hypothesis testing concerns decisions that result in acceptance or rejection of a hypothesis about an unknown state of nature. Whereas confidence intervals give the user a range of values within which a parameter such as the mean can be assumed with a high degree of confidence, in hypothesis tests we examine claims (hypotheses) about states of nature. In this chapter, the hypothesis concerns a single distribution parameter such as the mean or the standard deviation. Sample data either do or do not suggest statistically significant differences from that hypothesis.

Many business decisions are based on a sample of data. For example, an automobile dealership might initiate changes in their service department based on a sample of repair waiting times. Hypothesis testing provides a means to understand whether observed changes are due to sampling variation or to actual shifts in process performance.

This chapter introduces a number of important concepts, including null and alternative hypotheses, test statistics, and *p*-values. These concepts will be applied to the statistical methods presented throughout the remainder of this book.

8.1.1 Accept or Reject Decisions for the Mean: H_0 versus H_A

In simple hypothesis testing, the decision is between accepting either the null hypothesis (H_0) or the alternative hypothesis (H_A). There are situations with more complex hypotheses, but they are not treated here. For simplicity, the following discusses hypotheses about a mean.

The null hypothesis can be derived in several ways. It might represent the status quo, the process mean before changes have been made, a standard, or a contractual value. The alternative hypothesis is often referred to as the *research question*, because in many applications the objective is to prove that it is true. Depending on how the research question is posed, the alternative hypothesis can be left-sided, right-sided, or two-sided.

Whenever possible in this book, the null hypothesis is written as a simple hypothesis of the form H_0: Parameter = Parameter$_0$ (e.g., Mean = Mean$_0$). A simple hypothesis specifies a single value for the parameter such as the mean. This null hypothesis value (Mean$_0$) is the one entered in the testing hypothesis window of JMP and is the value with which the error probability α is associated. Mean$_0$ represents the value of the mean assumed or claimed under the null hypothesis.

Left-Sided Alternative Hypothesis

For testing a mean, the values hypothesized in a left-sided alternative are located to the left of the null hypothesis mean. The left-sided hypothesis is used to test that the actual mean is smaller than the mean claimed in the null hypothesis. It is useful to write the two hypotheses side by side. Left-sided hypotheses concerning the mean are written as follows:

H_0: Mean = Mean$_0$ versus H_A: Mean < Mean$_0$

Note: Some prefer to write the null hypothesis as the complement to the alternative hypothesis. This would lead to a statement of the null hypothesis H_0: Mean $\geq$ Mean$_0$. Here we use the strict equality, because the *p*-value is associated with Mean$_0$.

Suppose you are trying to show that process changes have reduced the mean order-processing time from the historical mean of 120 minutes. The null hypothesis would state that the mean of order processing times has not been reduced from its historical value, i.e., H_0: Mean = Mean$_0$ = 120 minutes. The alternative hypothesis would state that the mean order processing time has been shortened, i.e., H_A: Mean < Mean$_0$ = 120 minutes.

Right-Sided Alternative Hypothesis

For testing a mean, the values hypothesized in a right-sided alternative are located to the right of the null hypothesis mean. The right-sided hypothesis is used to test that the actual mean is larger than the mean claimed in the null hypothesis. Right-sided hypotheses concerning the mean are written as follows:

H_0: Mean = Mean$_0$ versus H_A: Mean > Mean$_0$

For example, a maker of automotive plastics needs to prove that the strength of a new material is higher than the comparable older plastic. The null hypothesis is that the mean strength of the new plastic material is no stronger than the old material's mean strength of 15 pounds per square inch (lbs/si). The alternative hypothesis is that the new plastic material has a mean strength that is higher than the mean strength of the old material, i.e., is greater than 15 lbs/si. The null hypothesis is H_0: Mean = Mean$_0$ = 15. The right-sided alternative is H_A: Mean > Mean$_0$ = 15.

Two-Sided Alternative Hypothesis

For testing a mean, the values hypothesized in a two-sided alternative are located on either side of the null hypothesis mean. The two-sided hypothesis is used to test that the actual mean is either smaller or larger than the single mean claimed in the null hypothesis. Two-sided hypotheses concerning the mean are written as follows:

H_0: Mean = $Mean_0$ versus H_A: Mean $\neq$ $Mean_0$

For example, a bakery famous for its delicious chocolate chip cookies needs to control the ingredients at a pre-specified proportion. The current best recipe states that on average there should be six chocolate chips per cookie. The null hypothesis is H_0: Mean = 6 chips per cookie. The alternative hypothesis includes any mean larger or smaller. H_A: Mean $\neq$ 6 implies that either Mean < 6 or Mean > 6.

Before you begin with a hypothesis test, clarify the following issues:

- What is your research question?

- How large a sample can you take?

- What is the consequence of a wrong decision?

Hypothesis Testing Steps

Hypothesis testing is a process that requires the following steps.

Step 1: Translate the research question into the *alternative hypothesis* H_A and formulate the *null hypothesis* H_0. Write down both hypotheses.

Step 2: Choose a method to collect the appropriate data for the test. This involves answering two questions:

- How should the data be collected (research design)?

- How many observations should be collected (sample size)?

Step 3: Collect the data. The usual assumption is that the data are collected as a random sample. Failure to choose random sampling might result in invalid conclusions.

Step 4: Use JMP to calculate the appropriate *test statistic* and its associated *p-value*. The *p*-value measures the strength of the data evidence in favor of the null hypothesis and is used to decide either to accept or to reject the null hypothesis.

Court Decisions and Hypothesis Testing

Decisions in criminal courts are related to hypothesis testing. Criminal courts must decide to acquit or convict an accused. The basic assumption in a court is that the accused is not guilty until proven otherwise. The prosecutor must present to the court sufficient evidence to prove the guilt of the accused. Only if the prosecutor presents the court with evidence that establishes guilt beyond a reasonable doubt can the court convict the accused. In a court, a guilty verdict depends only on the evidence presented. The verdict of guilty is a function of the actual state of guilt only to the degree that available evidence and the actual state of guilt are related. A non-guilty person might be convicted if the presented evidence points to guilt. Likewise, a guilty person might be judged not guilty if insufficient evidence is available to prove guilt. Both of these situations lead to errors in the decision-making process, as shown in Table 8.1.

Table 8.1 Hypothesis Testing and Court Decisions

Verdict	Actual Guilt	
	H_0 is true: Not Guilty	H_A is true: Guilty
Not Guilty	OK	Error
Guilty	Error	OK

Acquitting a non-guilty person or convicting a guilty person are both shown as "OK" decisions in this table. However, convicting a non-guilty person or acquitting a guilty person are both erroneous decisions. Criminal courts try to err on the side of giving an accused every benefit of the doubt by requiring evidence of guilt "beyond a reasonable doubt." Courts have rules of evidence. Only evidence meeting certain standards is admitted. Of course, the final decision must be based in the law itself.

Statistical hypothesis testing is very similar. The null hypothesis can be construed as the statistical analogue to the judicial presumption of innocence. In statistics, rejection of the null hypothesis depends only on sample evidence and is a function of the sample data, the rules concerning data collection, and the rules of statistical inference.

Like courts, statistical hypothesis testing distinguishes between two types of errors, as shown in Table 8.2. The probability of making the incorrect decision when the null hypothesis is true is called the α (alpha) probability or *significance level*. The probability of making the incorrect decision when the null hypothesis is false is called the β (beta) probability. The *power* of a test is the probability that the null hypothesis is correctly rejected and is measured by $1 - \beta$.

Criminal courts challenge the prosecutor to present evidence in sufficient quality and quantity that it is "beyond a reasonable doubt." "Beyond a reasonable doubt" is the threshold for conviction. Evidence below that threshold, although it might point toward the guilt of the accused, is deemed insufficient for conviction. Statistical hypothesis testing uses a similar threshold. The data, as summarized by the test statistic and p-value, need to point in the direction of the alternative hypothesis with sufficient strength to reject the null hypothesis. The common criterion is that any p-value smaller than the α error probability leads to a rejection of the null hypothesis. The α and β error probabilities are criteria in selecting the proper sample size. It should be emphasized that statistical hypothesis testing does not result in correct decisions all the time, but in correct decisions with calculable probabilities.

Table 8.2 Decisions and Errors in Hypothesis Testing

Decision	Actual State	
	H_0 is true	H_A is true
Accept H_0/Reject H_A	OK	Error probability β
Reject H_0/Accept H_A	Error probability α	OK

8.1.2 Significance Level α

The threshold that allows us to decide whether to accept or reject a null hypothesis is called the significance level α (alpha). It is also called the Type I error probability.

- It is the probability α of wrongly rejecting the null hypothesis when in fact it is true (see Table 8.2).

- It defines the probability of obtaining "extreme values of a sample statistic," given that the null hypothesis is actually true.

- It is the probability that a sample statistic falls into the rejection region when H_0 is true.

- It is designated α (alpha). Typical values are 0.01, 0.05, 0.10, and 0.25.

- It is the threshold for the p-value, with p-values smaller than α resulting in rejection of H_0.

In most business and social science applications, the gold standard is $\alpha = 0.05$. Critical engineering or medical applications might require a smaller α. When the cost of testing is very high, α might be as high as 0.25 and even higher.

8.1.3 Test Statistic

Test statistics summarize the data under H_0 and are used to decide which hypothesis to accept and to reject. The test statistic depends on the assumptions made and the quantity and quality of data available. Test statistics come in many forms. For example, the t-statistic, commonly used in tests for a single mean of a Normal distribution, consists of three components:

- The sample mean as the evidence.

- $Mean_0$ hypothesized in the null hypothesis as the standard to which the evidence is compared.

- The standard error of the (sample) mean as a measure of the precision of the sample mean. As has been shown earlier, the standard error of the sample mean can be reduced by increasing the sample size.

Other methods such as regression analysis require different test statistics.

8.1.4 *p*-Value

The *p*-value is the result of a hypothesis test and is derived from the observed value of the test statistic and its related distribution. The *p*-value measures the strength of the sample evidence in support of the null hypothesis. A large *p*-value lends stronger support to the null hypothesis than a small *p*-value. The range of *p*-values is from zero to one. Like probabilities, they are unit-less. Technically, they are calculated as follows: Given that H_0 is true, the *p*-value is the probability of obtaining a test statistic at least as extreme (towards H_A) as the observed test statistic. However, the *p*-value is strictly speaking not a probability, because it is calculated after the sample outcome has been observed.

Calculating *p*-values can be inconvenient, unless the calculations are performed by software like JMP. For the *t*-test, JMP labels *p*-values as Prob > |t| for two-sided, Prob <t for left-sided, and Prob >t for right-sided alternative hypotheses.

There are good reasons why p-values are commonly used in hypothesis testing:

- p-values are standardized to a range $0 \leq p \leq 1$. This allows a comparison of p-values from different applications. The value of a test statistic, however, is method specific and cannot be compared across applications.

- Computer packages give p-values.

Some prefer confidence intervals to hypothesis testing with p-values. The main reason for this preference of confidence intervals is that they directly express the range of plausible parameter values. Still, many problems are phrased as decisions, and that is why the p-value continues to have undiminished applicability.

8.1.5 Decision Rule to Accept or Reject H₀

The decision rule in a hypothesis test is as follows:

- Accept H_0, if the p-value $\geq \alpha$.

- Reject H_0, if the p-value $< \alpha$.

A p-value smaller than α indicates weak sample evidence in support of the null hypothesis, and, as a result, leads to its rejection. The value α is the threshold between acceptance (for $p \geq \alpha$) and rejection of the null hypothesis (for $p < \alpha$).

When the null hypothesis is rejected, the result is said to be "statistically significant at the α-level." Other ways of phrasing the rejection of the null hypothesis are to say "significant at the α-level" or "with statistical significance at the α-level." The phrase "at the α-level" is often suppressed when $\alpha = 0.05$ is used as the standard significance level. Research papers typically report either the p-value or an appropriate α-level of significance.

8.1.6 Example: Order Processing Times

At a company selling herbal tea by mail order, the mean processing time of orders has remained unchanged at 120 minutes during the past two years. Recently, important but expensive process changes were introduced. Management needs to demonstrate that the changes actually reduced mean order-processing time. If the changes did not reduce order-processing times, then they will revert to the cheaper old process. If the changes had no effect, you would assume that the mean time remains at 120 minutes. This value is the null hypothesis, written as a simple hypothesis specifying a single value:

H_0: Mean = $Mean_0$ = 120 minutes.

From the research question, decide what needs to be shown about the changes in the processing time.

Left-Sided Alternative Hypothesis

In the example, show that the changes reduced mean order-processing time. This will be the alternative hypothesis, written as follows:

H_A: Mean < $Mean_0$ = 120 minutes

This is a left-sided alternative hypothesis, because the mean is specified as less than (to the left of) the null hypothesis Mean = 120 minutes. Figure 8.1 shows that the alternative hypothesis does not state a specific value for the mean. It simply says that the mean is less than $Mean_0$. Yet only one of these infinitely many means can be the actual mean. The graded area to the left of H_0 is intended to indicate that the farther to the left the actual mean is from $Mean_0$, the more likely the test will reject H_0.

Figure 8.1 Left-Sided Alternative Hypothesis for H_0: Mean = 120 Minutes

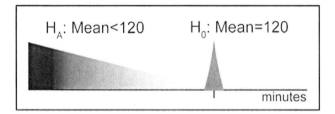

Repeat samples will show variability of the sample mean around the actual mean whatever its value is, because the sample mean is a random variable with a mean and a standard deviation (called the standard error of the mean). The central limit theorem suggests that in many applications sample means tend to be normally distributed. Figure 8.2 shows such a (solid line) distribution of the sample mean around the null hypothesis mean. Even if H_0 is true, the sample mean outcome does not equal the hypothesized mean every time. Instead, sample means are distributed around the hypothesized mean so that some sample outcomes trigger a rejection of the null hypothesis, even though it is true.

The actual mean could also be part of the alternative hypothesis, i.e., Mean < 120. In Figure 8.2, this situation is indicated by the dashed density on the left. The mean of the alternative hypothesis is not specified exactly. However, in order to calculate the β error probability, a specific mean value must be selected. In this left-sided test, the β error probability for a Mean = 110 minutes is larger than the one for a Mean = 100 minutes, everything else being the same. The α error probability is always calculated for the null hypothesis mean, here Mean = 120 minutes.

What if the actual mean is higher than the null hypothesis mean? This case is of little interest in a left-sided alternative hypothesis, but might nevertheless occur. Such situations have an even higher probability of accepting the null hypothesis than if the null hypothesis itself were exactly true. In such a case you would be even more prone to accept the null hypothesis. Accepting the null hypothesis is the correct decision, because the decision would not result in a claim that the mean order processing time has been shortened.

Figure 8.2 Distribution of Sample Mean under H_0 and H_A for Left-Sided H_A

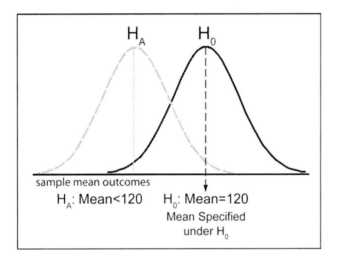

Right-Sided Alternative Hypothesis

You could also consider certain changes to the order process that increased mean order-processing time. Then the alternative hypothesis is

H_A: Mean > $Mean_0$ = 120 minutes

This is a right-sided alternative hypothesis because the H_A mean values are greater than (to the right of) the null hypothesis value $Mean_0$ = 120 minutes. Figure 8.3 shows a schematic for a right-sided H_A. The graded area to the right of H_0 indicates that the farther to the right the actual mean is from $Mean_0$, the more likely the test will reject H_0.

Figure 8.3 Right-Sided Alternative Hypothesis for H_0: Mean = 120 Minutes

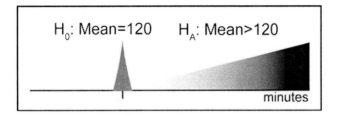

Figure 8.4 shows that assuming a mean from the right-sided alternative hypothesis, the sample mean distribution is to the right of the null hypothesis distribution.

Figure 8.4 Distribution of Sample Mean under H_0 and H_A for Right-Sided H_A

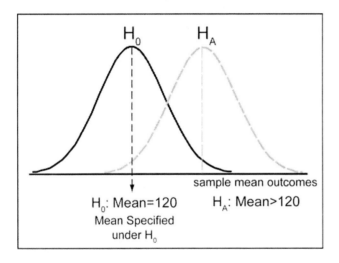

Two-Sided Alternative Hypothesis

Lastly, to demonstrate change without regard to the direction of the change (increase or decrease in order-processing time), a two-sided alternative hypothesis is required:

H_A: Mean $\neq$ 120 minutes = $Mean_0$.

The mean values of the alternative hypothesis can be on either side of the null hypothesis. Figure 8.5 shows a schematic for a two-sided H_A. The graded area is now on both sides of H_0. It indicates that the farther in either direction the actual mean is from $Mean_0$, the more likely the test will reject H_0.

Figure 8.5 Two-Sided Alternative Hypothesis for H_0: Mean = 120 Minutes

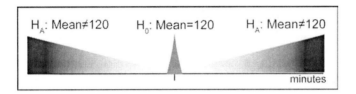

Figure 8.6 presents a graph of the two-sided alternative hypothesis. The distribution of mean values under the null hypothesis is in the center. Because H_A is two-sided, potential alternative hypothesis distributions could have a mean larger or smaller than the null hypothesis mean. Since the alternative hypothesis was not specified as two single values, any mean less than or greater than $Mean_0$ is part of the alternative hypothesis.

Figure 8.6 Two-Sided Alternative Hypothesis with H_0 in Center

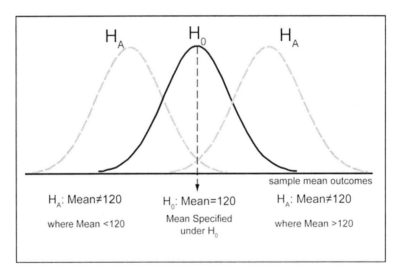

In order to select the correct *p*-value in JMP, write down the null hypothesis and the alternative hypothesis. In the herbal tea order-processing example, the null hypothesis is that process changes did not affect the mean order processing time of 120 minutes. The alternative hypothesis specifies the change in the mean processing time the recent process modifications were intended to have.

- Are we trying to show that processing times decreased? Then it is a left-sided H_A.

- Are we trying to show that processing times increased? Then it is a right-sided H_A.

- Are we trying to show that processing times either decreased or increased? Then it is a two-sided H_A.

In the order processing times example, we are trying to show that process changes shortened mean processing time. A left-sided H_A is selected. Refer to the statement of H_0 and H_A to select the proper *p*-value in the JMP output.

Data: After some process changes have been made, 10 order-processing times have been randomly selected and recorded: 129, 115, 125, 100, 109, 114, 108, 109, 102, 92. They are stored in *OrderProcessing.jmp*.

Problem: Do these 10 observations (the evidence) indicate that the mean processing has been reduced from an average of 120 minutes? In the notation of hypothesis tests we write H_0: Mean = 120 minutes versus H_A: Mean < 120 minutes. This is a left-sided alternative hypothesis. The sample mean is 110.3 minutes and the sample standard deviation is 11.2.

Just as in deciding the innocence or guilt of an accused, a threshold concerning the quality of the evidence is needed. For example, would we be willing to say that there was a reduction in processing time based on:

- 1 observation with a processing time of 110?

- 10 observations with a mean processing time of 110?

- 100 observations with a mean processing time of 110?

From what we know about the standard error of the sample mean and how the sample size reduces it, we can conclude that the sample mean based on 100 observations is more precise than the one based on ten observations, which in turn is more precise than the one based on only one observation. However, we cannot tell whether the sample mean of 110 is lower than $Mean_0 = 120$ with statistical significance.

Statistical hypothesis testing uses two related criteria, the test statistic and the associated *p*-value, to provide objective criteria for a proper conclusion. In a left-sided alternative hypothesis, the test statistic for the mean will result in a rejection of the null hypothesis if the value of the test statistic (t-Ratio) is sufficiently to the left of 0 so that the *p*-value is less than α. This will be explained next.

8.1.7 Test Statistic, Significance Level, Critical Value, and *p*-Value

Test statistics are quantities calculated from the sample data. Test statistics and the significance level α are used to define acceptance and rejection regions. From the distribution of the test statistic, find the α quantile for a left-sided alternative hypothesis, the $(1 - \alpha)$ quantile for a right-sided alternative hypothesis, and the $(\alpha/2)$ and $(1 - \alpha/2)$ quantiles for a two-sided alternative hypothesis. These quantiles define the acceptance and rejection regions of the null hypothesis for test statistics. The *p*-value of any test statistic in the rejection region is less than the significance level α.

The left panel of Figure 8.7 shows the distribution assuming the null hypothesis is true for a left-sided alternative hypothesis. It also shows the critical value and the acceptance and rejection regions. In practice, this process requires knowing the distribution of the test statistic and its appropriate quantile. With JMP, the values of the appropriate test statistics are calculated within each platform along with the *p*-values associated with the test statistics.

The right panel of Figure 8.7 shows the results of the order-processing example. We assume that the times are approximately normally distributed and that the 10 observations were randomly selected from a large (post-change) collection of completed orders. Because the sample standard deviation $s = 11.16$ is calculated from the data, the t-distribution needs to be used. (JMP chooses the proper distribution automatically, depending on the inputs specified.)

Figure 8.7 Critical Value, Acceptance, and Rejection Region in Left-Sided Alternative Hypothesis

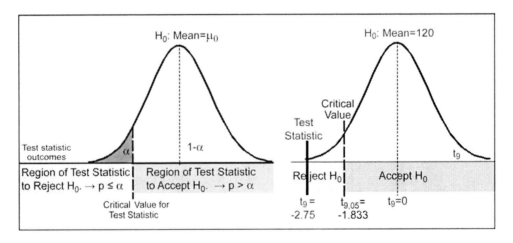

The test statistic appropriate for the above hypotheses is the t-statistic (or t-Ratio in JMP). In the order-processing example, it is as follows:

$$t_{n-1} = \frac{\bar{y} - \mu_0}{s/\sqrt{n}} = t_9 = \frac{110.3 - 120}{11.156/\sqrt{10}} = \frac{-9.7}{3.528} = -2.75$$

This test statistic is compared with the critical value represented by the 5% quantile of the t-distribution with 9 degrees of freedom $t_{9,0.05} = -1.833$. (The quantile can be obtained from JMP using the t-distribution quantile function.) Since $t_9 = -2.75$ is less than $t_{9,0.0} = -1.833$, it falls into the rejection region. The conclusion is to reject the null hypothesis, H_0: Mean = 120 minutes, in favor of the alternative hypothesis H_A: Mean < 120 minutes.

The t-statistic contains several components.

- $\bar{y} = 110.3$ is the sample mean and represents the evidence.

- $\mu_0 = 120$ is the mean of the null hypothesis (Mean$_0$).

- $s/\sqrt{n} = 11.156/\sqrt{10} = 3.528$ is the standard error of the (sample) mean and measures the precision of the sample mean.

- The sample size n determines the degrees of freedom $(n - 1)$, which in turn identifies the appropriate t-distribution. The sample size is 10 and the degrees of freedom are $10 - 1 = 9$.

How extreme is $t_9 = -2.75$, assuming H_0 is true? The *p*-value can be used to answer this question.

p-Value: In JMP, it is most convenient to base the accept/reject decision on the *p*-value, because (1) JMP calculates *p*-values, and (2) *p*-values are always on a scale from 0 to 1. The interpretation of the *p*-value does not depend on the test statistic. It can be said that with the *p*-value, "One rule fits all!": Accept H_0 if $p \geq \alpha$; reject H_0 if $p < \alpha$.

In the delivery times example, the *p*-value corresponding to the *t*-test statistic t = –2.75 is calculated from P[t9 < –2.75] = 0.0112, leading to a rejection of H_0 at $\alpha = 0.05$.

Critical Value: What is behind the test statistic and its critical value? In testing for the mean with the standard deviation estimated from the data, the corresponding *t*-test statistic follows a Student's t-distribution with $(n - 1)$ degrees of freedom. For a one-sided alternative hypothesis, the critical value for the *t*-test statistic is the $1 - \alpha$ quantile of the t-distribution.

- For left-sided hypotheses, reject H_0 if the t-test statistic $< -t(1 - \alpha)$. (Reject H_0 if $p < \alpha$)

- For right-sided hypotheses, reject H_0 if the t-test statistic $> t(1 - \alpha)$. (Reject H_0 if $p < \alpha$)

Notice that for these hypotheses, the rejection rules are the same for the *p*-value, but differ for the t-test statistic. Using JMP, t(0.95) for 9 degrees of freedom is 1.83. Reject H_0, because $t_9 = -2.75 < -1.833$. The correspondence between the t-test statistic, *p*-value, α-critical value and α is shown in Table 8.3.

Table 8.3 Correspondence between Test Statistic and *p*-Value

From Sample	In JMP Distribution Platform	Related to Significance Level
t-statistic	t-Ratio	α-critical value of t-distribution
p-value	Prob < t, Prob > t, Prob > |t|	α

We measure the distance of the data evidence from the null hypothesis by evaluating the area under the t-distribution density (for left-sided H_A) to the left of the test statistic, and call this area the *p*-value. The *p*-value measures how likely it is to obtain a sample mean and standard deviation that will produce a test statistic $t = -2.75$ or smaller, assuming that the null hypothesis Mean = 120 is true.

8.1.8 Example: Order Processing with JMP

The data in *OrderProcessing.jmp* consist of a single column containing sample order processing times. The Distribution platform is used to obtain basic results. Click the red triangle next to the variable name and select **Test Mean** from the options menu of the Distribution platform (left panel of Figure 8.8). This action produces the Test Mean specification dialog box (center panel of Figure 8.8).

Specify the null hypothesis value Mean = 120. Leaving the standard deviation field blank uses the sample standard deviation as a default. The resulting *p*-values are based on the t-statistic with $(n - 1)$ degrees of freedom, where *n* is the sample size. (Entering a standard deviation value would result in a test that assumes that the true standard deviation is entered, resulting in a z-statistic with *p*-values calculated from the standard Normal distribution.) Click **OK** and observe the results as shown in the right panel of Figure 8.8.

These results restate the hypothesized value and key values calculated from the data. They also give the test statistic as t = –2.7496. The last three rows are three different *p*-values (labeled as Prob) corresponding to two-sided, right-sided, and left-sided alternative hypotheses. JMP placed an asterisk (*) next to *p*-values less than 0.05. In the present example, we need to look at only the left-sided *p*-value (Prob < t) = 0.0112.

Figure 8.8 Distribution Options with Test Mean Results for Order Processing Times

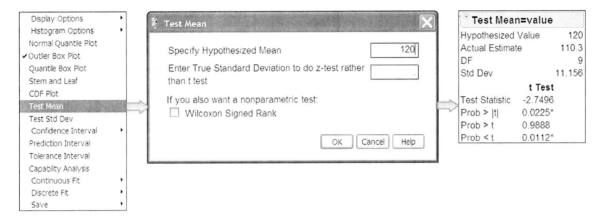

You need to specify H$_A$ before the proper *p*-value can be selected. Table 8.4 shows the *p*-values from JMP and discusses the choice of the left-sided *p*-value for the rejection of H$_0$.

Table 8.4 Test Statistic and *p*-Values in JMP

Statistic	Example value	Direction	Interpretation
Test Statistic	−2.7496	Non-directional	The test statistic is to the left of the critical value $t_{.05, 9} = -1.833$, which is from a Student's t-distribution with $n - 1 = 9$ degrees of freedom (df). The Test Mean window does not show the critical value. Use *p*-values instead.
Prob > |t|	0.0225	Two-sided *p*-value	The *p*-value is less than $\alpha = .05$, so H_0 could be rejected, but it is ignored because H_A is left-sided.
Prob > t	0.9888	Right-sided *p*-value	This is the wrong side for this left-sided hypothesis! Ignore it.
Prob < t	0.0112	Left-sided *p*-value	This left-sided *p*-value is less than $\alpha = .05$, so reject H_0! It is half of the two-sided *p*-value.

8.1.9 Confidence Intervals and Two-Sided Hypothesis Testing

A confidence interval for a parameter, e.g., a mean μ or a standard deviation σ, can be viewed as the set of all null hypotheses with two-sided alternative hypotheses that could not be rejected with the given data. You cannot reject a two-sided hypothesis at significance level α as long as the parameter specified in H_0 is within the $(1 - \alpha)100\%$ confidence interval.

Example: Herbal Tea Mail Order

Management of Herbal Tea Mail Order wants to know whether a recent change in the design of the Web site had any effect, positive or negative, on the daily value of orders. The data are contained in the column titled Daily Value $ of *Herbal Tea Mail Order.jmp*. The average daily value of orders for the year before the change in Web site was $5657. Test whether the Web site change had an impact on Daily Value $.

The null hypothesis H_0: Mean = \$5657 is that the mean of Daily Value \$ did not change because of the change in the Web site. The alternative hypothesis H_A: Mean $\neq$ \$5657 claims that the mean of Daily Value \$ changed. The basic results from JMP show that the Sample Mean = \$6244 and that the 95% confidence interval is from \$5701 to \$6787. Figure 8.9 shows that the hypothesized mean is outside the 95% confidence interval and allows the rejection of the null hypothesis using a two-sided α = 0.05. It appears that the Web site design change increased the average daily value of orders. The Test Mean procedure confirms this. The two-sided p-value Prob > |t| = 0.0346 is less than α = 0.05. Had the null hypothesis been anywhere from \$5701 to \$6787, the null hypothesis could not have been rejected.

Figure 8.9 Hypothesized Mean and 95% Confidence Interval Do Not Overlap. Reject H_0

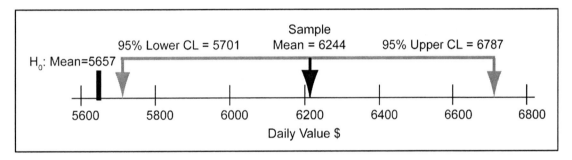

8.2 Sample Size Needed to Test H₀: Mean = Mean₀ versus Hₐ: Mean = Meanₐ

8.2.1 Introduction

In statistical hypothesis testing, the true state of nature is unknown and decisions are based on sample information. Sample information can be misleading, pointing towards one hypothesis when in fact the other hypothesis is true. This fact results in Type I and Type II errors, with error probabilities α and β respectively. Using statistics, we cannot decide with certainty, but we can calculate the probabilities with which we make each of these erroneous decisions. Likewise, we can find a sample size that will result in a test with the desired α and β error probabilities. To calculate sample sizes for simple hypothesis testing, several quantities need to be specified:

- H_0: Mean = Mean_0 and H_A: Mean = Mean_A, two specific values for the null and alternative hypotheses.

- α, the error probability of rejecting H_0: Mean = Mean_0 when H_0 is in fact the true state of nature.

- β, the error probability of rejecting H_A: Mean = Mean_A (accepting H_0) when H_A is exactly true. Since H_A might have many values, a specific value (Mean_A) has to be selected to perform calculations.

- σ, a value for the standard deviation. Sometimes a reasonable guess might have to be made in order to perform sample size calculations. In applications such as polling, the worst-case value can be used for sample size determination.

Figure 8.10 shows the relationship of means and error probabilities for a right-sided alternative hypothesis. The null hypothesis and alternative hypothesis are shown as the center of each distribution of the sample mean. The horizontal axis represents outcomes for the sample mean $\bar{y}$, which is in this case the test statistic. If H_0 is true, the mean of the $\bar{y}$ distribution is to the left in this right-sided test. If H_A is true, it is to the right. In between H_0 and H_A is the critical value that defines the accept and reject regions. For specific H_0 and H_A, the sizes of the regions define the α and β error probabilities.

Figure 8.10 Error Probabilities α and β and Difference to Detect in a Right-Sided Hypothesis

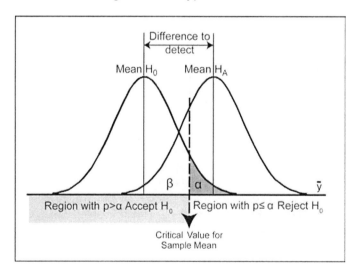

If H_0 is true, you want to accept it with a high probability $(1 - \alpha)$. If H_A is true, you want to accept it with a high probability $(1 - \beta)$. This probability $(1 - \beta)$ is called the **power** of the test. Figure 8.11 shows this for a right-sided hypothesis test.

Figure 8.11 Probabilities to Accept H_0 and H_A When They Are True

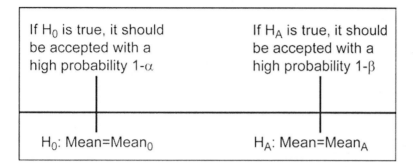

There is a strict relationship between α and β error probabilities, the distance between $Mean_0$ and $Mean_A$, and the standard deviation σ. In terms of distributions, the relationship can be visualized using two Normal distributions with different means representing the different hypotheses, assuming the standard deviations to be the same. In calculating sample sizes, you need to specify the Difference to detect. This is a difference $|Mean_0 - Mean_A|$ between H_0 and H_A. It is important to the investigator that the probability to accept H_0 is $1 - \alpha$ if H_0 is true; likewise, the probability to reject H_0 is $1 - \beta$ if a *specific* H_A is true. The sample size n is chosen to satisfy these two requirements.

As a rule, a relatively large difference between $Mean_0$ and $Mean_A$ requires a small sample size, while a relatively small difference requires a large sample size. Relatively small error probabilities require large sample sizes and vice versa. Similarly, a relatively small standard deviation results in small sample sizes and vice versa. (Note that the distributions depicted in Figure 8.10 are of the sample mean rather than of individual outcomes.)

8.2.2 Example: Sample Size Calculations for Mama Mia's Pizza Parlor

Situation: Mama Mia's Pizza Parlor has been known for pizzas with lots of cheese. Mama Mia rightly boasted that every one of her pizzas is topped with two ounces of imported Romano cheese. Recently, Mama Mia's daughter Ethel took over this famous establishment. Sure enough, some of her regular customers got into an argument about whether the quantity of cheese has been reduced since then. They suggest sampling the new pizzas.

This problem will be discussed in the remainder of the chapter. Next, the sample size will be calculated using JMP. Then an approximate formula will be used to calculate the sample size. Lastly, power curves will be constructed for this problem.

Problem: How many pizzas do you have to sample? Your friends want to be correct with probability 0.95 if Ethel continues to use two ounces of cheese as her mother did. If she reduced the amount of cheese by 0.25 ounce, then they want to be correct in identifying this lower value with probability 0.90. Experience shows that the standard deviation of cheese on pizzas varies with $\sigma = 0.25$.

Requirements for Sample Size Calculations: The following values are needed to calculate sample size for discovering a process improvement:

- a value of the process standard deviation $\sigma = 0.25$

- the mean of the existing process, which is $Mean_0 = 2.0$

- the potential change in mean, which is to $Mean_A = 1.75$

- the difference to detect, which is $| Mean_0 - Mean_A | = |2.0 - 1.75| = 0.25$

- values for the error probabilities $\alpha = 0.05$ (failure to find $Mean_0$ when it is true) and $\beta = 0.10$ (failure to find $Mean_A$ when it is true)

Note that JMP asks for the power instead of β. The power $= 1 - \beta = 1 - 0.10 = 0.9$.

JMP can be used to obtain exact sample size calculations, or there are approximate formulas based on the Normal distribution.

8.2.3 Sample Size in JMP

The Sample Size and Power platform, selected from the **DOE** menu shown in Figure 8.12, can be used to calculate the required sample size.

Figure 8.12 Sample Size and Power in JMP

For the current example, click the **One Sample Mean** button in the left panel of Figure 8.13. This action produces another window in which the necessary values are specified (right panel of Figure 8.13).

Figure 8.13 Sample Size Menu and One Sample Mean Specifications

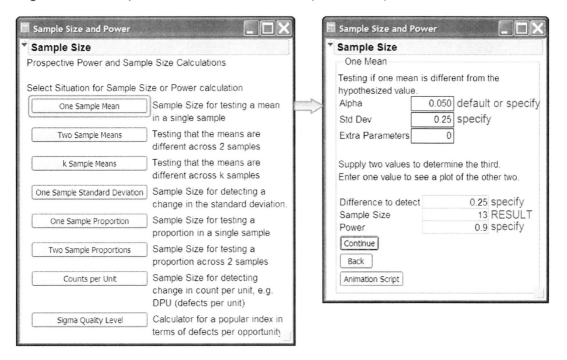

Specify all values, but leave the sample size field blank. Alpha = 0.05 by default, but can be altered if desired. The Error Std Dev (of individual values) is assumed to be 0.25 in this example. Do not specify Extra Params. The Difference to detect is $|2 - 1.75| = 0.25$, and the desired Power = $1 - \beta = 1 - 0.1 = 0.9$.

Click the **Continue** button to obtain the required sample size of 13. Fractional sample sizes are not possible, because only whole pizzas are sampled. When fractional sample sizes are obtained, JMP rounds up to the next largest value.

8.2.4 Approximate Formulas for Sample Size

For a two-sided alternative hypothesis, the sample size formula requires $\alpha/2$ and β, because the α errors occur on either side of the null hypothesis, whereas the β error occurs only on one of the two sides. The two-sided approximate sample size formula is

$$n = \frac{\left(z_{1-\alpha/2} + z_{1-\beta}\right)^2 \cdot \sigma^2}{\left(\text{Mean}_0 - \text{Mean}_A\right)^2}$$

For $\alpha = 0.05$ and $\beta = 0.10$, the two-sided sample size formula requires $z_{0.975} = 1.960$ and $z_{0.9} = 1.282$ from the Standard Normal quantiles. The approximate required sample size for the two-sided example is

$$n = \frac{\left(1.960 + 1.282\right)^2 \cdot 0.25^2}{\left(2.00 - 1.75\right)^2} = 10.51 \approx 11$$

This approximate sample size is smaller, because it uses Normal distribution quantiles rather than the t-distribution quantiles in JMP.

For a one-sided alternative hypothesis, the sample size is very similar, except that both error probabilities are all assigned to one side. The formula for the approximate sample size is

$$n = \frac{\left(z_{1-\alpha} + z_{1-\beta}\right)^2 \cdot \sigma^2}{\left(\text{Mean}_0 - \text{Mean}_A\right)^2}$$

For $\alpha = 0.05$ and $\beta = 0.10$, the one-sided approximation requires $z_{0.95} = 1.645$ and $z_{0.90} = 1.282$ from the Standard Normal table.

8.2.5 Power Curves

The power of a test is $(1 - \beta)$. The power of a test depends on α and on the sample size n. A power near 1 is desirable when the alternative hypothesis is true, because it indicates that the test is able to reject H_0 given that a value of H_A is the true value. Similarly, a power near 0 is desirable when the null hypothesis is true.

For a given α, power curves show how the power of a test changes as a function of either the sample size with a fixed difference between the means of the null and the alternative hypothesis (Figure 8.14), or as a function of this difference with a fixed sample size (Figure 8.15). The power curve shows how well the test design achieves its purpose, and should be used before the data are collected.

The power curve of Figure 8.14 shows power as a function of various sample sizes for a fixed Alpha, Error Std Dev and Difference in Means. It is constructed by leaving blank the fields for Sample Size and Power. Only the Difference is specified at 0.25. The result is a curve with Power on the vertical axis and Sample Size on the horizontal axis. With this curve you can find the power of a test for a given sample size or, conversely, the sample size required for a given power. The intersecting vertical and horizontal reference lines illustrate one of these relationships. In Figure 8.14, the vertical reference line is marked for a sample size of $n = 13$. This line intersects with the power curve when the power is approximately 0.90, as shown by the horizontal reference line. The power is actually higher because the calculated sample size $n = 12.58$ was rounded to 13. The curve indicates that raising the sample size to about $n = 20$ will increase the power to a value nearer to 1. On the other hand, a reduction of the sample size to $n = 6$ gives a power of approximately 0.5. Such a test would find a Difference = 0.25 only with probability 0.5.

Figure 8.14 Power Curve as a Function of Sample Size

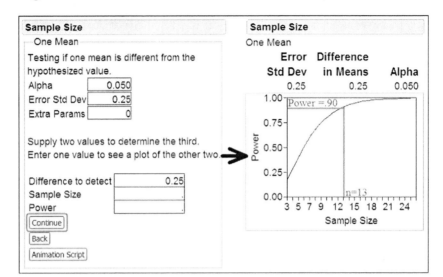

A second power curve is shown in Figure 8.15. This power curve presents power as a function of the difference to detect, while keeping Alpha, Error Std Dev and Sample Size fixed. Specify a value for sample size, and leave Difference to detect and Power blank, as shown in the left panel of Figure 8.15. In this example, the sample size is $n = 13$ and, as before, Alpha = 0.05 and the Error Std Dev = 0.25. The resulting power curve has an S-shape.

This power curve shows that for a sample size $n = 13$, small differences between H_0 and H_A have only a small power. For a difference of 0.15, the power is approximately 0.5. When the H_A is true with an actual mean 1.85 ounces of cheese per pizza, the test has an even probability of accepting H_0 or H_A. The vertical and horizontal lines mark the case for a difference equal to 0.25 ounces. In this case, the test will accept H_A that the actual mean of cheese per pizza is 1.75 ounces with a probability of more than 0.9.

Figure 8.15 Power Curve as a Function of Difference

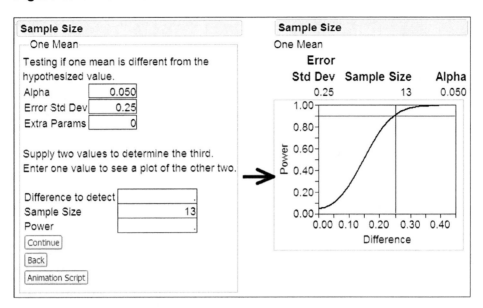

Steep power curves are a consequence of larger sample sizes. The steepness indicates how drastic the transition is from preferably accepting H_0 to preferably accepting H_A. Likewise, smaller differences can be detected with high power. To verify this claim, repeat the example of Figure 8.15, but with a sample size $n = 100$. Keep the horizontal axis in the range from 0 to 0.40.

Power curves are a popular tool in designing sample collection protocols in quality management, because samples are often repeatedly drawn from different work shifts or batches of production. The power curve shows how sensitive the sampling procedure is in discovering deviations from the norm as represented by the parameter value postulated in H_0.

8.2.6 Sample versus Process Standard Deviation (s versus σ)

For testing a hypothesis concerning a mean, H_0: Mean = $Mean_0$ versus H_A: Mean < or > or ≠ $Mean_0$, use either a process standard deviation that has been estimated over a long time period or the sample standard deviation calculated from the data. A standard deviation estimated over a long time period or accumulated from a large sample is considered to be a KNOWN standard deviation σ.

Standard deviation estimates from a single small to moderate sample size do not provide precise enough information. Consequently, the standard deviation σ is assumed UNKNOWN. These two cases are treated differently in JMP and are presented here.

Standard Deviation σ Is Unknown—Use Sample Standard Deviation s

When the standard deviation is not known, it needs to be estimated from the sample data. In JMP this case is invoked by leaving the Enter True Standard field blank. JMP uses a default test statistic as follows:

$$t_{n-1} = \frac{\bar{y} - \text{Mean}_0}{s/\sqrt{n}}$$

This test statistic is assumed to have a t-distribution with $(n-1)$ degrees of freedom.

For this example, the friends wanted to buy the indicated thirteen pizzas, but found they had cash for only 10 pizzas. A smaller sample reduces the power of the test, but can still give useful results. The pizzas yielded the following weights of cheese in ounces: 1.74, 1.68, 2.03, 2.04, 1.78, 1.84, 1.98, 1.87, 1.81, and 1.88. For the analysis in JMP, we need to construct a data table with one column labeled "Cheese per pizza" and 10 rows. This data table is contained in *MamaMiaCheese.jmp*.

The Distribution platform from the Analyze menu analyzes a single variable (Cheese per pizza) as in Figure 8.16. The results show that the average amount of cheese on the 10 pizzas sampled was 1.865 oz. and that the standard deviation was only 0.1211 ounces (which is of course less than the value used in the sample size calculations!).

Figure 8.16 Basic Output of *MamaMiaCheese.jmp*

Quantiles			Moments	
100.0%	maximum	2.0400	Mean	1.865
99.5%		2.0400	Std Dev	0.1211289
97.5%		2.0400	Std Err Mean	0.0383043
90.0%		2.0390	upper 95% Mean	1.9516504
75.0%	quartile	1.9925	lower 95% Mean	1.7783496
50.0%	median	1.8550	N	10
25.0%	quartile	1.7700		
10.0%		1.6860		
2.5%		1.6800		
0.5%		1.6800		
0.0%	minimum	1.6800		

Does this confirm your friends' suspicion that Mama Mia's pizzas contain less cheese? To answer this question, test the null hypothesis that the mean equals 2.0 ounces using the **Test Mean** pull-down menu.

The dialog box on the left in Figure 8.17 shows null hypothesis Mean = 2.0, but the value for the **Enter True Standard Deviation** field is left unspecified, because the standard deviation σ is unknown. The right panel of Figure 8.17 shows all three possible p-values of the p-test. In this example, you are interested only in the left-sided p-value labeled Prob < t. This one-sided p-value = 0.0032 is much below the significance threshold α = 0.05. Therefore, reject the null hypothesis. The statistical evidence shows that on average the amount of cheese per pizza has decreased.

Figure 8.17 Specify Null Hypothesis and Test That Mean = 2.0 Using Std Dev = 0.1211

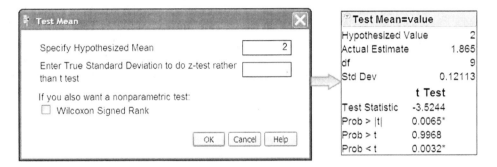

Standard Deviation σ Is Known

This case applies when the standard deviation is known from historical experience or otherwise. Specify σ so that its value can be included in calculating the following test statistic:

$$z = \frac{\bar{y} - \text{Mean}_0}{\sigma / \sqrt{n}}$$

This test statistic is assumed to have a Normal distribution with Mean = 0 when the null hypothesis is true. JMP uses this version of the test statistic whenever a value is entered for the true standard deviation, as indicated in Figure 8.18. The value specified in the **Enter True Standard Deviation** field appears as "Sigma given" in the result table on the right.

Figure 8.18 Test Mean Window and Results with Sigma Known

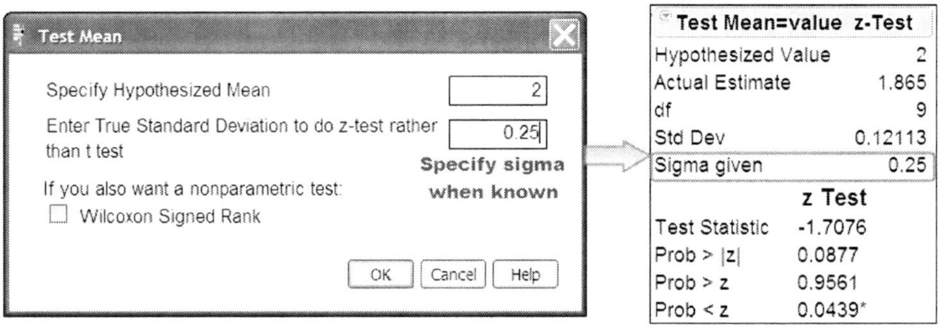

Using a known standard deviation $\sigma = 0.25$ causes JMP to perform the z-test based on the Normal distribution instead of the t-distribution. The results in Figure 8.18 show that the "Test Statistic" is based on the "z Test" instead of the " t Test" of Figure 8.17. The z-test uses the "Sigma given" = 0.25 instead of the "Std Dev" = 0.1211. The left sided *p*-value associated with the z-test and standard deviation = 0.25 is Prob < z = 0.0439. This *p*-value is below the $\alpha = 0.05$ significance threshold. Therefore, reject the null hypothesis. This *p*-value is larger than the corresponding one from the t-test, because the data are assumed more variable with a standard deviation = 0.25 instead of the Std Dev = 0.1211 for the t-test. It might be useful to investigate how precise the standard deviation estimate is.

Confidence Interval for the Standard Deviation σ

For the Mama Mia cheese example, the 95% confidence intervals for the mean and the standard deviation are available from the Confidence Interval option, as shown in the left panel of Figure 8.20. The point estimate s = 0.121. The results of Figure 8.19 show the 95% confidence interval for sigma is $0.083 < \sigma < 0.221$. This interval does not include the assumed value $\sigma = 0.25$. This seems to be a case where the original assumption about σ was not quite correct. Note that the confidence interval for the standard deviation is an example of the point estimate (Std Dev) not being in the center of the confidence interval. The calculations for this confidence interval are based on the asymmetric Chi-Square distribution.

Figure 8.19 95% Confidence Limits for Mean and Standard Deviation

Confidence Intervals				
Parameter	Estimate	Lower CI	Upper CI	1-Alpha
Mean	1.865	1.77835	1.95165	0.950
Std Dev	0.121129	0.083317	0.221134	

8.2.7 Hypothesis Test for the Standard Deviation σ

The hypotheses are H_0: $\sigma = \sigma_0$ versus H_A: $\sigma <, >, \neq \sigma_0$. The test statistic $\chi^2_{n-1} = (n-1)s^2/\sigma_0^2$ requires a Chi-Square distribution with $(n-1)$ degrees of freedom. In JMP, the test is carried out by the Test Std Dev option. The menus and results are shown in Figure 8.20.

Figure 8.20 Summary of Test for Hypothesized Standard Deviation

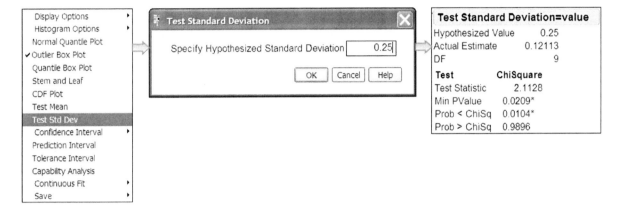

In this case, we test H_0: $\sigma = 0.25$ versus H_A: $\sigma < 0.25$. Only the left-sided *p*-value is of interest. It is labeled as Prob < ChiSq and is 0.0104. Because p = 0.0104 < α = 0.05, the null hypothesis can be rejected at the 0.05 level of significance. The actual standard deviation is less than the hypothesized σ = 0.25.

Normal Quantile Plot

The previous hypothesis test for the standard deviation assumes that the observations are normally distributed. If that assumption is not met, the *p*-value might be wrong and the conclusions might be invalid. Check the normality assumption with the Normal Quantile Plot option from the Distribution platform. The normal quantile plot for the data in *MamaMiaCheese.jmp* is shown in Figure 8.21. The horizontal axis measures cheese per pizza. The inner vertical axis is on a normal quantile scale, while the outer vertical scale from 0 to 1 gives the cumulative probability. Normally distributed data are expected to fall near the straight solid line that is fitted to the mean and standard deviation of the data. The horizontal dashed line at the 0.50 quantile of the vertical axis intersects with the solid line at the mean (1.865). The slope of the solid line is determined by the standard deviation (0.12113). In a perfect fit, all values from a distribution with mean 1.865 and standard deviation .12113 would fall on the solid line. The two dashed flared lines, referred to as Lillefors bounds, specify 95% confidence bounds. As long as observations are within the lower and upper Lillefors bound, conclude that the data are normal; at least there is no evidence that they are not normal. The width of the bounds depends on the number of observations. For large samples, the bounds can be so tight as to be only visible with difficulty.

Figure 8.21 Normal Quantile Plot of *MamaMiaCheese.jmp*

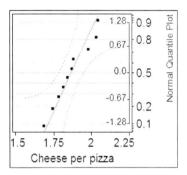

In Figure 8.21 the 10 data points are well within the 95% bounds, and near the fitted solid line. It can be concluded that the data are normally distributed. This conclusion also affirms the validity of the 95% confidence interval and the hypothesis test for the standard deviation.

8.3 Summary

- Hypothesis testing is analogous to making decisions in a criminal court where the accused is pronounced either guilty or not guilty based on the evidence presented to the court.

- Hypothesis testing concerns two states of a population characteristic. These two assumed states are called the null hypothesis and the alternative hypothesis.

- The assertion that the population mean equals a certain value is a simple null hypothesis.

- Alternative hypotheses often contain more than one value of the population characteristic.

- Alternative hypotheses are either left-sided, right-sided, or two-sided.

- Hypothesis testing results in a decision to accept or reject the null hypothesis, or to reject or accept the alternative hypothesis.

- Hypothesis testing may result in wrong decisions. The probability that the null hypothesis is rejected when in fact it is true is called α (alpha). The probability that the alternative hypothesis is wrongly rejected is called β (beta).

- The significance level of a hypothesis test is α. The significance level is used as an accept/reject threshold for the null hypothesis.

- The *p*-value is a number between 0 and 1. It measures how much the test statistic supports the assumption that the null hypothesis is true. A *p*-value less than the significance level α results in a rejection of the null hypothesis. Using the *p*-value in the decision rule is a method to accept or reject the null hypothesis.

- Instead of the *p*-value and α, you can use the test statistic and the critical value in a decision rule to accept or reject the null hypothesis.

- Critical values define acceptance and rejection regions for the test statistic. These correspond in one-sided alternative hypotheses to this decision rule: Accept H_0 if $p \geq \alpha$. Reject H_0 if $p < \alpha$.

- Hypothesis tests with two-sided alternative hypotheses can be performed using confidence intervals. For a 5% significance level, use a 95% confidence interval. If the hypothesized population characteristic is within the confidence interval range, accept the null hypothesis; otherwise reject it.

- Sample sizes that yield hypothesis tests with error probability α and β need to be calculated before data are collected.

- The required sample size also depends on how small the *difference to detect* is. A small difference, relative to the population standard deviation, requires a large sample.

- Sample sizes can be calculated in JMP using the DOE platform.

- Power curves are used to examine the trade-offs between sample size, difference to detect, alpha, and power.

8.4 Problems

1. For each of the following situations, set up the null and alternative hypotheses. Using a bell-shaped curve with the null hypothesis value for the mean, sketch the critical region and show the associated α as the area under the bell-shaped curve.

   a. The chlorine level in a swimming pool must be on average 4.0 units. It is suspected that the chlorine level has increased. Assume $\alpha = 0.05$.

   b. Actual daily travel expenses are budgeted to be on average $150 per day. Management needs to find out if the average has changed. Assume $\alpha = 0.02$.

   c. The average response time to calls made to a customer service center has been 30 seconds. A new call protocol aims at shortening that time to on average below 30 seconds. Assume $\alpha = 0.10$.

2. The supervisor of a stock room keeps track of the inventory level of a critical spare part. He is planning to conduct a test of hypothesis that the mean number of critical spare parts on hand is 25. Write down the research questions that would be addressed by each of the three alternative hypotheses—left-sided, right-sided, and two-sided.

3. At a bottling plant, a soft drink is automatically filled in bottles that have an 8.5-ounce capacity. The nominal amount in each bottle is 8 ounces. Bottles that overflow cause product to be wasted, while bottles that are under filled are not salable. Neither situation is desirable. By keeping the filling process mean at 8.1 ounces, most bottles will be properly filled. The quality assurance engineer is concerned that the process mean might have shifted. A random sample of 20 bottles is measured. State the null and alternative hypotheses. Characterize this test as left-, right- or two-sided.

4. The mean delivery time for an order from an online bookseller has been 3.2 days. The bookseller would like to reduce the delivery time by using a different delivery service. It is proposed that 500 packages be randomly selected and shipped via another delivery service. Write down the null and alternative hypotheses that can be used to determine if the bookseller should use a different delivery service. Characterize this test as left-, right- or two-sided.

5. New drivers hired by a trucking firm attend a one-week orientation class that culminates in a written examination and a road test. As a condition of their employment, the drivers must pass the road test and score at least 85 (out of 100 points) on the written examination. Historically the mean score on the exam has been 90, but there has been concern that the scores have dropped. The scores for the most recent orientation class are recorded in *trucking_exam.jmp*.

   a. Conduct a test of hypothesis to determine if there has been a decline in the mean written exam score.

   b. What assumptions must be met in order to assure the validity of the hypothesis test?

   c. If this test of hypothesis needs to be repeated in the future, what changes for the data-collection process would you recommend to better satisfy the assumptions?

6. An independent contractor sells roasted peanuts and popcorn at a baseball stadium. He has a standing order with his supplier for 20 cases of popcorn and 15 cases of peanuts per game during baseball season. If the demand is higher than the pre-ordered quantity, the contractor can speed-order additional merchandise from a store next to the baseball stadium at a much higher price. The file *popcorn_peanuts.jmp* contains a random sample from last year's sales per game. Based on this data, would you recommend that the contractor change his standing order? If so, how? Discuss how you reached your conclusion.

7. Two medical researchers are analyzing data from a clinical trial testing a new formulation of a well-known pain reliever. The current formulation is advertised as being effective for 12 hours. The new formulation is designed to provide pain relief beyond 12 hours. Both researchers are analyzing the same data. One of the researchers states that based on his analysis the average time of pain relief with the new formulation did not exceed the current formulation with statistical significance. The other researcher disagrees, claiming that his analysis shows that average pain relief was longer with statistical significance. Analyze the data. What is your conclusion? How can you explain the difference in the two researchers' conclusions? The data can be found in *new_formulation.jmp*.

8. A school psychologist needs to ascertain that the mean IQ score of fourth graders in a school district is 100. She plans to randomly sample 100 fourth graders and administer IQ tests. The results are given in the file *IQscores.jmp*.

    a. Set up the appropriate statistical test of hypothesis.

    b. Find the *p*-value for this test. Give an explanation of the *p*-value for this test.

9. A computer help desk is trying to improve service times to resolve customer inquiries. Currently, it takes an average of 5 minutes to resolve a customer call. A random sample of 20 call times is given in *helpdesk_call_times.jmp*.

    a. Give a statistical summary of the data.

    b. Use these data to test at the 5% significance level whether or not the help desk has improved the mean time to resolve customer calls.

    c. Find a 95% confidence interval on the true mean resolution time. Explain whether in this problem the hypothesis test and confidence interval are equivalent.

10. The manager of a grocery store is concerned that a local dairy's actual mean filling amount for half gallon cartons is less than the nominal amount of milk. He wishes to substantiate his claim by taking a random sample of 50 cartons and measuring the amount of milk contained in each carton. The data are recorded in the file *MilkCartons.jmp*.

    a. Set up a test of hypothesis that is appropriate for this situation.

b. Do the data support the grocery manager's suspicion that the dairy on average fills less than the nominal half gallon? Give your reasons for your conclusion.

c. Do the fill amounts of individual cartons have to be normally distributed to test the hypothesis for the mean fill amount? Explain why or why not.

d. State the 95% confidence interval for the mean amount of milk. Explain the meaning of this confidence interval.

11. A consumer research group is ready to recommend a brand of champagne to its subscribers. As part of their recommendation, they want to give an average price. They obtain the following prices (in dollars) of the champagne at six randomly selected stores. The prices can be found in *champagne_prices.jmp*.

a. Find a 95% confidence interval and interpret it.

b. Does the data support the premise that champagne price averages $30? Show your statistical test of hypothesis, using $\alpha = 0.05$.

c. Exclude the apparent outlier and test the hypothesis again. Does the outlier affect your conclusion? If so, why? Show how you arrived at your answer.

12. Internal auditors are preparing to conduct an annual audit of invoices and need to determine how many invoices must be examined. Historically the errors in invoices have had a mean of $25 and a standard deviation of $20.

a. Prepare a table that shows the sample sizes required to estimate the mean error for α values of 0.01, 0.025, 0.05, and 0.10. Assume the difference to detect = $10 and $\beta = 0.10$.

b. Generate a power curve that shows how sample size changes with power for a difference to detect of $10 and $\alpha = 0.05$.

c. Generate a power curve that shows how sample size changes with difference to detect for $\beta = 0.10$ and $\alpha = 0.05$.

d. Give a recommendation for the sample size to be used for the audit. State the corresponding power, α, and the difference to detect. Justify the values you chose.

13. A medical researcher is testing a new pain reliever and wishes to estimate the number of hours of pain relief for patients who have bone fractures. Prior testing shows the mean time of pain relief is 13 hours with a standard deviation of 2.5 hours. The study protocol requires that the hypotheses be tested with $\alpha = 0.05$ and Power = 0.95. The researcher wishes to detect the difference in the mean to within 1.0 hour. There are 15 patients available for testing the new medication.

a. Is this sample size adequate given the requirements of the study? If not, how many additional patients would be required to meet the requirements of the study?

b. What difference in the mean could be detected for the 15 available patients?

14. A manufacturer claims that a new fiber optic cable has a mean breaking strength $\mu = 15$ kg with a standard deviation of 0.5 kg. To test the null hypothesis that $\mu = 15$ kg against the alternative that $\mu < 15$ kg, a random sample of lengths from 50 different rolls of cable will be tested. The test will be conducted with $\alpha = 0.05$. The data for a fiber optic cable-breaking strength test can be found in *fiber_cable_strength.jmp*.

    a. Sketch a bell-shaped curve with the null hypothesis mean in the center and mark the rejection region. Define Type I error α (alpha) and explain what it means in the context of drawing a conclusion from sample evidence.

    b. What are the consequences of reducing $\alpha = 0.05$ to $\alpha = 0.01$ for a test of hypothesis?

    c. Set up and perform a test of the null hypothesis that $\mu = 15$ kg against the alternative that $\mu < 15$ kg.

    d. Test whether the assumption that the standard deviation is 0.5 kg is supported by the data.

15. Formulate a question that could assist with a business decision in your job, career, or personal life. Find a source of data that can be analyzed to answer the question. Analyze the data and summarize your problem and results in no more than one page. Your summary should include the following:

    a. Describe the business situation and the problem to be addressed.

    b. Cite the source of your data. Discuss why you selected this data source and the quality of the data. Do you have any reservations about using this data?

    c. Describe the data.

    d. Set up and test a statistical hypothesis to answer the question.

    e. Discuss how the analysis can be used in this situation and what business actions are indicated.

8.5 Case Study: Traffic Speed Limit Change

A controversy has emerged in a town regarding a section of roadway with a speed limit of 35 miles per hour. Citizens living along the road contend that the speed limit should remain at 35 due to the number of residences along the road. They believe safety is a key issue for residents pulling out of their driveways and children playing in the yards. They believe the town is not providing sufficient enforcement of the speed limit in this area. Other citizens who commute along this stretch of highway claim that the speed limit is too slow and should be raised to at least 40 miles per hour. When police monitor this roadway, their current practice is to stop only those motorists traveling more than 5 miles over the speed limit.

Business Problem

The town must decide whether the speed limit should remain at 35 mph, or should be raised. Further, the town must determine whether the police should change their speed limit enforcement practices.

Strategy

The town will hold a public hearing that is expected to be contentious and well attended by both sides of the controversy. The town has hired a traffic-planning firm to collect a small amount of data and perform a preliminary analysis that will be ready in time for the hearing.

Tasks

The preliminary data are contained in the file *TrafficSpeedAnalysis.jmp.*

1. Give a statistical summary of the data. Include in your summary descriptive statistics along with the percentage of vehicles exceeding the speed limit and the percentage of vehicles that would be stopped by law enforcement.

2. Based on this data, is there sufficient evidence that the average speed exceeds the posted speed limit of 35 mph? Set up and perform a statistical test of hypothesis.

3. Reevaluate the data, this time excluding the speeds of those vehicles that would be stopped by the police. Set up and perform a statistical test of hypothesis.

4. Discuss the results of the tests performed in steps 2 and 3 as it pertains to drawing a conclusion about the problems with the roadway. What would your recommendation to the town be? Explain how you arrived at your conclusion.

5. Write a brief report or prepare four slides in presentation format that contain the following: Slide 1—Problem statement, Slide 2—Statistical summary of preliminary speed data, Slide 3—Summary of relevant statistical tests, Slide 4—Conclusions and recommendations.

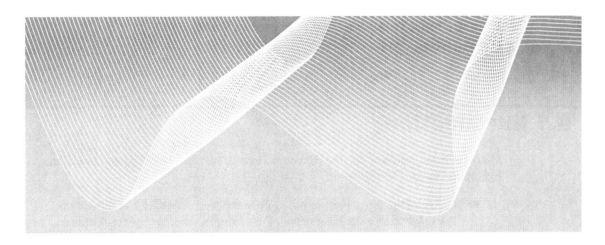

Chapter 9

Comparing Two Means

9.1 Introduction 224

    9.1.1 Hypotheses for Comparing Two Means 224

    9.1.2 Paired *t*-Test versus Two-Sample *t*-Test 225

9.2 Two-Sample *t*-Test 227

    9.2.1 Detail Example: Comparing Processor Speeds of Two Brands 228

    9.2.2 Two-Sample *t*-Test in JMP 230

    9.2.3 Interpretation of Two-Sample Results 232

9.3 Paired *t*-Test 233

    9.3.1 Detail Example: Advertising Effect in Test Markets 234

    9.3.2 Paired *t*-Test in JMP 236

    9.3.3 Interpretation of Paired *t*-Test Results 237

9.4 Paired *t*-Test versus Two-Sample *t*-Test on the Same Data 238

    9.4.1 Example: Abrasion Resistance 238

    9.4.2 Summary of Differences between the Paired and Two-Sample *t*-Test 240

9.5 Summary 241

9.6 Problems 241

9.7 Case Study: Westville Meat Processing Plant 244

9.8 References 246

9.1 Introduction

Many business problems require the comparison of two processes or two populations. Examples include the following:

- Compare the mean performance of two processes with each other without reference to an external standard. For example, compare the clock-speed of two brands of computer processors.

- Perform a "Before and After" comparison of mean performance. For example, compare the effectiveness of a new advertising campaign using same store sales at two points in time, before and after advertising.

- Select the better of two processes, if one is indeed better.

The questions here are similar to those asked in the one-sample problem of the previous chapters. The key difference is that the comparisons are based entirely on sample information and do not resort to external standards. In comparing the clock-speed of two computer processor brands, both brands are sampled and compared. In evaluating the effectiveness of a new advertising campaign, sales are measured before and after the advertising has had its effect on customers.

Instead of relying on past experience for information on the old process, new sample information is collected for both the old and the new processes. By not resorting to external reference values, both groups (levels, processes, products, brands, treatments, etc.) are treated equally. Each has to provide the sample evidence on which the comparison is based. This makes for more valid comparisons, because data on both processes are collected under similar conditions.

9.1.1 Hypotheses for Comparing Two Means

In comparing the means of two processes or populations, a typical research question asks if the means of the two groups or levels are different. One hypothesizes that the difference between the two means is different from zero. This leads to hypothesis statements for comparing the means of two processes A and B as shown in Table 9.1.

Table 9.1 Versions of Null and Alternative Hypotheses

Hypothesis	Version 1	Version 2
Null hypothesis	$\text{Mean}_A - \text{Mean}_B = 0$	$\text{Mean}_A = \text{Mean}_B$
Left-sided H_A	$\text{Mean}_A - \text{Mean}_B < 0$	$\text{Mean}_A < \text{Mean}_B$
Right-sided H_A	$\text{Mean}_A - \text{Mean}_B > 0$	$\text{Mean}_A > \text{Mean}_B$
Two-sided H_A	$\text{Mean}_A - \text{Mean}_B \neq 0$	$\text{Mean}_A \neq \text{Mean}_B$

The null hypothesis states that the two means are equal, or that the mean difference is zero. There are again three possible alternative hypotheses: Mean_A is less than Mean_B (left-sided), Mean_A is greater than Mean_B (right-sided), and Mean_A does not equal Mean_B (two-sided). The hypotheses can be written in two equivalent versions as shown in Table 9.1. Version 1 emphasizes the difference between the two means, whereas Version 2 looks more at equality or inequality of means. They both express the same hypotheses.

In comparing two means, the alternative hypothesis often is two-sided, because you might not have a preconceived notion as to which of the two means might be larger or smaller. For example, in qualifying one of two suppliers, a business does not know which if either of the two suppliers is superior. The alternative hypothesis might also be one-sided. For example, a business might consider switching to a new supplier only if the new supplier is superior to the current one. The new supplier's material has to be proven stronger than that of the current supplier.

9.1.2 Paired *t*-Test versus Two-Sample *t*-Test

Depending on how the data are collected, either the two-sample *t*-test or the paired *t*-test may apply. Certain situations might preclude one or the other data collection method. On occasion, either data collection design is applicable.

The first case is the *paired t-test*. An example is the evaluation of the effect of an advertising campaign in a department store chain in which sales are recorded before and after the ads. A number of stores are randomly selected to participate. The number of stores *n* is the sample size. The data consist of pairs of observations (before and after) with one pair from each store. The same sample of stores is measured by their difference in sales performance at two different time points, hence the frequently quoted "same store sales" for this type of comparison. The left-panel of Figure 9.1 labels before (B) and after (A) the advertising campaign in the same order because it is always the same. Different stores might have widely differing sales. The paired approach reduces the variation due to different store sales. Rather than taking one random sample of stores before and then another sample of stores after the advertising campaign, it is much more efficient to examine sales of the same stores before and after the advertising campaign, and record the difference in sales.

Figure 9.1 Before-After and Randomized Paired Comparisons

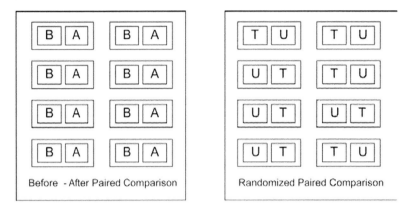

The right panel of Figure 9.1 shows paired situations with the two treatments in random order. An example of such a situation is when similar (or related) pairs are used to test the effectiveness of a treatment. For each pair, one test specimen is left untreated (U), while the other is treated (T). In those cases, it is preferable to randomize the treatment assignment for each pair. Pairing or matching observations reduces the variability that arises from selecting different sampling units. On each sampling unit, a direct comparison is made: before versus after, untreated versus treated. Pairing allows us to find actual differences in means that would be hidden with other sampling designs, e.g., the two-sample approach.

The second case, the *two-sample t-test*, requires two independent samples. The comparisons are between the results of two samples, so it is less direct than the paired comparisons. The two-sample approach is used when natural pairing between processes is impractical or impossible, because of the destruction of the test pieces or for many other reasons.

The sample sizes of the two different groups can be unequal. Suppose you compare the performance of two brands of computer processors. With the two-sample approach, there is no specific requirement that the number of processors tested for each brand is the same. A sample of brand IND processors is an independent random selection from brand IND processors. Similarly, brand ANT processors are randomly selected from their production. They were produced at a different location with different production processes. The two samples are different and independent. Figure 9.2 shows a two-sample design in a comparison of two brands of processors. In this two-sample case, the number of observations is 10 in each group. Ideally, the 20 processors should be tested in random order under similar conditions.

Figure 9.2 Two-Sample Design of 10 Processors in Each Sample

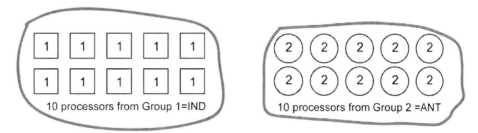

The two-sample *t*-test is discussed first, followed by the paired *t*-test.

9.2 Two-Sample *t*-Test

The two-sample *t*-test is used to test the difference between the means of two normally distributed populations. The two-sample *t*-test requires several assumptions:

- The two samples of size n_A and n_B are independent. This usually means that subjects or test specimens are randomly assigned to one of the two treatment groups. If that is not possible, they should be at least randomly selected from the populations whose means are to be compared.
- The order in which subjects or specimens are tested is random.
- The variances of the two populations are equal. If they are unequal, a different procedure needs to be used.

Random assignment is an integral part of the two-sample *t*-test. In randomized clinical studies in which two treatments (Control versus Treatment, Placebo versus Treatment) are compared, a subject who consents to participate will be randomly assigned to one of two treatments. Random assignment of test specimens to one of two groups is not always possible. For example, when you compare two different grades of material derived from two different sources or technologies, the origin of the material is fixed. In that case, you randomize the run order in which the individual tests on the specimens are conducted. You need to avoid testing all of one grade of material first, followed by testing of the other grade, since observed differences might be due to the run order.

Randomization of the run order is performed for the following reasons:

- to eliminate external systematic differences other than the ones deliberately introduced into the experiment
- to ensure the effective independence of observations

One of the most common external influences on results is unfamiliarity with running the test. Differences in temperature, humidity, personnel, and equipment are other reasons why randomization is required to meet statistical assumptions.

9.2.1 Detail Example: Comparing Processor Speeds of Two Brands

Situation: A computer assembler needs to compare the clock-speed of two processor brands ANT and IND, both with nominal clock-speed of 2800 MHz. Although these processors are rated by the respective manufacturers, they have been shown to be variable and to deviate from the design value.

Problem: Compare the mean processor speeds of the two brands. Determine whether they are different.

This problem statement leads to the following hypothesis: H_0: $Mean_{ANT} = Mean_{IND}$ versus H_A: $Mean_{ANT} \neq Mean_{IND}$. The alternative hypothesis is two-sided, because the problem statement only asks whether they are different. Any statistically significant difference in sample means leads to a rejection of the null hypothesis.

Data Requirements: Ten randomly selected processors from each brand are to be tested. This makes two distinct and independent samples of equal size with $n_{IND} = n_{ANT} = 10$. (Because the samples are taken from two independent populations, the number in each group could be unequal.) From each brand, only those with design speed 2800 MHz are to be chosen. The test specimens for both samples are taken from production.

Each observation needs to be classified as coming from one group or the other. Thus, each observation contains the continuous Y measurement of interest and a second nominal X-variable (or X, Factor). The two levels of the X, Factor identify the group or population from which each observation is sampled. In JMP two columns are required, one for the Y, Response, the other to identify the X, Factor. Figure 9.3 shows the data file *ProcSpeedTwoSam.jmp*. It has two required variable columns, an optional third column, and 20 observation rows:

1. The X, Factor Brand identifies an observation as being ANT or IND. It must be a nominal or ordinal variable. If the data type of the group identifiers is numeric, make sure to change the modeling type to nominal by clicking the variable name and choosing **Column Info**.

2. The Y, Response Clock-Speed contains the measurements of interest. It needs to be a continuous variable. This variable contains the clock-speed measurements of 20 different processors with a design clock-speed of 2800 MHz.

The third column, Run Order, is an optional variable used to test assumptions. The numbers in this column indicate the sequence in which the 20 processors were tested. The Run Order should be generated randomly, so that external influences on the results can be reduced. The data in Figure 9.3 are sorted by run order, because they were entered into the data file after each measurement.

Figure 9.3 Data File *ProcSpeedTwoSam.jmp*

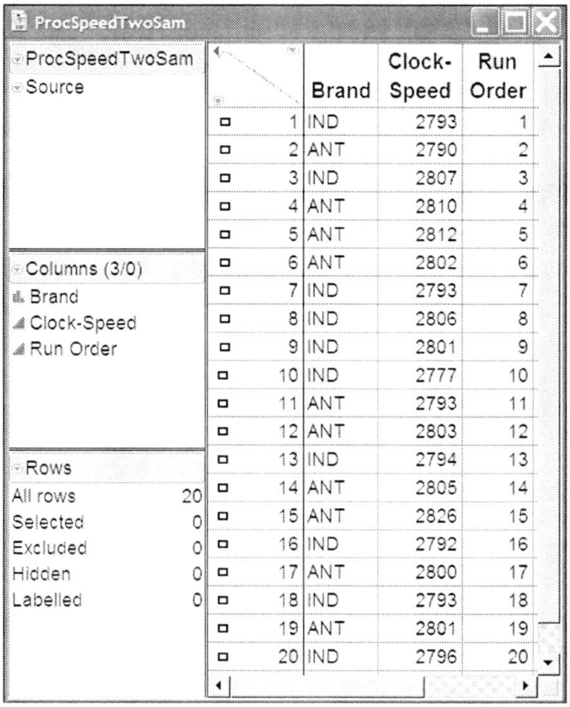
ProcSpeedTwoSam
ProcSpeedTwoSam
Source
Columns (3/0)
Brand
Clock-Speed
Run Order
Rows
All rows 20
Selected 0
Excluded 0
Hidden 0
Labelled 0

	Brand	Clock-Speed	Run Order
1	IND	2793	1
2	ANT	2790	2
3	IND	2807	3
4	ANT	2810	4
5	ANT	2812	5
6	ANT	2802	6
7	IND	2793	7
8	IND	2806	8
9	IND	2801	9
10	IND	2777	10
11	ANT	2793	11
12	ANT	2803	12
13	IND	2794	13
14	ANT	2805	14
15	ANT	2826	15
16	IND	2792	16
17	ANT	2800	17
18	IND	2793	18
19	ANT	2801	19
20	IND	2796	20

Anticipated Results: There is no reason to believe that the companies associated with the two brands would misrepresent the performance of their main product offering. One cannot *a priori* assume that one brand is superior. Nor is there any indication that one manufacturer is superior to the other. A quick look at the plot of the test results (Figure 9.4) shows that the lines representing the group means are very close to the overall mean.

Figure 9.4 Plot of Clock-Speed of Processor by Brand

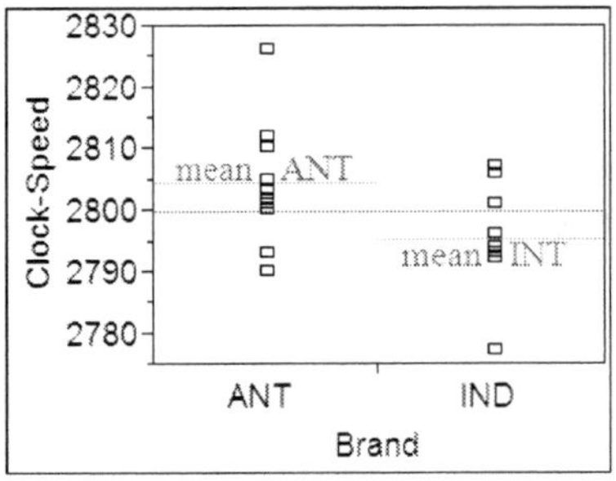

In Figure 9.5, the means of ANT, IND, and the overall mean (Mean) are shown at a scale that is determined by the minimum and the maximum value of all 20 processors. Relative to the largest and smallest, the IND and ANT means are very close. Using the two-sample *t*-test and its *p*-value, we can determine whether they are significantly different.

Figure 9.5 Overall Mean and Brand Means on a Scale from Minimum to Maximum

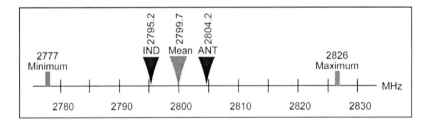

9.2.2 Two-Sample *t*-Test in JMP

The analysis is performed using the Fit Y by X platform. It can be selected from the **Analyze** menu as shown on the left in Figure 9.6.

Figure 9.6 Fit Y by X Variable Selection

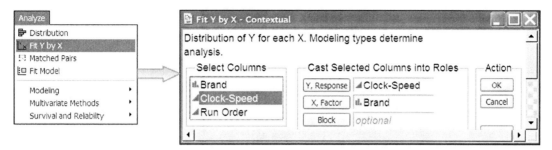

This platform requires a Y, Response and an X, Factor. In Figure 9.6, only the **Y, Response** and the **X, Factor** fields are filled in. Other fields allow analyses of more specialized data sets and can be left empty. Clicking **OK** produces a simple plot of the Y by the X, Factor already shown in Figure 9.4. In order to get the two-sample *t*-test, click the red Options triangle and select **Means/Anova/Pooled t** as shown on the left of Figure 9.7. JMP will produce the desired results for the two-sample *t*-test marked as a), b), and c). Additional output can be ignored.

The Summary of Fit a) shows the overall Mean Speed in MHz of all 20 processors to be 2799.7 MHz. The Root Mean Squared Error (RMSE) is 9.366607 MHz. The RMSE is the square root of a weighted average of the two level variances. It can be used assuming that the two actual but unknown standard deviations are the same.

Result b) shows individual group means and standard errors of each mean based on the pooled standard deviation (RMSE). The individual confidence intervals use the pooled standard deviation instead of the individual standard deviations. As has been mentioned, this procedure assumes equal true standard deviations in the two groups.

Figure 9.7 Fit Y by X Results for Two-Sample *t*-Test

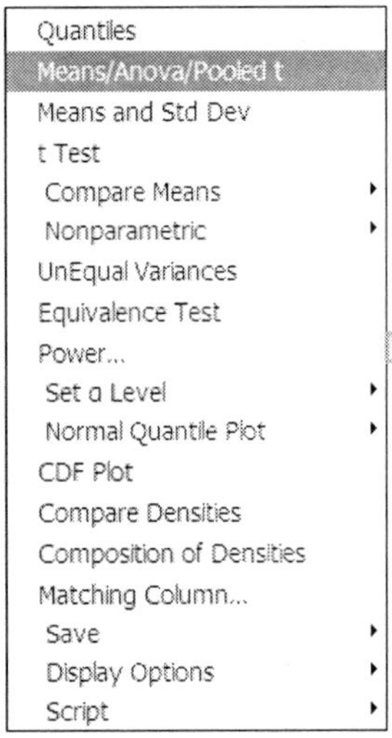

a) **Summary of Fit**

Root Mean Square Error	9.366607
Mean of Response	2799.7
Observations (or Sum Wgts)	20

b) **Means for Oneway Anova**

Level	Number	Mean	Std Error	Lower 95%	Upper 95%
ANT	10	2804.20	2.9620	2798.0	2810.4
IND	10	2795.20	2.9620	2789.0	2801.4

Std Error uses a pooled estimate of error variance

c) **t Test**

IND-ANT

Assuming equal variances

Difference	-9.000	t Ratio	-2.14855
Std Err Dif	4.189	DF	18
Upper CL Dif	-0.200	Prob > |t|	0.0455*
Lower CL Dif	-17.800	Prob > t	0.9772
Confidence	0.95	Prob < t	0.0228*

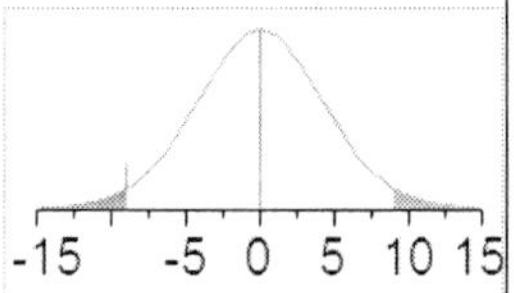

Results marked c) contain the two-sample *t*-test. They show that the estimated difference between the two group means is Difference = –9.000 MHz. This is the difference between the sample mean of IND minus the sample mean of ANT. The 95% confidence limits for the mean difference are labeled Lower CL Dif = –17.800 MHz and Upper CL Dif = –0.200 MHz. This confidence interval does not include the "no difference" value 0. In this example, the two-sided *p*-value Prob > |t| =0.0455 is appropriate, since the question only asked for differences in mean speed without indicating which direction the difference should take. The diagram on the right shows the density of the t-distribution with 18 degrees of freedom (df). The shaded area represents the two-sided *p*-value of 0.0455.

9.2.3 Interpretation of Two-Sample Results

Two-Sample Inference: The interpretation of the results of a two-sample *t*-test is similar to that of one-sample test results. The observed difference of –9.0 MHz between the processor means of IND and ANT is statistically significant, because the two-sided *p*-value for testing H_0: $Mean_{ANT} - Mean_{IND} = 0$ versus H_A: $Mean_{ANT} - Mean_{IND} \neq 0$ is Prob > |t| = 0.0455. This *p* -value is smaller than the threshold $\alpha = 0.05$ for rejecting the null hypothesis. The decision is to reject H_0. There is sufficient sample evidence to support the conclusion that mean processor speeds are different with $\alpha = 0.05$ significance. Another indicator to reject H_0 is that the 95% confidence interval from Lower CL Dif = –17.800 to Upper CL Dif = –.200 does not include 0.

Confidence Intervals by Level: The mean clock-speed of all 20 processors is 2799.7 MHz (Summary of Fit). The individual means are 2804.2 MHz for ANT processors and 2795.2 MHz for IND processors (Means for Oneway Anova). The individual 95% confidence interval for the mean clock-speed of ANT processors is from 2798.0 to 2810.4 MHz. This confidence interval indicates that with 95% confidence the mean clock-speed of ANT is known to within ±6.2 MHz of the sample mean 2804.2. (ANT's individual mean of 2804.2 MHz is the midpoint between the two confidence limits.) The confidence interval uses the Root Mean Square Error (RMSE) as the standard deviation. The 95% confidence interval for the mean clock-speed of IND processors is 2789.0 to 2801.4 MHz. It is of equal width as the ANT interval, because it also uses the RMSE. Its midpoint is the IND clock-speed mean of 2795.2 MHz. Both individual confidence intervals overlap, yet the two-sample confidence interval for the mean difference does not include zero. This shows that it is insufficient to look at variables individually and expect to come up with valid conclusions.

Conclusion: With p = 0.0455, the ANT mean clock-speed is different from the INT mean. The confidence interval tells us that the mean difference could be very small. It remains for the experts to declare the magnitude of the difference practically important.

9.3 Paired *t*-Test

The easiest way to determine when to use a paired *t*-test is to ask whether a direct comparison is possible with regard to the two groups or treatments. The paired *t*-test should be used whenever two readings are taken on the same sampling unit, as in the following cases:

- before and after weight measurements on participants in a weight loss study
- number of loans written by a retail branch of a national banking chain before and after an advertising campaign
- test specimens prepared in duplicate, with one part treated and the other part not treated

Sometimes, the paired *t*-test is also used when test subjects can be arranged into related or like pairs. The best-known example for this is the study of twins. Two related or similar test specimens are paired but randomly assigned to a different treatment method. The results are treated as matched pairs.

Although pairing is usually preferable to two-sample randomization, it might not always be possible to form pairs. For example, in comparing the effect of two diets on blood pressure in humans, only one randomly selected group of subjects can be committed to diet A, and a second group to diet B. It is physically inadvisable to feed both diets to the same subjects, since it would not produce results useful in a comparison.

9.3.1 Detail Example: Advertising Effect in Test Markets

Situation: A national finance corporation wants to determine the effectiveness of an advertising campaign for a particular type of loan. It needs to measure the effect of a new promotional campaign on the number of loans written in a particular month.

Problem: Determine the effectiveness of the advertising campaign by comparing the mean number of loans written in September with those of November.

This problem statement leads to the following hypothesis: H_0: $Mean_{Nov} - Mean_{Sep} = 0$ versus H_A: $Mean_{Nov} - Mean_{Sep} > 0$. The alternative hypothesis is right-sided, because the problem statement asks whether the advertising campaign results in an increase of written loans in November over September. A statistically significant increase in the number of loans written will lead to a rejection of the null hypothesis. (Note that the right-sided hypothesis can be changed to left-sided by simply reversing the direction of the difference, i.e., by writing $Mean_{Sep} - Mean_{Nov}$.)

Data Requirements: The market research director decides to run the advertising campaign in eight test markets during November and to use September loans as baseline data. In the JMP data table, each row is assigned to an individual sampling unit, a test market. Two columns are required, one for loans written in September, before the advertising campaign, and one for loans written in November after the advertising campaign. The data are recorded in units of 10000 loans written.

Figure 9.8 Difference in Loans Before and After Campaign Is Observation of Interest

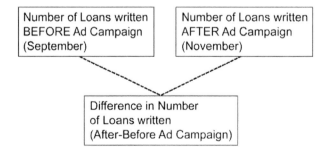

An optional third column is the difference between November and September loans. Figure 9.8 shows this diagrammatically. The Difference column is the difference between the number of loans written in November and September in each test market. It is the

column that is actually being analyzed. JMP has a special platform called *Matched Pairs* with which paired data can be analyzed. The Matched Pairs platform, which does not require this difference column, will be used in the following abrasion resistance example.

The data are contained in *Test Mrkt-paired.jmp*. The column labeled Difference is created in the formula editor as shown in Figure 9.9. After you click the **Apply** button, the calculated differences appear in the Difference column as shown on the right of Figure 9.9.

Figure 9.9 Formula Window for Difference Variable for Text Mrkt-paired.jmp

Anticipated Results: It is instructive to plot the paired differences contained in the Difference column. They are shown in Figure 9.10. The plot shows that all but one of the differences is positive. Some differences are very small. It is best not to guess as to whether these data support an assertion of a statistically significant difference between September and November loans.

Figure 9.10 Differences between September and November Loans in 10000 Units

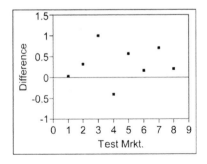

9.3.2 Paired *t*-Test in JMP

For pedagogical reasons, we present first the approach using the Distribution platform, which requires the Difference column. This approach shows that the data to be analyzed are the pairwise differences, rather than the individual measurements.

To obtain the desired test results, analyze the Difference column using the Distribution platform. Specify the variable **Difference** in the **Y, Columns** field as shown in Figure 9.11. For this problem, we look only at the Moments of Difference table of the standard Distribution output and ignore the histogram, box plot and quantiles.

The sample mean of the difference in the number of loans written is 0.31125. Recall that the number of loans written is represented in units of 10000 loans. Therefore, the mean difference is 3112.50 loans. The standard deviation of the difference is 0.43209 (4320.90 loans). The two-sided 95% confidence interval contains the zero mean difference as a possibility. However, the alternative hypothesis is right-sided, so this confidence interval should not be used for testing.

In order to test a hypothesis within the Distribution platform, left-click the red triangle next to the variable name (**Difference**) to obtain the Distribution options menu. Select **Test Mean** as the desired procedure. There is no need to change the default null hypothesis of zero. The lower right panel of Figure 9.11 shows the results of the Test Mean procedure with a null hypothesis $Mean_{Nov} - Mean_{Sep} = 0$. The test statistic for the *t*-test is 2.0374 with 7 degrees of freedom. The resulting right-sided *p*-value is Prob $> t = 0.0405$. Because this *p*-value is less than $\alpha = 0.05$, the null hypothesis is rejected.

Figure 9.11 Variable Selection for Difference and Paired *t*-Test with Distribution Platform

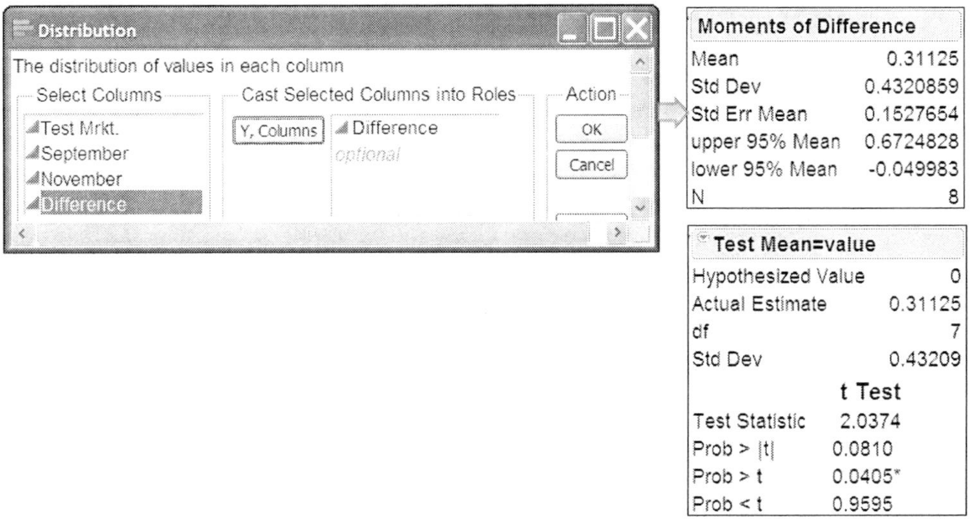

Moments of Difference	
Mean	0.31125
Std Dev	0.4320859
Std Err Mean	0.1527654
upper 95% Mean	0.6724828
lower 95% Mean	-0.049983
N	8

Test Mean=value	
Hypothesized Value	0
Actual Estimate	0.31125
df	7
Std Dev	0.43209

	t Test
Test Statistic	2.0374
Prob > |t|	0.0810
Prob > t	0.0405*
Prob < t	0.9595

9.3.3 Interpretation of Paired *t*-Test Results

Is the mean of 0.31125 greater than 0 with statistical significance? The research question asks for a right-sided alternative hypothesis, because only an increase in the number of loans is of interest. The right-sided *p*-value, Prob > t = 0.0405, shows that advertising increases the mean of written loans at the $\alpha = 0.05$ significance level. Note that this conclusion is based on only eight test markets. Is the change due to advertising? This is a question of cause and effect and statistics does not supply the answer. Other factors, such as increases of sales in expectation of coming holidays, could also have accounted for the statistically significant change. In order to answer that question, the analyst needs to understand more than the numbers and investigate other potential contributors to changes in the number of loans written.

9.4 Paired *t*-Test versus Two-Sample *t*-Test on the Same Data

The paired *t*-test and the two-sample *t*-test have the same null hypothesis and alternative hypotheses. In practice, the choice of the appropriate *t*-test is based on the data collection method. The data files are arranged differently for the two methods. Nevertheless, it is possible to erroneously arrange the two-sample outcomes as paired, as long as the number of observations in each group is the same. Paired data also can be (wrongly) arranged as two-sample data. Using the wrong *t*-test, based on the data collection method, leads to flawed conclusions and business decisions.

9.4.1 Example: Abrasion Resistance

This example shows that you must be careful in selecting the proper analysis method to get correct results. The paired *t*-test is the correct choice, while two-sample *t*-test is wrong.

Situation and Problem: A material needs to be abrasion resistant. In this example, 10 duplicate test pieces are obtained from different production batches. One is randomly selected to be coated with a thin film that supposedly enhances abrasion resistance. The other piece is left untreated. Abrasion resistance is recorded according to a standard industry protocol. Higher values indicate more resistance to abrasion and vice versa. The paired *t*-test approach is the correct approach, because test pieces were collected in pairs. The alternative hypothesis for this example is one-sided, because you are primarily interested that the treatment increases abrasion resistance, and not that it has an adverse effect:

$$H_0: \text{Mean}_{\text{Treated}} - \text{Mean}_{\text{Untreated}} = 0 \text{ versus } H_A: \text{Mean}_{\text{Treated}} - \text{Mean}_{\text{Untreated}} > 0$$

Figure 9.12 Data of 10 Pairs of Specimens (Paired) and 20 Individual Specimens (Two-Sample)

AbrasionResPaired.jmp				
AbrasionResPaired			Untreated	Treated
		1	12.1	14.7
Columns (2/0)		2	10.9	14
Untreated		3	13.1	12.9
Treated		4	14.5	16.2
		5	9.6	10.2
		6	11.2	12.4
Rows		7	9.8	12
All rows	10	8	13.7	14.8
Selected	0	9	12	11.8
Excluded	0	10	9.1	9.7
Hidden	0			
Labelled	0			

AbrasionResStacked.jmp				
AbrasionResStacked			Treat Type	Resistance
Source		1	Untreated	12.1
		2	Treated	14.7
Columns (2/0)		3	Untreated	10.9
Treat Type		4	Treated	14
Resistance		5	Untreated	13.1
		6	Treated	12.9
Rows		7	Untreated	14.5
All rows	20	8	Treated	16.2
Selected	0	9	Untreated	9.6
Excluded	0	10	Treated	10.2
Hidden	0	11	Untreated	11.2
Labelled	0	12	Treated	12.4

Data Requirements: O. L. Davies, in *Design and Analysis of Industrial Experiments*, 2d ed. (1956), provides the data for this test. The data are in two columns (Untreated, Treated). Each row represents a test piece. One column is for the abrasion results of the untreated sections and the other for those of the treated sections of the same test piece. The proper data table *AbrasionResPaired.jmp* is shown on the left in Figure 9.12.

JMP Analysis with Matched Pairs Platform: The two columns Treated and Untreated of *AbrasionResPaired.jmp* are both selected for the **Y, Paired Response** in the variable selection window. The numerical results are shown on the left of Figure 9.13. The right-sided *p*-value Prob > t = 0.0030 is less than $\alpha = 0.05$. Conclude that with statistical significance the coating treatment produces material that is more abrasion resistant than if left untreated.

Figure 9.13 Abrasion Resistance Results for Matched Pairs and Two-Sample *t*-Test

Matched Pairs			
Difference: Treated-Untreated			
Treated	12.87	t-Ratio	3.56511
Untreated	11.6	DF	9
Mean Difference	1.27	Prob > |t|	0.0061*
Std Error	0.35623	Prob > t	0.0030*
Upper 95%	2.07585	Prob < t	0.9970
Lower 95%	0.46415		
N	10		
Correlation	0.84045		

Two-Sample t-Test (Fit Y by X)			
t Test			
Treated-Untreated			
Assuming equal variances			
Difference	1.2700	t Ratio	1.457205
Std Err Dif	0.8715	DF	18
Upper CL Dif	3.1010	Prob > |t|	0.1623
Lower CL Dif	-0.5610	Prob > t	0.0811
Confidence	0.95	Prob < t	0.9189

Incorrect Two-Sample Test with Fit Y by X Platform: For comparison, the data for the incorrect two-sample *t*-test are contained in *AbrasionResStacked.jmp*, shown on the right of Figure 9.12. The two-sample approach assumes 20 independent samples. Ten randomly selected test pieces are left untreated. Ten test pieces undergo the coating. The JMP data table requires 20 rows, 10 for each treatment. The two columns are Treat Type (nominal) and Resistance (continuous). The data are analyzed with the Fit Y By X platform by selecting the Resistance column as **Y, Response** and the Treat Type column as the **X, Factor**. The results in the right panel of Figure 9.13 show the same mean Difference = 1.27. However, the one-sided *p*-value Prob > t = 0.0811 is not significant at $\alpha = 0.05$. One cannot reject the null hypothesis. The two-sample approach does not find a statistically significant difference in mean abrasion resistance between treated and untreated test pieces. The conclusion obtained by the two-sample *t*-test is different from the paired *t*-test approach. The correct paired *t*-test rejects the null hypothesis with p = 0.0030. The incorrect two-sample approach accepts the null hypothesis with p = 0.0811.

9.4.2 Summary of Differences between the Paired and Two-Sample *t*-Test

In practice, the choice of analysis method is not arbitrary, but is based on the experimental setup, i.e., the way the data were collected. Because of the direct comparison, tests based on paired data are more sensitive in discovering smaller differences than the two-sample *t*-test. In many applications, the two-sample *t*-test is the only one practically possible. For example, in destructive testing, specimens are destroyed as a result of the test. Pairing a destroyed specimen would be very difficult if not impossible. Given a choice for collecting data by either method, the paired *t*-test is typically preferred.

9.5 Summary

- The two-sample *t*-test and the paired *t*-test are two methods to compare two means.

- The choice of either of the methods depends on how the data were collected. Only one method will apply to a given data set.

- The two-sample *t*-test requires two independent samples, possibly with an unequal number of observations in each.

- The paired *t*-test requires pairs of observations taken on the same or related sampling units. These observations could be of the Before-After type, or they could be two treatments applied to the same or related sampling units.

- The paired approach tends to be more sensitive in detecting smaller mean differences, because paired observations compare directly and reduce sampling as a source of variation.

- The paired approach cannot always be used, especially in situations requiring destructive testing or when the similarity of test units is difficult to establish.

- In the two-sample approach, units are randomly selected and assigned to one of the treatments.

- In the two-sample approach, the run order of measurements should be random to avoid unintended effects relating to time and learning.

- In the paired approach, pairs should be randomly selected. If possible, the order of the two treatments should be randomized.

9.6 Problems

1. Recommend the most appropriate data collection and analysis method—paired or two independent samples—for each of the following situations. Explain your selection.

   a. An accounting department will award a contract to one of two payroll service companies. The contract will be awarded after a trial run by each company, and will in part be based on the mean number of errors made in 1000 paychecks and associated records.

   b. A pharmaceutical company is testing a flu vaccine that is delivered by nasal spray as compared to a formulation given by injection. A group of volunteers who received these vaccinations is monitored throughout the flu season. If group members contract the flu, the duration of the illness is recorded.

    c.   A manufacturer of walking shoes is comparing the durability of a new material for soles with the material presently used. The comparison is based on the miles walked before a hole appears in the shoe sole.

    d.   A running club wants to evaluate the effectiveness of a new training program on members' times in long distance races.

2. A medical doctor has developed the North Coast Diet and wants to determine if the diet produces statistically significant weight loss in the initial two weeks. Fifteen test subjects begin the diet and their weights are recorded, in pounds, at the end of two weeks. The data are given in *NorthCoastWeightLoss.jmp*.

    a.   Set up the null and alternative hypotheses that address the research question.

    b.   Is the doctor's claim that statistically significant weight loss occurs in the first two weeks supported by the results? Explain how you arrived at your answer.

    c.   Give a 95% confidence interval on the mean weight loss. Give a brief interpretation of this confidence interval.

    d.   What data collection method was used for this test? Explain why this method was appropriate for this case.

    e.   Eliminate the apparent outlier and retest the hypothesis without it. Does this alter the conclusion?

    f.   Comment on the data. Do you have any reservations about drawing conclusions from this data?

3. The Martello Toy Manufacturing Co. would like to recommend the battery brand with the longest life for use with their toys. After an initial screening of eight different brands, they decide on testing the Zodiak and Softcell brands. The test setup specifies that 15 batteries of each brand are to be tested on Martello's Dancing Monkey until the batteries are exhausted. The time in hours to exhaustion is recorded as the life of the battery. These are the results:

Zodiak	5.3	5.7	5.2	6.3	7.5	5.4	6.6	4.7	7.6	5.7	7.5	5.8	6.6	6.6	6.6
Softcell	5.3	2.5	3	4.8	2.1	6.3	5.1	4.2	8	.7	8.1	4.9	2.8	2.9	3.9

    a.   Can you recommend one brand over the other using $\alpha = 0.05$?

    b.   Calculate a 95% confidence interval that includes the range of possible mean differences.

    c.   (Optional) How many monkeys should be used in testing the batteries? Consider one monkey for testing all 30 batteries, two monkeys with one monkey used for Zodiak and the other for Softcell, 15 monkeys with each monkey testing one battery of each brand, or 30 monkeys with one monkey per battery? What are the advantages and pitfalls of each design?

4. A consumer testing group evaluates two types of leaf blowers, Dudelbag and Bagobust. On each individual test, ten bushels of unpacked oak leaves are scattered over a lawn area of 50 by 50 feet. The effectiveness of the leaf blower is measured by the time (in minutes) it takes to collect all leaves in a target circle of 10 feet diameter in the corner of the lawn area. The results of the test are contained in the file *LeafBlowers.jmp*.

   a. Set up the appropriate hypothesis test to determine whether the mean time to complete the test differs by brand.

   b. Using JMP, test the hypothesis that there is a mean difference in mean performance between the blowers.

   c. Find a 95% confidence interval for the difference in mean performance.

   d. Find a 95% confidence interval for the mean time to complete the job for Dudelbag.

5. A nutritionist has developed the West River Diet and is testing it on ten randomly selected clients before publishing the diet in a trade magazine. In particular, the nutritionist would like to be able to claim that the West River Diet will cause a mean weight loss of 10 pounds in the first month. The data are as follows:

Client	1	2	3	4	5	6	7	8	9	10
Before	185	205	253	162	145	315	196	195	154	221
After	180	190	258	151	135	302	192	187	141	206

   a. Give a short statistical summary of the results.

   b. Does the nutritionist have statistical evidence to support the claim of an average 10-pound weight loss in the first month? Show how you arrived at your answer.

   c. Give a 95% confidence interval on mean weight loss. Write a brief interpretation of this confidence interval.

   d. Client 3 gained weight on this diet. Reanalyze the data without Client 3. Compare the mean weight loss with and without Client 3. Does your conclusion change? Should Client 3 be excluded from the analysis because this client gained weight?

   e. Discuss whether you believe the weight loss claim from this study will apply to a broad population. Explain your reasoning.

6. In order to substantiate an advertising claim about the content of saturated fat, a manufacturer of cooking oil tested its own brand Corn Mash Xtra against the leading competitor, Saffron Light. The test was performed by an independent laboratory and involved 15 randomly selected bottles of both brands. The results can be found in the file *CookingOil.jmp*.

   a. Can you conclude from these data that both oils contain about the same percentage of saturated fats? Use $\alpha = 0.10$.

b. Find a 95% confidence interval for the difference in saturated fat content between the two brands of oil. Would you draw the same conclusion with the confidence interval as with the hypothesis test? Why or why not?

7. A manufacturer of bathtub enclosures produces a plastic polymer as a base compound and reinforces it with a filler to achieve strength. The fillers are supplied by one of two vendors, Polyfiber and Neptune Plastics. Some of the plant technicians have claimed that Polyfiber produces better filler than Neptune Plastics. A preliminary test is set up as follows: Six batches of the plastic polymer are separated in sub batches and either mixed with Polyfiber or Neptune Plastics filler. Test specimens measuring 4" x 4" are subjected to a standard breaking strength test that measures the force required to break through a particular test specimen. The results were as follows:

Batch	Polyfiber	Neptune Plastics
1	1045	1015
2	1215	1005
3	1207	990
4	930	966
5	960	940
6	1060	920

Do these data show with statistical significance that Polyfiber is a better filler than Neptune Plastics? Use any α you think is appropriate.

9.7 Case Study: Westville Meat Processing Plant

This year the Westville meat processing plant has introduced a new gourmet line of meat, the success of which is of particular interest to the board of directors, who would like to see Westville's market share increase. Westville provides packaged meat to regional grocery chains and its product line has included traditional cuts of beef and pork for the consumer market. The new gourmet meat is marinated in one of two flavors and comes ready for the grill with no additional preparation required. The gourmet meat is packaged in smaller quantities than traditional packages.

Business Problem

The general manager of the Westville meat processing plant is preparing for the annual meeting with the board of directors and wishes to present objective evidence to demonstrate the success of the new product line. Here are the essential questions to be addressed:

1. Has the addition of the gourmet meat line increased sales for all Westville products offered at groceries? In other words, is the new product generating increased sales or are consumers substituting the new product for other Westville products?

2. Has there been demand for the new product that has gone unsatisfied due to distribution or production problems?

Strategy

The market research department has been asked to identify, collect, and analyze relevant data to be used in developing the presentation to the board.

Task

The general manager would like a briefing on the results of the data analysis, typically a one-page summary for each issue. There is no prescribed format; however, each briefing sheet should concisely and clearly describe the issue being addressed, summarize the results, and provide conclusions or recommended actions. A few sentences and well-chosen graphical or numerical results for each issue are sufficient.

The market research department has provided you with the following data and information that you should use to prepare the briefing sheets.

Sales: Data were collected from 15 randomly selected groceries in the region that carry the full line of Westville products. The total weekly sales for Westville meat was obtained for a one-week period, one month before the introduction of the new product line. The total weekly sales was again obtained for a one-week period, one month after the new product had been available. Weekly sales are recorded in thousands of dollars. The data set is found in *WestvilleMeatSales.jmp*.

Distribution and Production: The distribution center that ships Westville meat to regional grocers has on some occasions been unable to fill the demand for the two flavors. This failure might require adjustment to the production schedule. Determine if there is a statistically significant difference between the two flavors in terms of unsatisfied demand. Also, give a confidence interval on the mean difference in unsatisfied demand between the two flavors. The pertinent data can be found in *WestvilleTwoFlavors.jmp*, which contains unfilled weekly demands, in number of cases, for the two flavors. The data are taken from the distribution center's inventory database.

9.8 References

Davies, O. L. 1956. *Design and Analysis of Industrial Experiments.* 2d ed. New York: Hafner Publishing.

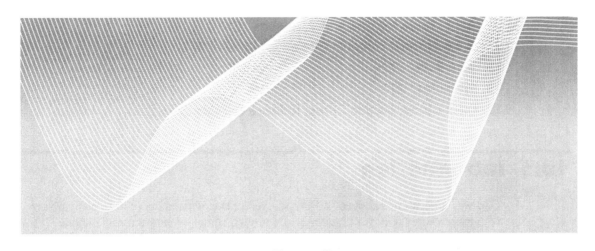

Chapter 10

Comparing Several Means with One-Way ANOVA

10.1 Introduction 248

    10.1.1 Fixed-Effect One-Way ANOVA 249

    10.1.2 One-Way ANOVA Model 250

    10.1.3 One-Way ANOVA Hypotheses 250

    10.1.4 Sources of Variation 252

10.2 Detail Example: Training Method and Time to Learn 253

10.3 One-Way ANOVA in JMP 254

    10.3.1 One-Way ANOVA with Fit Y by X 254

    10.3.2 Interpretation of Results 258

    10.3.3 Means Comparisons 259

10.4 Checking Assumptions of ANOVA Model 265

    10.4.1 Residuals 265

    10.4.2 Example: Stock Indices by Industry 267

    10.4.3 Side-by-Side Box Plots 269

10.4.4 Tests for Unequal Variances 270
10.4.5 Normal Quantile Plot of Residuals 271
10.5 Summary 272
10.6 Problems 273
10.7 Case Study: Carpal Tunnel Release Surgery 276

10.1 Introduction

One-way analysis of variance (ANOVA) extends the two-sample *t*-test to comparisons with more than two means. Examples where such comparisons are useful include determining the following:

- whether the type of gasoline affects fuel consumption
  Y = fuel consumption (miles per gallon)
  X = gasoline type (regular, intermediate, premium)

- whether a teaching method affects learning speed
  Y = time to learn a task (minutes)
  X = teaching method (self-study, class-room learning, computer-based learning)

- whether a dishwasher brand affects performance
  Y = score on standard test established by the Advertising Council
  X = brand (GE, Maytag, Miele, Whirlpool)

- whether a type of medicine affects cholesterol levels
  Y = total cholesterol level (mg/dl)
  X = type of treatment (statins, bile acid sequestrants, gemfibrozil)

One-way ANOVA requires one Y-variable and one X-factor. It comes in two versions: fixed-effect ANOVA and random-effect ANOVA. Fixed-effect ANOVA is about comparing means of Y from a fixed number of X-factor levels (for example, the mean learning speed of three different teaching methods). Fixed-effect ANOVA is used to test the null hypothesis that the factor level means of Y are equal for each X-factor level. The alternative hypothesis claims that at least two factor level means are different. These ANOVA hypotheses answer the question, "Are there any differences in Y means between factor levels?" An additional statistical procedure answers the question, "Which means, if any, are different?"

A random-effect model is about identifying factors that contribute to the overall variation. It tests the hypothesis that an X-factor, as a potential source of variation, contributes with statistical significance to the overall variation in Y. Random effects are used in measurement analysis. Even though JMP easily accommodates random-effect models, they are not treated in this text.

10.1.1 Fixed-Effect One-Way ANOVA

Fixed-effect one-way analysis of variance (ANOVA) is used for testing the equality of several means from several independent samples. If the number of factor levels is two, one-way ANOVA is reduced to the two-sample *t*-test. The data are structured accordingly into a Y-variable and an X-factor.

- The *Y-variable* is the continuous performance variable or the characteristic of interest.
- The *X-factor* is nominal and represents the groups, factor levels, or treatments across which the Y-variable is to be compared.

A level is a particular numerical value of a quantitative factor or a category of a qualitative factor. All the X-factors in the preceding examples are *qualitative (*or *nominal) factors*. Quantitative factor levels are *selected* values from a continuous variable. They need to be converted to nominal levels. An example of a *quantitative factor* is advertising exposure in seconds to determine the best length of a commercial. The variable time in seconds is actually a continuous variable. However, for practical reasons only commercials of a particular length can be broadcast. One selects several specific lengths of exposure to a commercial, such as 15 seconds, 30 seconds, or 60 seconds, as they are available from advertising media. The continuous factor is recoded as a nominal factor, possibly by using qualitative descriptions such as short, medium, or long exposure time. (Alternatively, such quantitative X-factors can be treated as continuous X-variables in regression analysis.)

Ideally, when experimental units are collected for testing, they are randomly assigned to the treatment levels of a single factor. For example, when levels represent various drug treatments, the study participants are randomly assigned to one of the treatments. However, when levels represent different suppliers or different technologies of a product, random assignment to a factor level is impossible, because each supplier or technology supplies its own inputs, such as raw materials. In this latter case, it is very important that testing be performed in random order.

One-way ANOVA provides great flexibility with respect to how many samples are observed at each factor level. As with the two-sample *t*-test, the number of observations at each treatment level can be unequal.

10.1.2 One-Way ANOVA Model

The one-way ANOVA model compares separate factor level means. Each observation is written as the mean of level t, $Mean_t$, and the random error ε_{ti}. The fixed-effect ANOVA model is

$$y_{ti} = Mean_t + \varepsilon_{ti},$$ where t = 1, ..., k, denotes the X-factor levels, and i = 1, ..., n_{lt} labels the observation within each level. Random errors ε_{ti} are usually assumed to be normally distributed with zero mean and standard deviation σ, written as N(0,σ).

The fixed-effect model is compared to the simpler null-hypothesis (or base) model $y_i = Mean + \varepsilon_i$, where Mean represents the overall mean and ε_i is the random error. The difference between a level mean and the overall mean $A_t = Mean_t - Mean$ is called the *effect of treatment t* or the deviation of $Mean_t$ from the overall Mean.

The usual ANOVA assumptions in the fixed-effect ANOVA model are that:

1. standard deviations σ across groups or levels are equal

2. individual observations are statistically independent

3. random errors ε_{ti} follow a Normal distribution N(0, σ)

10.1.3 One-Way ANOVA Hypotheses

The *null hypothesis* for comparing k means is as follows:

H_0: $Mean_1 = Mean_2 = ... = Mean_k$ (all k means are equal), or H_0: $A_t = 0$, for t = 1, ..., k, (all k factor effects are 0).

The *alternative hypothesis* is H_A: At least one of the k means is different from one of the other (k − 1) means, or at least one of the A_t is non-zero. The ANOVA alternative hypothesis says that one, two, or several means are different, or that all means are different from each other.

Figure 10.1 labels three levels numerically (1, 2, and 3) and shows the null hypothesis and two possible scenarios for the alternative hypothesis. The left panel shows the distribution of outcomes when the null hypothesis is true, because all three means are equal. The center panel represents one possible situation of the alternative hypothesis in which Mean1 and Mean2 are the same, but are smaller than Mean3. The panel on the right shows a third alternative hypothesis with all three means different from each other.

Figure 10.1 Null Hypothesis and Two Possible Alternative Hypotheses in ANOVA with Three Levels

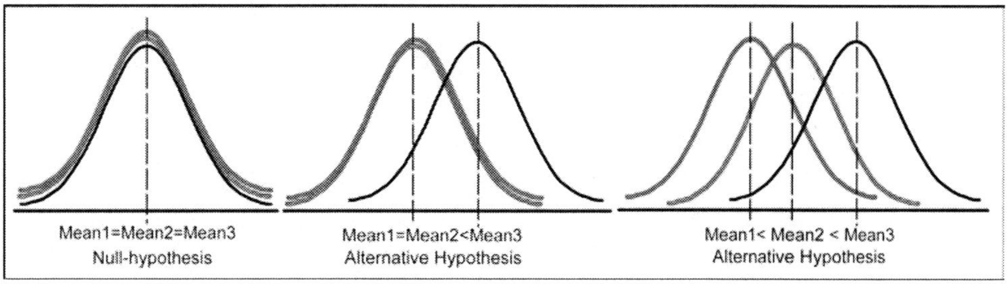

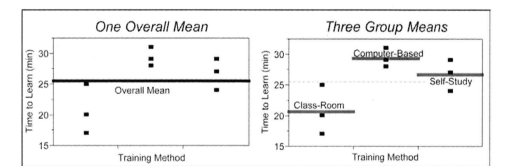

Sample data allow us to evaluate which of the hypotheses should be accepted. We compare the base model involving only the overall mean versus the one-way ANOVA model involving the level means. In the example of the next section, three different training methods are evaluated for their effect on the Y-response Time to Learn. The three training methods are Class-Room, Computer-Based, and Self-Study. If the null hypothesis is true, one overall mean is sufficient to represent all of the data. If the alternative hypothesis is true, the level means are a better representation. The two models are shown in Figure 10.2, with one overall mean in the left panel and three different group means in the right panel.

Figure 10.2 Deciding Which Model or Hypothesis Fits Better

Based on the closeness of fit, the graph seems to indicate that three group means represent the data better than one overall mean. This impression needs to be confirmed with a hypothesis test and additional analysis as outlined in the following discussion.

10.1.4 **Sources of Variation**

Analysis of variance got its name because it separates the variability of the Y-values into different sources of variation. Total variation is the variation around a single overall mean (using the base model). In one-way ANOVA, the total variation of Y is measured by the Total Sum of Squares (or C Total in JMP). Differences in level means are said to cause Between Levels Variation. The total variation of the Y-response is decomposed into separate sums of squares:

- Variation due to the X-factor levels is measured by the *Model Sum of Squares* or *Between Levels Sum of Squares*.

- *Unexplainable* experimental error is measured by the *Error Sum of Squares*.

The decomposition of Total Sum of Squares into Between Levels Sum of Squares and Error Sum of Squares is shown in Figure 10.3. The proportional size of Error and Between Levels Sum of Squares is helpful in deciding in favor of the null hypothesis or the alternative hypothesis. One assumes that if differences between level means are negligible and H_0 is true, then the Between Levels Sum of Squares is going to be relatively small. If differences between level means are relatively large and H_A is true, then the Between Levels Sum of Squares is going to be large relative to the Error Sum of Squares. This fact is used in the F-Ratio, the test statistic for testing the null hypothesis H_0: $Mean_1 = Mean_2 = ... = Mean_k$.

Figure 10.3 Decomposition of Total Sum of Squares into Two Components

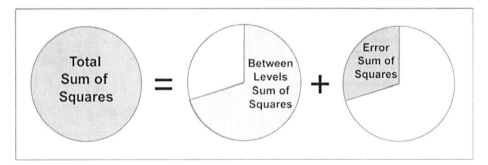

10.2 Detail Example: Training Method and Time to Learn

Situation: A business needs to train a large number of employees in reporting skills. The goal is to select a training method that, on average, has the shortest learning times. The Human Resources Department suggests three training methods: a self-study text, class-room instruction, or an interactive computer-based DVD. Nine employees are randomly selected and are randomly assigned to one training method. A typical learning task is selected to test the three methods. The time that it takes employees to learn the task is recorded in minutes.

Problem: Do the three methods have the same mean learning times? If not, which of the three methods has the shortest learning time? This problem statement leads to the following null hypothesis:

$$H_0: \text{Mean}_{\text{Class-Room}} = \text{Mean}_{\text{Self-Study}} = \text{Mean}_{\text{Computer-Based}}$$

The alternative hypothesis H_A claims that at least one of these three means is different from the other two. H_A allows the possibility that all three are different. The standard ANOVA H_A is always two-sided. More detailed mean comparisons can be conducted with JMP and will be discussed later.

Data Requirements: As with the two-sample *t*-test, one-way ANOVA data come from independent samples. The sampling units are unrelated, and each unit is reported as one observation. Figure 10.4 shows three employees assigned to each level, although the number of employees could vary by level.

The one-way ANOVA data table requires two columns, one column for the continuous Y-variable and one for the nominal X-factor that identifies the factor level of each observation.

Figure 10.4 One-Way ANOVA Data Design: Several Independent Samples

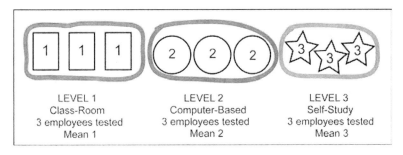

LEVEL 1	LEVEL 2	LEVEL 3
Class-Room	Computer-Based	Self-Study
3 employees tested	3 employees tested	3 employees tested
Mean 1	Mean 2	Mean 3

Figure 10.5 shows the data for the training method and learning time example in *TrainingTime.jmp*. Column 1 is the X-factor called Training Method. Its three levels are Self-Study, Class-Room, and Computer-Based. For each level, there are three rows representing the three independent observations for that level. Altogether, there are nine rows for the nine observations. Column 2 contains Time to Learn in minutes as the Y.

Figure 10.5 Data Table *TrainingTime.jmp*

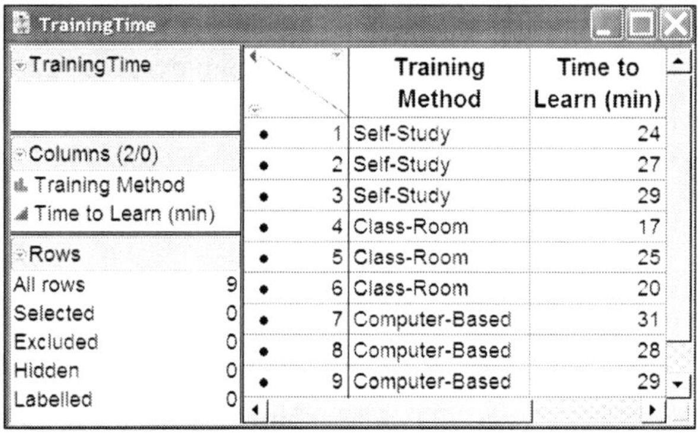

	Training Method	Time to Learn (min)
1	Self-Study	24
2	Self-Study	27
3	Self-Study	29
4	Class-Room	17
5	Class-Room	25
6	Class-Room	20
7	Computer-Based	31
8	Computer-Based	28
9	Computer-Based	29

Anticipated Results: The simple plot of the times to learn on the right in Figure 10.7 shows that the three methods resulted in different learning times. Are these differences statistically significant? One-way ANOVA and means comparisons will be required to demonstrate statistical significance.

10.3 One-Way ANOVA in JMP

For one-way ANOVA, both the Fit Y by X platform and the Fit Model platform can be used. Fit Y by X gives specific output designed for the one X-factor model.

10.3.1 One-Way ANOVA with Fit Y by X

From the **Analyze** menu, select **Fit Y by X**. Select **Time to Learn (min)** as the **Y, Response**, and **Training Method** as the **X, Factor**, as shown in Figure 10.6.

Figure 10.6 Variable Selection in Fit Y by X

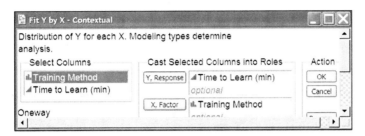

Click **OK** to obtain the basic plot of Time to Learn (min) by Training Method shown in Figure 10.7. In order to obtain basic ANOVA results, select **Means/Anova** from the options menu as shown on the left. This option produces a number of default outputs:

1. the original data plot by level augmented by mean diamonds (Figure 10.8)

2. a table of estimates and 95% confidence intervals for the level means (Figure 10.9)

3. a Summary of Fit table with the overall mean and Root Mean Square Error (Figure 10.10)

4. the Analysis of Variance table (Figure 10.11)

Figure 10.7 Scatter Plot of Time to Learn in Minutes by Training Method and Options Menu

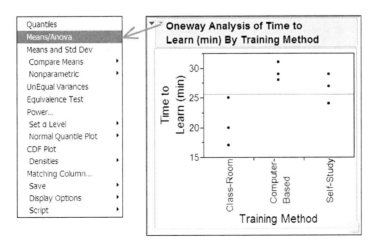

In Figure 10.8, the plot of the data on the right in Figure 10.7 is augmented with mean diamonds. The plot has been manually annotated with the numerical values to clarify the meaning of various corners. (JMP does not add numbers.)

- The horizontal line across the entire plot represents the overall mean (Mean of Response = 25.56).

- Each factor level has its own mean diamond.

- The horizontal center line of each mean diamond represents the level mean. The level mean for Class-Room is 20.67, the one for Computer-Based is 29.33, and that for Self-Study is 26.67 minutes.

- The lower corner (vertex) of each diamond represents the lower 95% confidence limit for the level mean. The upper vertex represents the upper 95% confidence limit. For the Class-Room level, the 95% confidence interval is from 16.59 to 24.74 minutes, for the Computer-Based mean from 25.26 to 33.41 minutes, and for Self-Study from 22.59 to 30.74 minutes.

- The horizontal width of the diamond at the mean line is an indicator of the relative sample size at that level. A wider diamond indicates a larger sample size relative to other levels. In the Training Method example all levels have $n = 3$ observations, so they are all equally wide.

Figure 10.8 Plot of Time to Learn (min) by Training Method with Mean Diamonds

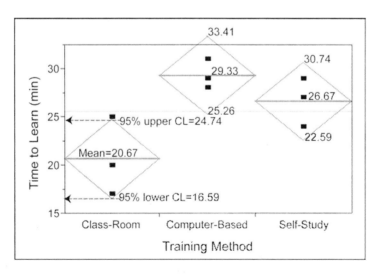

The numerical values for the means and confidence intervals in Figure 10.8 were taken from the Means table in Figure 10.9.

Figure 10.9 Means of Time to Learn (min) for Each Training Method with Confidence Limits

Means for Oneway Anova					
Level	Number	Mean	Std Error	Lower 95%	Upper 95%
Class-Room	3	20.667	1.667	16.59	24.74
Computer-Based	3	29.333	1.667	25.26	33.41
Self-Study	3	26.667	1.667	22.59	30.74
Std Error uses a pooled estimate of error variance					

The numbers are rounded so that they can be read with ease. The table has a separate row for each level containing the number of observations, mean, standard error of the mean, and lower and upper 95% confidence limits. A note at the end of the table reminds us that the Std Error column and the confidence intervals are all based on the pooled estimate of the standard deviation (RMSE in the Summary of Fit of Figure 10.10). This estimate assumes that the level standard deviations are equal.

The Summary of Fit (Figure 10.10) contains the overall Mean of Response = 25.56 minutes of all nine observations, the Root Mean Square Error (RMSE) = 2.89 minutes (rounded), and the total number of observations.

Figure 10.10 Excerpt of Summary of Fit Report

Summary of Fit	
Root Mean Square Error	2.886751
Mean of Response	25.55556
Observations (or Sum Wgts)	9

The Analysis of Variance table in Figure 10.11 shows the following:

- the decomposition of the Total Sum of Squares = 168.22 into Training Method Sum of Squares = 118.22 and Error Sum of Squares = 50.00 (see also Figure 10.3 and Figure 10.12)

- the Mean Square Error (MSE) = 8.3333 representing the variance estimate around the fitted means. MSE is the square of the RMSE (see Summary of Fit)

- the overall p-value Prob > F = 0.0263 indicating the strength of the data evidence in favor of the null hypothesis that there are no differences between level means

Figure 10.11 Analysis of Variance Table

Analysis of Variance					
Source	DF	Sum of Squares	Mean Square	F Ratio	Prob > F
Training Method	2	118.22	59.1111	7.0933	0.0263*
Error	6	50.00	8.3333		
C. Total	8	168.22			

The *p*-value Prob > F = 0.0263 is a crucial piece of information, because it results in a rejection of the null hypothesis at $\alpha = 0.05$. Conclude that at least one of the means is different from the others. With H_0 rejected, a means comparison as shown next is advised.

10.3.2 Interpretation of Results

ANOVA can be approached as a two-step process. In step 1, the overall null hypothesis that the level means are equal is tested. If the null hypothesis is rejected, step 2 determines which of the means are different from each other. Step 1 examines the breakdown of the Total Sum of Squares into components and uses the F-Ratio to test the overall significance. Step 2 uses confidence intervals to determine which of the means are statistically different. Other useful analysis, such as contrasts, could complement confidence intervals. These methods are not treated here, but are available in JMP.

Total Sum of Squares Decomposition: Figure 10.12 shows the basic idea of the ANOVA process. You can use the X-variable (Training Method) to explain the variation in the Y-variable. The Total Sum of Squares is the overall Y variation without accounting for any differences due to the X-variable. Variation due to differences in means of the X-factor is measured by the Between Levels (Training Method) Sum of Squares. It represents the amount *explained* by the ANOVA model. The Error Sum of Squares is the variation in Y *unexplained* by the ANOVA model. It is the pooled variation around the level means.

In the ANOVA table of Figure 10.11, the column Sum of Squares shows a relatively large Between Levels Sum of Squares for Training Method (118.22) and a correspondingly small Error Sum of Squares (50). This indicates that the model explains a relatively large amount of the original Total Sum of Squares (C Total = 168.22) and points to a rejection of the null hypotheses that all means are equal.

Figure 10.12 Decomposition of Total Sum of Squares for Time to Learn Example

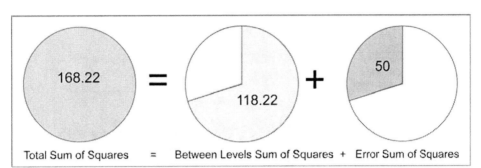

| Total Sum of Squares | = | Between Levels Sum of Squares | + | Error Sum of Squares |

F-Ratio Test: The F-Ratio test is used to decide whether to accept or reject the overall null hypothesis. The F-Ratio test uses the ratio of two Mean Square Errors, i.e., MS_{Model}/MS_{Error}, as the test statistic. JMP uses the fact that $(k-1)$ degrees of freedom are associated with the Between Levels Sum of Squares and $(N-k)$ degrees of freedom with the Error Sum of Squares to calculate the p-value based on the F-Ratio.

The ANOVA table (Figure 10.11) contains the F-Ratio and its p-value, Prob > F. In this example, the F-Ratio = 7.0933 results in a p-value Prob > F = 0.0263, which is less than $\alpha = 0.05$. Thus, reject the null hypothesis that there are no differences in mean learning times between training methods, and conclude that some level means are different. The F-Ratio and its p-value tell us only that some means are different, but not which of the means are different. This is answered by means comparisons in a second analysis step.

10.3.3 Means Comparisons

Mean comparisons are confidence intervals and hypothesis tests that examine the differences between level means or combinations of means. In this section, we deal with two-sided 95% confidence interval procedures. Specific hypothesis tests for linear combinations of means could be carried out using the JMP Contrasts option, but are not shown here.

Compare Means in Fit Y by X: To find out which means are different with statistical significance, use one of the four methods of the Compare Means option (see Figure 10.13). The four options are described in the following sections.

Figure 10.13 Compare Means—Each Pair, Student's t, if Prob > F Is Less Than $\alpha = 0.05$

Quantiles		
Means/Anova		
Means and Std Dev		
Compare Means ▸	Each Pair, Student's t	Student's t tests for all possible individual comparisons, no adjustment for multiple tests.
Nonparametric ▸	All Pairs, Tukey HSD	
UnEqual Variances	With Best, Hsu MCB	
Equivalence Test	With Control, Dunnett's	

Each Pair, Student's t Confidence Intervals

Each Pair, Student's t is a confidence interval procedure and hypothesis test involving pairwise differences of level means with the shortest confidence intervals. It finds individual confidence intervals and *p*-values. This option can be used only if the overall *p*-value (Prob > F in JMP) is less than α. The training method and time to learn example qualifies for this Compare Means test, because Prob > F = 0.0263 is less than $\alpha = 0.05$.

The Each Pair, Student's t is useful when only a few differences are of interest. Even though each individual confidence interval has a 95% confidence level, the overall confidence level that all confidence intervals are correct simultaneously is less than or equal to the stated 95%. The fewer mean comparisons made, the closer the overall confidence level is to the 95% mark.

The Compare Means—Each Pair, Student's t option produces several tables. Figure 10.14 shows the differences table with point estimates of each pairwise difference and 95% confidence intervals for the three possible mean differences. It also gives the *p*-values assuming a null hypothesis that the two means are equal against a two-sided alternative hypothesis.

Figure 10.14 Each Pair, Student's t Mean Comparisons—Differences Table

Level	- Level	Difference	Lower CL	Upper CL	p-Value
Computer-Based	Class-Room	8.67	2.90	14.43	0.0104*
Self-Study	Class-Room	6.00	0.23	11.77	0.0438*
Computer-Based	Self-Study	2.67	-3.10	8.43	0.3011

The estimated difference between the means of Computer-Based and Class-Room is 8.67 minutes. The corresponding 95% confidence interval is from 2.90 to 14.43 minutes. This is a statistically significant difference between the means, because the confidence interval does not contain the 0 difference or because the two-sided *p*-value = 0.0104 is below $\alpha = 0.05$. Self-Study also requires significantly more time to learn the task than Class-Room instruction, with a two-sided *p*-value = 0.0438. However, the difference between

Computer-Based and Self-Study is not statistically significant at $\alpha = 0.05$, because the *p*-value is 0.3011.

A different view of mean comparisons is the Connecting Letters Report displayed in Figure 10.15. The results show the estimated level means in the Mean column. Level means labeled with the same capital letter are not different with $\alpha = 0.05$ significance, whereas level means without a common letter are statistically different. Computer-Based and Self-Study share the letter "A" and are not different. The Class-Room mean is the only level to show the letter "B" and therefore is different from the other two level means at $\alpha = 0.05$.

Figure 10.15 Connecting Letters Report for Compare Means

Level		Mean
Computer-Based	A	29.33
Self-Study	A	26.67
Class-Room	B	20.67
Levels not connected by same letter are significantly different.		

Other Compare Means in Fit Y by X

Each Pair, Student's t is a reasonable method when you are interested in a few mean differences. The Fit Y by X platform in JMP enables three other multiple comparison procedures.

The *All Pairs, Tukey HSD* comparison for differences between means is used when confidence intervals for all possible differences are needed. All Pairs, Tukey HSD confidence intervals come at a price, because they are much wider than Each Pair intervals.

With Best, Hsu MCB compares the largest or the smallest mean with other means. This is especially useful in selection problems when you need to be sure that the best is really the best and not merely part of a "best group." A question answered by the With Best, Hsu MCB is this: Do the data support a statistically significant #1 ranking for the largest (or smallest) mean? For example, in evaluating the performance of four suppliers, you might need to know whether the best scoring supplier is better than the second best with statistical significance or whether the observed difference might be due to random outcomes.

The *With Control, Dunnett's* procedure is useful for comparing several treatment level means to a single control level. Because it limits the number of comparisons, Dunnett's confidence intervals tend to be shorter than the Tukey HSD procedure. The Dunnett procedure was developed for testing drugs, but is useful in other applications as well.

Compare Means in Fit Model

The Fit Model platform can also be used for one-way ANOVA. Some output is identical to Fit Y by X; some is slightly different. Here we focus on confidence intervals for the pairwise differences of level means, because these apply to two-way ANOVA (see Chapter 11) as well. Fit Model offers the following methods:

1. LSMeans Student's t is similar to the Each Pair, Student's t.

2. LSMeans Tukey HSD is similar to All Pairs, Tukey HSD.

3. LSMeans Contrast allows testing of hypotheses concerning specific linear combinations of means such as H_0: $Mean_A = (Mean_B + Mean_C)/2$.

4. LS Means Student's t and LSMeans Tukey HSD each offer several presentations of mean comparisons.

Methods 1 and 2 include the Crosstab report, Connecting Letters report (similar to Connecting Letters in Fit Y by X shown in Figure 10.15), Ordered Differences report (similar to Mean Differences and Confidence Interval in Fit Y by X shown in Figure 10.14), and the Detailed Comparisons Report (Student's t only). The next section shows the Crosstab report.

Crosstab Report of Fit Model

Begin by selecting **Fit Model** from the **Analyze** menu. The Y-variable is entered in the field for Y, while the X-factor is entered in the Construct **Model Effects** field of the Model Specification window of Fit Model (Figure 10.16). Clicking the **Run Model** button produces extensive output. Fit Model uses a more general approach to estimate means and calls them LSMeans for Least Square Means. LSMeans are the preferred means in ANOVA, because they adjust for unequal sample sizes in higher order ANOVA. LSMeans are interpreted like regular means.

Figure 10.16 Selecting Y and Model Effects in Fit Model

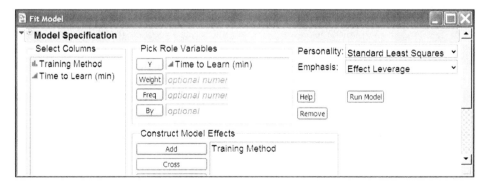

LSMeans comparisons can be obtained from the submenu of the X-factor (Training Method) analysis window. Click the red triangle and select **LSMeans Student's t** as in Figure 10.17. This method is the same as Each Pair, Student's t of Fit Y by X.

Figure 10.17 Options for Model Effect (Training Method in Fit Model)

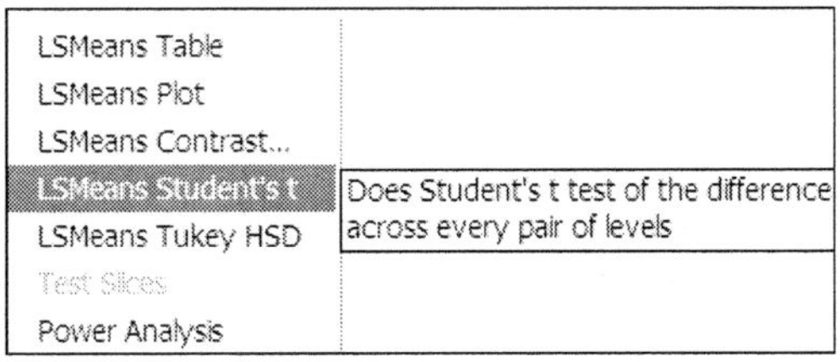

LSMeans Table	
LSMeans Plot	
LSMeans Contrast...	
LSMeans Student's t	Does Student's t test of the difference across every pair of levels
LSMeans Tukey HSD	
Test Slices	
Power Analysis	

Choosing **LSMeans Student's t** results in the default Crosstab report of all differences between Means (Figure 10.18). Each cell contains four numbers.

1. Mean[i] – Mean[j]: The difference between Row [i] mean and Column [j] mean

2. Std Err Dif: The standard error of the mean difference

3. Lower CL Dif: Lower 95% confidence limit for the difference between Row mean and Column mean

4. Upper CL Dif: Upper 95% confidence limit for the difference between Row mean and Column mean

For the cell defined by the row [i] labeled Class-Room and the column [j] labeled Computer-Based, the following results are shown:

- Mean[i] – Mean[j] = –8.6667
- Std Err Dif = 2.35702
- Lower CL Dif = –14.434
- Upper CL Dif = –2.8992

Figure 10.18 Crosstab Report for Pairs of Training Method Means

LSMeans Differences Student's t			
α= 0.050 t= 2.44691			
		LSMean[j]	
Mean[i]-Mean[j] Std Err Dif Lower CL Dif Upper CL Dif	Class-Room	Computer-Based	Self-Study
Class-Room	0	-8.6667	-6
	0	2.35702	2.35702
	0	-14.434	-11.767
	0	-2.8992	-0.2326
Computer-Based	8.66667	0	2.66667
	2.35702	0	2.35702
	2.89924	0	-3.1008
	14.4341	0	8.43409
Self-Study	6	-2.6667	0
	2.35702	2.35702	0
	0.23257	-8.4341	0
	11.7674	3.10076	0

For a two-sided alternative hypothesis, differences are statistically significant at α = 0.05 when 0 is not included in the corresponding 95% confidence interval. When both the Lower CL and Upper CL Dif are either less than 0 or greater than 0, the 95% confidence interval does not contain the null hypothesis difference 0, and the null hypothesis that the mean difference is zero should be rejected. When the confidence interval includes 0, the mean difference is not significantly different from zero.

The results of Figure 10.18 agree with those of Figure 10.14. At α = 0.05, the Class-Room teaching method is different from the other two methods with statistical significance, because none of the 95% confidence intervals for the differences involving Class-Room include 0. The 95% confidence interval for the difference in means of Time to Learn between Class-Room and Computer-Based is from –14.434 to –2.899, between Class-Room and Self-Study is from –11.767 to –0.233 minutes, and between Self-Study and Computer-Based ID from –8.434 to 3.101 minutes. This 95% confidence interval for the mean difference in Time to Learn includes 0, and the null hypothesis cannot be rejected. The Crosstab report contains the same results above and below the major diagonal, except that the sign of the difference and the confidence interval are changed. The main diagonal should be ignored.

10.4 Checking Assumptions of ANOVA Model

This section uses the time to learn example and a new example of changes in stock prices in different industries to demonstrate how to check common ANOVA assumptions. When certain assumptions are not met by the data, more advanced analysis might be required. Diagnostic checks on original data and residuals verify that key criteria are valid and that the method of analysis is appropriate for the data. For example, with data violating the assumption of equal variances between levels, *p*-values might be misstated and pairwise comparisons might require different methods.

10.4.1 Residuals

What are residuals? A residual is that part of an observation that is unexplained by the model. A residual is the difference between the observed value and its corresponding level mean:

Residual of Y = Observed Y − Estimated Average of Y

Residuals are used to check the equality of variances, to verify randomness, and to find outliers or poorly fitting observations. Figure 10.19 shows the **Save Residuals** option, which enables you to save residuals to the data table. This creates a new column containing the residuals.

Figure 10.19 Saving Residuals to a Data Table Using Fit Y by X

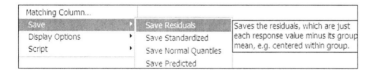

In the training method and time to learn example, the residual for Observation 1 (Self-Study) can be calculated as follows: The observed value is 24. The Self-Study Mean is 26.67. The residual is 24 − 26.67 = −2.67.

Assumptions in ANOVA: The following paragraphs discuss three key assumptions of ANOVA and methods to evaluate them. The three assumptions are that (1) the residuals are independent, (2) the variance of the residuals is the same for all levels, and (3) the residuals are normally distributed.

1. **The residuals are independent:** The random sample should contain independent observations. Non-randomness can invalidate any statistical

interpretation of results. In practice, this assumption is often violated, especially by historical data.

A plot of residuals versus time order might show trends in the data or changing variability over time. Increasing or decreasing residuals might indicate that other factors influence the outcomes. Increasing or decreasing values over time might reveal such phenomena as learning during the test or fatigue by the respondents. If values show time trends, you must question how stable and representative the sample results are. The left panel in Figure 10.20 plots the residuals of Time to Learn against time order. The residuals do not exhibit any particular pattern. Therefore, it is reasonable to assume that the observations are mutually independent.

2. **The variance of the residuals is the same for all levels:** The variances within each level are **assumed** to be equal. A violation of the assumption of equal variance leads to misstated p-values and possibly to wrong conclusions. The center plot of Figure 10.20 plots residuals versus the three training methods. The time order of the data collection has been used to label the residuals. The plot shows differences in the residual variability. For example, Class-Room shows the highest spread, but not to a degree that is worrisome for the validity of ANOVA. In fact, the Levene test for the equality of the three variances, discussed in the next section, reveals no statistically significant difference of the variances.

3. **The residuals are normally distributed:** A violation of this assumption is not as crucial for the analysis of mean values. However, non-normality in the residuals does affect predictions of future outcomes.

Figure 10.20 Residuals Plots versus (1) Time Order, (2) Training Method, (3) Normal Quantiles

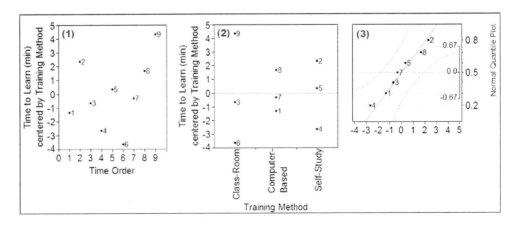

The normality of the residuals can be confirmed with a normal quantile plot from the Distribution platform (discussed at the end of Chapter 8). Use the JMP Distribution platform to obtain a normal quantile plot of the previously saved residuals. As an indication that the assumption of normality is met, the plot should show residuals falling along the straight line.

The plot on the right in Figure 10.20 shows the residuals closely following the straight line, and all residuals are within the flared Lillefors bounds, indicating that there is no evidence to assume that the residuals are not normally distributed.

Several other plots are available for examination of residual patterns or trends:

- A plot of residuals versus predicted values can be used to examine whether variability changes with increasing predicted values. This is a default plot in the Fit Model platform.

- A plot of residuals versus a variable that was not included in the analysis might indicate improvements to the model.

10.4.2 Example: Stock Indices by Industry

This example shows analysis of data in which the observations show significantly different standard deviations (variances) for the different levels.

Situation: Forty stocks were randomly selected, 10 each from one of four industry groups (levels): Manufacturing, Chemical, Electronics, and Service. For each stock, the change in price over a 9-quarter period was calculated. The 9-quarter change of each stock price is given as an index. An index value of 1 means no change, an index less than 1 means a decrease in value, and an index greater than 1 means an increase in value over the 9-quarter period.

Problem: Determine whether one or more of the industry groups have superior average performance with regard to stock price changes. Also, check the assumptions of ANOVA.

Data Requirements: Figure 10.21 shows an excerpt of the file *StocksBYIndustry.jmp*. The file contains three data columns. Column 1 contains the ID of each stock by sector. The values run from 1 to 10. Column 2 contains the Industry X-factor with the levels Chemical, Electronics, Service and Manuf. The column Priceindex shows the change of the stock price since the beginning of the first quarter, at which time the price was indexed to 1. Thus, a value 1.067 indicates a 6.7% increase in stock price.

Figure 10.21 Excerpt of Data File *StocksBYIndustry.jmp*

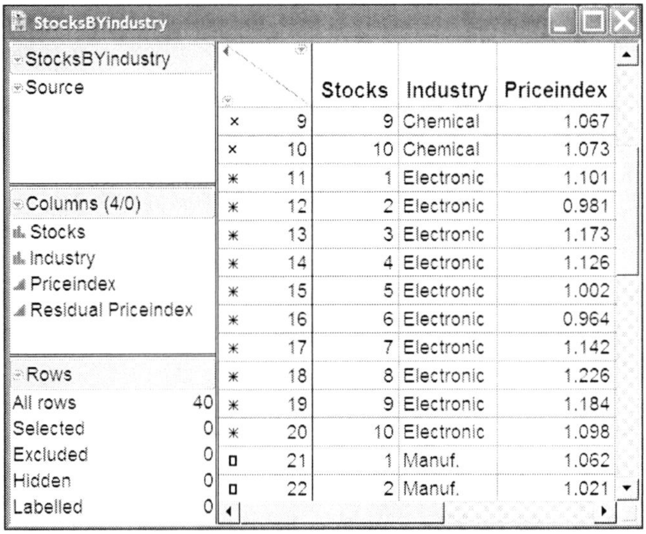

StocksBYIndustry

- StocksBYIndustry
- Source

Columns (4/0)
- Stocks
- Industry
- Priceindex
- Residual Priceindex

Rows	
All rows	40
Selected	0
Excluded	0
Hidden	0
Labelled	0

		Stocks	Industry	Priceindex
×	9	9	Chemical	1.067
×	10	10	Chemical	1.073
*	11	1	Electronic	1.101
*	12	2	Electronic	0.981
*	13	3	Electronic	1.173
*	14	4	Electronic	1.126
*	15	5	Electronic	1.002
*	16	6	Electronic	0.964
*	17	7	Electronic	1.142
*	18	8	Electronic	1.226
*	19	9	Electronic	1.184
*	20	10	Electronic	1.098
□	21	1	Manuf.	1.062
□	22	2	Manuf.	1.021

Use the Fit Y by X platform to obtain industry means and individual standard deviations from the **Means and Std Dev** option of the menu as shown in Figure 10.7. The **Means and Std Dev** option yields the results in Figure 10.22. This table contains the number of observations in each level, the level means, the individual level standard deviations, and the standard errors based on the level standard deviations. The standard error of the mean is the standard deviation divided by the square root of the sample size. The standard deviations and standard errors are calculated from level-specific data, whereas the Means/Anova option uses the RMSE.

Figure 10.22 Table of Means and Standard Deviations by Level

Means and Std Deviations of Priceindex by Industry

Level	Number	Mean	Std Dev	Std Err Mean	Lower 95%	Upper 95%
Chemical	10	1.0648	0.0557	0.0176	1.0250	1.1046
Electronics	10	1.0997	0.0900	0.0285	1.0353	1.1641
Manuf.	10	1.2251	0.1830	0.0579	1.0942	1.3560
Service	10	1.0430	0.0270	0.0085	1.0237	1.0623

From the **Moments** list of the JMP Distribution platform, obtain the overall mean price index of all 40 stocks as Mean = 1.10815 and the overall standard deviation Std Dev = 0.124814. We need to answer whether these are statistically significant differences in the average price by industry, and if differences are found, find the industries that offer the highest returns.

10.4.3 Side-by-Side Box Plots

Side-by-side box plots are an effective graphical tool to illustrate group differences. To obtain side-by-side box plots in the Fit Y by X platform, select the **Box Plots** option from the **Display Options** submenu (Figure 10.23) and box plots will be added to the scatterplot. Each level has its own box plot. All box plots are on the same vertical scale.

The side-by-side box plots show differences between group medians and differences in spread. Manufacturing price indices show considerably more variation than those of the other groups. This group has the largest box representing the central 50% of the data in each group and the longest whiskers. The horizontal lines at the end of the whiskers represent the smallest and largest values of each group. For a detailed explanation of the box plot and other characteristics of this graph, try using the JMP Help tool.

Figure 10.23 Box Plots by Industry with Connected Means

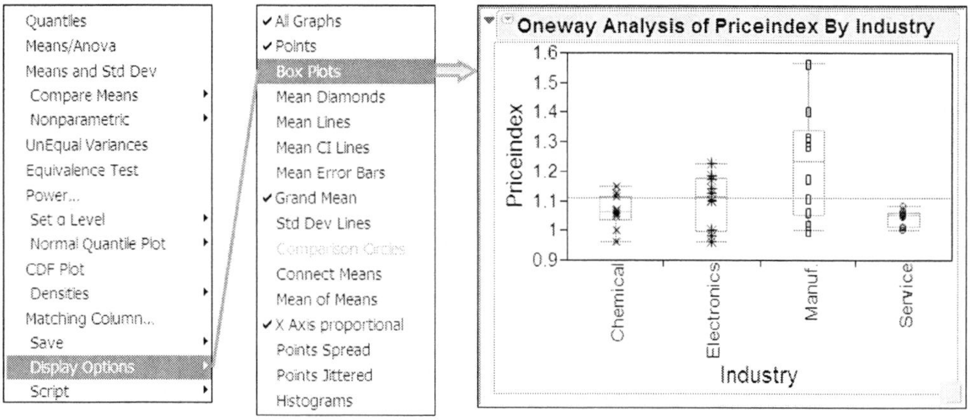

In one-way ANOVA, the predicted values are the level means. In this example, the level means are the Priceindex means by industry sector (Figure 10.22). The box plots in Figure 10.23 show that the price index of Manufacturing varies the most. Also, the Manufacturing median is much larger than the other medians.

10.4.4 **Tests for Unequal Variances**

Figure 10.22 and Figure 10.23 indicate that the variation of returns in the Manufacturing sector may be larger than in the other industries. One-way ANOVA assumes that the group variances are all equal. If they are not equal, then the *p*-value of the F-test might not be correct and the pairwise confidence intervals would require different formulas. Fit Y by X gives the option to test for unequal variances. Select this option as in Figure 10.24.

Selecting the **Unequal Variances** option leads to the table in Figure 10.25 with four different tests for unequal variances. We use the Levene test, because it is a versatile test for unequal variances in ANOVA and widely used in Six Sigma quality programs.

Figure 10.24 Fit Y by X Option Menu for Unequal Variance Test

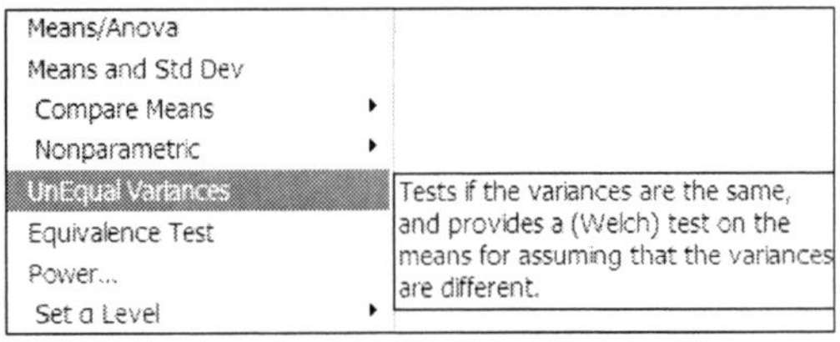

Levene Test for Unequal Variances: The Levene test is used in this example to show that the group variances (and standard deviations) are different. The null hypothesis of the Levene test states that the standard deviations in each group are the same. The alternative hypothesis claims that at least one of the standard deviations is different. In Figure 10.25, the *p*-value of the Levene test is p < .0001. The null hypothesis that all variances are the same must be rejected. This is a clear indication that the variances within the industry groups are different. The Manufacturing standard deviation is much larger than the other three standard deviations, as can be seen in Figure 10.23.

Figure 10.25 Levene Test for Equality of Group Variances

Test	F Ratio	DFNum	DFDen	Prob > F
O'Brien[.5]	5.1782	3	36	0.0045*
Brown-Forsythe	10.4120	3	36	<.0001*
Levene	10.6500	3	36	<.0001*
Bartlett	9.4692	3	.	<.0001*

Welch F-Test for ANOVA When Variances Are Unequal: The consequences of unequal variances are that the standard ANOVA cannot be trusted to give the right *p*-values and confidence intervals. The Welch ANOVA method is the appropriate analysis.

JMP provides the *p*-value for the overall Welch ANOVA as shown in Figure 10.26. This Unequal Variances *p*-value Prob > F = 0.0221 is still below α = 0.05. The null hypothesis that the means are equal can be rejected. However, the *p*-value = 0.0221 is much higher than using the standard equal variances approach (p = 0.0024).

Figure 10.26 Welch ANOVA F Test When Group Variances Are Unequal

Welch Anova testing Means Equal, allowing Std Devs Not Equal			
F Ratio	DFNum	DFDen	Prob > F
4.1346	3	17.392	0.0221*

In JMP, the Welch ANOVA test results in an acceptance or rejection of the overall null hypothesis that all means are equal. It does not give detailed pairwise comparisons of means.

10.4.5 Normal Quantile Plot of Residuals

Residuals should be normally distributed. This can be confirmed through a normal quantile plot from the Distribution platform. The plot should show residuals falling along the straight line as an indication of their normality.

A normal quantile plot provides several insights:

- An approximate straight-line relationship between normal scores and residuals indicates that the normality assumption is reasonable. A curved or asymmetric plot might suggest a transformation of the data.
- Outlying residuals at either end of the data might indicate special problems with these data points, i.e., inexperienced operators.

Figure 10.19 shows how to save residuals to the data table. After the residuals are saved, they are analyzed with the Distribution platform as a single variable. Figure 10.27 shows a histogram, box plot, and normal quantile plot of residuals of the stocks indices by industry example.

In a normal quantile plot, if all residuals fall on the straight line (more or less), they can be considered normally distributed. The curved lines on either side of the line are the Lillefors confidence bounds. If all residuals fall within these bounds, you would expect the assumption of normality to hold. In Figure 10.27, all residuals are within the Lillefors bounds. However, the industry markers show that the extreme residuals, represented by the rectangles, come from the Manufacturing group. The price indices of the

Manufacturing group show a wider spread. The data set might be too small to point to any problems with normality.

Figure 10.27 Histogram with Fitted Normal Density and Normal Quantile Plot of Residuals

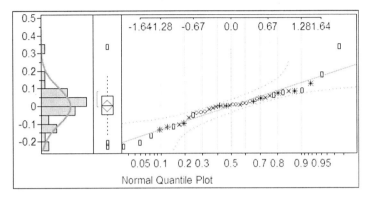

10.5 Summary

- Fixed-effect one-way ANOVA is used to compare several means.

- One-way ANOVA requires several independent samples. The number of observations in each sample might be different.

- Sampling units are randomly *selected from* or randomly *assigned to* one of the levels.

- The ANOVA model explains the data in terms of group means. The ANOVA model is compared to the base model with a single overall mean.

- The level means are used to explain some of the variation of the Y-variable. The decomposition of the Total Sum of Squares shows how much of the variation around a single mean has been explained by factor levels.

- Use side-by-side plots (mean diamonds or box plots) to examine the data for reasonableness.

- The F-Ratio tests that there are significant differences in the level means.

- If the *p*-value of the F-Ratio is less than α, use the Compare Means—Each Pair, Student's t 95% confidence intervals and test individual mean differences either with the ordered pairs report or the connected letters report.

- The ANOVA assumptions include independent and normally distributed residuals with constant variance across level means.

- Check the independence of residuals by plotting them against time or other variables.

- Check for equality of variance using the Levene test and with a plot of residual versus predicted Y.

- If the group standard deviations are different with statistical significance, use the Welch ANOVA test to determine the significance of the differences in group means.

- Check for normality using a normal quantile plot of the residuals.

10.6 Problems

1. Which of the following situations are amenable to analysis using one-way analysis of variance? Explain why ANOVA is or is not the appropriate method of analysis. For each case where one-way ANOVA is appropriate, identify the Y-variable and the X-variable. Indicate the conditions required or list the assumptions to ready each case for ANOVA, if possible.

   a. An optometrist compares reimbursements by four insurance plans for eye glasses.

   b. A plastics engineer compares the strength of a material by three different production lines.

   c. An insurance company is analyzing the amount of money paid on automobile claims based on the age of the claimant.

   d. A medical school is conducting a clinical study to determine whether there is a gender-based difference in the duration of pain relief for a new analgesic pain medication.

   e. A library is considering raising some of its fines for overdue books and media. To assist in making this decision, the library is comparing the number of days overdue for books, audio recordings, and videos for one month.

   f. A manufacturer of LEDs used in lighting tests the light level of their own brand against two other leading manufacturers.

2. The retailing manager of a food chain wishes to determine whether product location has any effect on the sale of pet toys. Three different aisle locations are considered: front, middle, and rear. A random sample of 18 stores was selected, with six stores randomly assigned each aisle location for the pet toy display. The size of the display area and the price of the product were held constant for all stores. At the end of the one-week trial period, the sales volume (in thousands of dollars) for each store was recorded. The results are given below.

Front Sales	Middle Sales	Rear Sales
8.6	3.2	6
7.2	2.4	4.6
6.2	2	4
5.4	1.4	2.8
5	1.8	2.8
4	1.6	2

a. Enter the data into a JMP data table in the proper format needed to carry out a one-way ANOVA.

b. Determine whether at least one of the aisle locations has a different sales mean than the others. Write out the appropriate hypotheses. Use the Fit Y by X platform to obtain an analysis of variance table. Find the F-Ratio and *p*-value used to test this hypothesis. What is your conclusion?

c. If at least one of the mean sales volumes differs, determine specifically which groups differ. From the one-way analysis of variance options, select **Compare Means → Each pair, Student's t**. Specifically, which groups differ?

d. Interpret the 95% confidence intervals for the pairwise differences in mean sales by location.

3. A study of graduates from three professional schools aims at determining the extent of differences in salaries four years after graduation. A pilot study yields the salaries (in thousands of dollars) contained in *Professional_Salaries.jmp*. Use the data to determine whether there are differences in the means of salaries four years after graduation between the three professions. Present your answer in sufficient detail with proper explanations.

4. The business manager of a summer classical music festival is preparing the schedule for next season. The goal is to have the concerts last as close to two hours as possible, excluding intermission, with each program containing one symphony. The symphony is selected first and then other shorter works are chosen to bring the total program length to approximately two hours.

   The music director postulated the theory that the later (chronologically) a symphony was written, the longer it was and thought this would be a helpful rule-of-thumb when choosing programs. The business manager randomly selected 24 symphonies from programs performed in previous years to test the theory. The symphonies are categorized into three historical periods: classical, romantic, and post-romantic. The classical period is the earliest, followed by the romantic, and then the post-romantic. The file *symphonies.jmp* contains the length of each symphony recorded in minutes and the associated historical period.

   a. Give a short statistical summary of the results.

   b. Do the data support the music director's theory? Show how you arrived at your answer.

   c. Beethoven's 9th Symphony seems to be uncharacteristically long, compared to other symphonies of the classical period. Exclude this symphony from the analysis. Does this change the conclusion?

5. Killson & Killson, a pharmaceutical manufacturer, would like to be able to claim that a new headache relief medication is better than the existing competitive brand. The new medication is tested in the form of two slightly different compounds, here labeled as "Active 1" and "Active 2". These compounds are compared to the rival brand and a control group in which patients receive a placebo (i.e., a non-effective medication such as sugar pills).

   The response variable is scored on a scale from 0 to 50, where 0 is no relief from pain and 50 is complete relief. The scores are reported 30 minutes after having taken the medication. Sixty-four subjects were randomly assigned to one of the four groups. This resulted in the data contained in the file *Headache.jmp*.

   a. Can Killson & Killson claim that their new medication is better at relieving headaches than the rival brand? Explain the specific reasons behind your answer, and write down all statistics or cite proper graphs that support your conclusion.

   b. Explain the meaning or use of the Root Mean Squared Error, here RMSE = 6.9792. How is RMSE different from the Standard Error of the Mean? What is the use of the F-Ratio, here $F = 6.1432$ with $p = 0.0010$?

c. Which of the medications are significantly different from the others at $\alpha = 0.05$? Is there one medication that seems to give the strongest relief?

10.7 Case Study: Carpal Tunnel Release Surgery

Repetitive stress injuries have become increasingly prevalent with the widespread use of computers. Carpal tunnel syndrome (CTS) is a repetitive stress injury characterized by pain, numbness, and tingling in the affected hand due to compression of the median nerve. A remedy to CTS is surgery. Recent medical studies suggest better outcome if surgery is performed early in the course of the condition.

The surgical procedure to correct CTS, called carpal tunnel release, involves enlarging the tunnel to relieve nerve compression. There are two types of surgery: open carpal tunnel release (OCTR) and endoscopic carpal tunnel release (ECTR). ECTR has two different techniques— single-portal and double-portal. OCTR, in which a larger incision affords the surgeon a better view of the nerve anatomy, has long been the standard procedure. Complications are rare and the long-term prognosis is excellent, with a second surgery seldom required. The primary disadvantage of OCTR is a longer recovery time compared to ECTR, a less invasive microsurgery allowing faster functional recovery and less post-operative discomfort. Complications rates are slightly higher for ECTR (particularly single-portal) as compared to OCTR. The long-term success of ECTR is not yet fully understood.

Business Problem

An insurer who provides workers' compensation coverage would like to evaluate recent carpal tunnel surgeries and their associated costs and lost work time. The cost given is the cost of the surgery and does not include subsequent follow-up or rehabilitation.

Task

The file *CarpalTunnelSurgery.xls* contains cost and return-to-work data. Based on this data, develop a recommendation for the preferred surgical procedure from the perspective of the insurer. Summarize your findings in a brief report. Be sure to include data description, all required assumptions, and the criteria used to make the recommendation.

Chapter 11

Two-Way ANOVA for Comparing Means

11.1 Introduction 278

11.2 Two-Way ANOVA without Replications 280

    11.2.1 Model and Sources of Variation 280

    11.2.2 Hypotheses for Each Model Term 282

    11.2.3 Detail Example: Prices of Wireless Mouse Devices by Brand at Different Stores 282

    11.2.4 JMP Analysis of the Wireless Mouse Example Using the Fit Model Platform 284

11.3 Two-Way ANOVA with Equally Replicated Data 292

    11.3.1 Model and Sources of Variation 292

    11.3.2 Hypotheses for Each Model Term 294

    11.3.3 What Is an Interaction Effect? 295

    11.3.4 Detail Example: Profitability of Futures Trading Customers of a Financial Institution (Equal Replications) 296

    11.3.5 JMP Analysis of Profitability of Futures Trading Customers 298

11.4 Two-Way ANOVA with Unequal Replications 306

11.5 Summary 306

11.6 Problems 307

11.7 Case Study: Fish Catch near Oil Rig 312

11.1 Introduction

Many business problems can be modeled as two or more independent variables or X-factors influencing a single Y-variable. Examples include the following:

- In comparing several brands of dishwashers for their effectiveness in removing soil from dishes, a company tests four brands (including their own). Each brand is in turn tested with each of the three detergents. Soil Removal Rating is the Y-variable, and Brand and Detergent are the two X-factors.

- Prices of wireless tracking devices (computer mouse devices) with comparable functionality might vary according to brand and the store where they are bought. Price is the Y-variable, and Brand and Store are the two X-factors.

- The profitability of customers engaged in futures trading with a financial institution might depend on the type of actions customers take and the demands on management they request. Profit per Customer per Year in Dollars is the Y-variable, and Customer Actions and Demands on Management are the two X-factors.

- The occupancy of hospital beds might depend on the care level a hospital provides and its ownership type. Occupancy in Percent is the Y-variable, and Care Level and Ownership are the two X-factors.

Two-way ANOVA is a method to explain the variation in Y with two X-factors. Two-way ANOVA is the appropriate technique for analyzing data that have the following characteristics:

Y = one continuous response variable

X_1, X_2 = two nominal explanatory factors

A single observation records the factor level combination (X_1, X_2) and the response Y. For this, three data columns are necessary.

Two-way ANOVA is a useful tool because the significance of each of the two X-factors can be ascertained along with the strength of each factor's contribution to explaining the Y-variation. As with many statistical methods, the way data are collected determines which two-way ANOVA model is applicable. There are three different designs to consider.

Two-way ANOVA without replications: Each of the two X-factors has a finite number of factor levels. Each factor level combination has exactly one Y observation. Such data are called unreplicated and allow conclusions about mean differences of levels for each factor.

Two-way ANOVA with equal replications: Each factor level combination has the same number of replicated observations. The replicated observations allow the addition of an interaction term to the model. The interaction effect can be viewed as a combination effect of the two factors in the sense that the effect of one factor depends on the level of the other factor. Interactions might be present because the responses are affected by synergisms or antagonisms between two (or more) factor levels. The interaction term allows a more differentiated inference of how each X-factor and each factor level combination affect the mean of the Y responses.

Two-way ANOVA with unequal replications: Data with unequal replications per factor level combination occur frequently in practical applications, especially when data are gathered without any specific data collection design or when test specimens are lost. ANOVA with unequal replications can be analyzed easily when each factor level combination is present in the data. With unequal replications, you must use the LSMeans (Least Squares Means) instead of the simple means, because only LSMeans account for the unequal sample sizes in the factor level combinations. (Although not treated in this text, a section on two-way ANOVA with unequal replications including a case study is available as a download.)

Depending on the data available, two-way ANOVA allows several conclusions, including:

- the importance and magnitude of each X-factor in explaining the values of the Y-variable
- an estimate of the size of factor effects and, with a two-way ANOVA with replications, the interaction effect between the two factors

11.2 Two-Way ANOVA without Replications

In two-way ANOVA without replications, each factor level combination is observed exactly once. The advantages of two-way ANOVA without replication are several:

- Analyzing two factors simultaneously decreases the Error Sum of Squares compared to analyzing one factor at a time, and this increases the chance of detecting statistically significant factor effects. In fact, neglecting a factor that explains a noticeable amount of variation might lead to statistically non-significant results for the other factor. In the example that follows comparing prices of wireless mouse devices by brand at different stores, the factor Brand is statistically non-significant unless the factor Store is included in the model.

- Each factor can have any number of factor levels. In the following example comparing prices of wireless mouse devices, the factor Brand has three levels and the factor Store has four levels.

- Given that data are properly collected, two-way ANOVA can tolerate modest deviations from its assumptions.

- It is important to account for all known major sources of variation by including in the analysis even factors that are not of primary interest, but might contribute to the variation in Y.

11.2.1 Model and Sources of Variation

The example discussed in this section concerns a price comparison of wireless computer tracking devices with Brand and Store as the two X-factors. The factor levels for Brand are the names of the three brands: Souris, Topo, and Maus. The factor levels for Store are the names of the four stores: CostLess, ServMax, BestDeal, and CompMart. With three brand and four store levels, there are altogether 3*4 = 12 factor level combinations. Since each level combination is observed exactly once, there are 12 observations of price.

The goal of a statistical ANOVA model is to explain as much of the variation in the Y-variable as possible using the available X-factors. In a two-way ANOVA model for unreplicated data, factor A has c levels and factor B has r levels. Each observation Y_{ti} is represented by the additive effects of the overall mean, the effect of level *t* of factor A, and the effect of level *i* of factor B. In addition, there is random variation. The model is written as

$$Y_{ti} = \mu + A_t + B_i + \varepsilon_{ti}$$

where μ is the overall mean.

A_t represents the fixed effect factor A at level t, with t = 1, ..., c. The effect A_t is the difference between the mean of level t of factor A and the overall mean μ. The factor A level effects sum to zero.

B_i represents the fixed effect factor B at level i, with i = 1, ..., r. The effect B_i is the difference between the mean of level i of factor B and the overall mean μ. Factor B level effects also sum to zero.

ε_{ti} is the residual error of the factor level combination (A_t, B_i). The errors follow a Normal distribution with mean 0 and constant standard deviation $\sigma(\sim N(0,\sigma))$.

In this model, the mean of a factor level combination is represented as an adjustment to the overall mean. The mean of level t of factor A is written as $\text{Mean}_{A(t)} = \mu + A_t$. Similarly, the mean of level i of factor B is written as $\text{Mean}_{B(i)} = \mu + B_i$. One assumes that all observations are statistically independent and the standard deviations are constant.

The assumptions of this model are concentrated on the residual error ε_{ti}. It assumes that residuals are normally distributed with mean 0 and standard deviation σ. Minor deviations from the Normal distribution assumptions can be tolerated except for prediction of future outcomes. Further, the observations at various factor level combinations are independent of each other. Selecting subjects randomly and conducting the test in random order are steps to ensure independence. The standard deviation σ is assumed to be the same for all factor level combinations, although one observation per factor level combination makes it difficult to evaluate this assumption.

Figure 11.1 Decomposition of Total Variation in Two-Way ANOVA without Replications

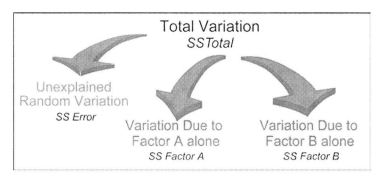

According to this model, the Total Variation of Y, represented by the Total Sum of Squares (SS_{Total} or C. Total) can be decomposed into two components, the Model Sum of Squares (SS_{Model}) and Error Sum of Squares (SS_{Error}). SS_{Model} is also referred to as explained variation, while SS_{Error} is referred to as unexplained variation. In JMP, this decomposition is shown in the ANOVA table. The Effect Tests table further decomposes the SS_{Model} into a Sum of Squares due to factor A ($SS_{Factor A}$) and Sum of Squares due to factor B ($SS_{Factor B}$). The decomposition is shown schematically in Figure 11.1. The arrows point from SS_{Total} to the three components generated by the decomposition. Each source of variation corresponds to a term in the two-way ANOVA model. The decomposition can be represented by an equation:

$$SS_{Total} = SS_{Model} + SS_{Error} = SS_{Factor A} + SS_{Factor B} + SS_{Error}$$

11.2.2 Hypotheses for Each Model Term

The hypotheses for both X-factors are similar to one-way ANOVA. For each factor, the null hypothesis states that there is no difference between the level means or equivalently that the factor effects are all zero. Each alternative hypothesis says that at least one level mean is different from the other means of that factor, or that some factor level effects are non-zero:

Hypotheses for Factor A: H_0(Factor A): $Mean_{A(1)} = Mean_{A(2)} = \ldots = Mean_{A(c)}$ versus H_A(Factor A): $Mean_{A(m)} \neq Mean_{A(n)}$ for at least two Factor A level means with m ≠ n.

Hypotheses for Factor B: H_0(Factor B): $Mean_{B(1)} = Mean_{B(2)} = \ldots = Mean_{B(r)}$ versus H_A(Factor B): $Mean_{B(k)} \neq Mean_{B(l)}$ for at least two Factor B level means with k ≠ l.

These two hypotheses can be tested independently using the *p*-values of the F-Ratios in the Effect Tests table.

11.2.3 Detail Example: Prices of Wireless Mouse Devices by Brand at Different Stores

Situation: A consumer magazine wants to give a best-buy recommendation to its readers on wireless mouse devices for computers. They collect data on the price of the brands of such devices. A single unit of each brand was purchased from each of four different stores on the same day. The store names are CostLess, ServMax, BestDeal, and CompMart. The three brands are Souris, Topo, and Maus. The purchased models were selected because they had about the same user functionality.

Research Question: The magazine wants to know whether the three brands all have the same average price. The magazine wants to incorporate any price differences in a best-buy recommendation. Price differences between stores are noted, but are not of primary interest to the magazine editors.

The research question leads to the hypotheses stated in the previous section. It also asks for a more detailed analysis of the results. Specifically, it asks to identify which of the means are different from each other with statistical significance. Since the objective is a possible best-buy recommendation, the magazine editors look for a mean lowest price.

Data Requirements: The data were entered into a spreadsheet. The format of the spreadsheet data is shown in Table 11.1. Each combination of Brand and Store shows exactly one non-discounted price. The primary task is to examine price differences by brand. Price differences between stores are included only because they might contribute to variability.

Table 11.1 Spreadsheet Data of Price of Wireless Mouse Devices by Brand and Store

Store \ Brand	Souris	Topo	Maus	Mean price by store
CostLess	27	35	31	31
ServMax	40	44	42	42
BestDeal	23	31	27	27
Compmart	22	26	24	24
Mean price by brand	28	34	31	Overall Mean = 31

The spreadsheet format is unsuitable for analysis in JMP. The data file for a two-way ANOVA example requires three columns. *WirelessMouse.jmp* is in this format (Figure 11.2):

- Column 1 contains Store, a nominal X-factor with four levels: CostLess, ServMax, BestDeal, and Compmart.
- Column 2 contains Brand, a nominal X-factor with three levels: Souris, Topo, and Maus.
- Column 3 contains Price, the continuous Y-response in dollars per single device.

There are exactly 12 rows; each row represents a single observation per factor level combination.

Figure 11.2 Data File *WirelessMouse.jmp* in Three-Column Format

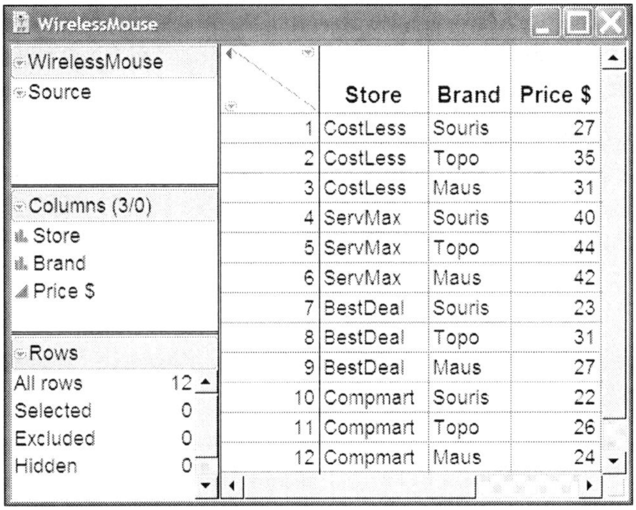

	Store	Brand	Price $
1	CostLess	Souris	27
2	CostLess	Topo	35
3	CostLess	Maus	31
4	ServMax	Souris	40
5	ServMax	Topo	44
6	ServMax	Maus	42
7	BestDeal	Souris	23
8	BestDeal	Topo	31
9	BestDeal	Maus	27
10	Compmart	Souris	22
11	Compmart	Topo	26
12	Compmart	Maus	24

Anticipated Results: It is expected that all three brands have different mean prices, but that there might also be considerable price differences between stores.

11.2.4 JMP Analysis of the Wireless Mouse Example Using the Fit Model Platform

A model with more than one X-variable requires the Fit Model platform. The Fit Y by X platform is unsuitable. Select **Fit Model** from the **Analyze** menu as shown in Figure 11.3. In the Model Specification window, assign each variable its appropriate role. The Y-variable is Price, while the X-factors, called model effects in the Fit Model platform, are Store and Brand.

Figure 11.3 Fit Model Variable Specification for Wireless Mouse Example

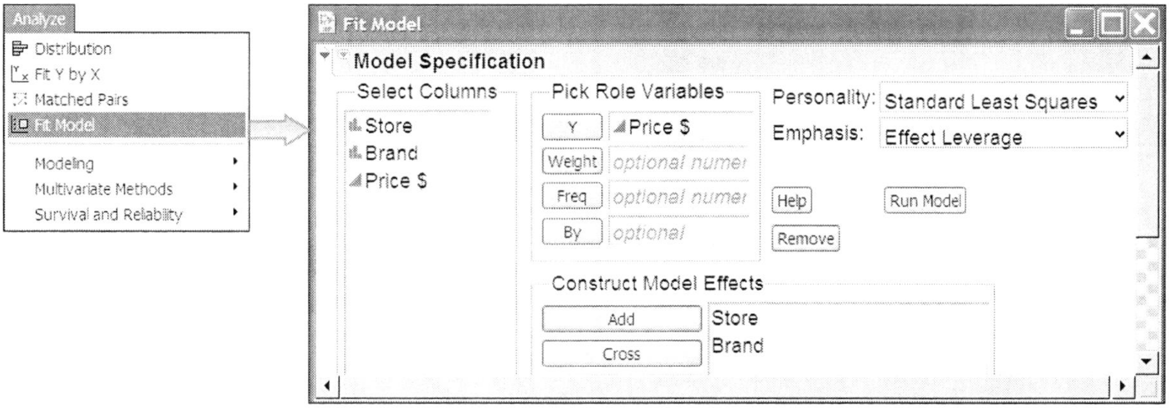

The Fit Model platform of JMP provides extensive output, not all of direct use in two-way ANOVA. The basic output is shown in Figure 11.4, Figure 11.5, and Figure 11.6. Figure 11.4 shows three quantities of interest from the Summary of Fit table:

- The Root Mean Square Error, or RMSE, gives the estimated standard deviation around the fitted two-way ANOVA model. In this example the rounded RMSE = 1.15 dollars.

- The Mean of Response = 31 dollars is the overall mean price of all 12 observations.

- The number of observations is 12.

Figure 11.4 Excerpt of Summary of Fit

Summary of Fit	
Root Mean Square Error	1.154701
Mean of Response	31
Observations (or Sum Wgts)	12

The Analysis of Variance table of Figure 11.5 decomposes the Total Sum of Squares $SS_{Total} = 638$ (in JMP labeled C. Total for corrected total) into two components, $SS_{Model} = 630$ and $SS_{Error} = 8$. It uses this decomposition to calculate the F-Ratio and its associated p-value for the overall model. The p-value of the overall F-Ratio, Prob > F is < .0001. (The asterisk indicates that this p-value is less than $\alpha = 0.05$.) Conclude at $\alpha = 0.05$ that the two-way ANOVA model is a better representation of the data than fitting the base model of a single overall mean.

Figure 11.5 Analysis of Variance Table of Wireless Mouse Price Example

Analysis of Variance				
Source	DF	Sum of Squares	Mean Square	F Ratio
Model	5	630.00	126.000	94.5000
Error	6	8.00	1.333	Prob > F
C. Total	11	638.00		<.0001*

The decomposition of SS_{Model} into individual factor sum of squares allows significance tests of individual model effects. The Effect Tests table in Figure 11.6 decomposes SS_{Model} into SS_{Brand} and SS_{Store}. The two F-Ratios test the significance of each component independently of the other. Both *p*-values in the Prob > F column are below the $\alpha= 0.05$ significance threshold. The *p*-value for Store is < .0001 and for Brand is 0.0010. For both factors, reject the null hypotheses. There is statistically significant evidence that some Brand means are different from other Brand means, and that some Store means are different from other Store means.

Figure 11.6 Effect Tests Table of Wireless Mouse Example

Effect Tests				
Source	DF	Sum of Squares	F Ratio	Prob > F
Store	3	558.00	139.5000	<.0001*
Brand	2	72.00	27.0000	0.0010*

Figure 11.7 shows a pie chart decomposition of the SS_{Total} based on $SS_{Total} = 638$, $SS_{Model} = 630$, $SS_{Store} = 558$, $SS_{Brand} = 72$, and $SS_{Error} = 8$.

Figure 11.7 Decomposition of SS_{Total} in Wireless Mouse Example

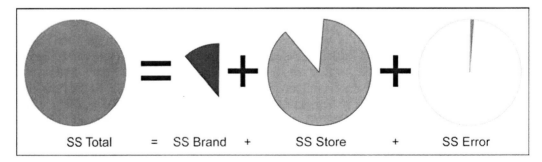

The pie on the left is SS_{Total}. The right side shows the three components. The largest source of variation is due to Store, while SS_{Error} is the smallest. Not including the X-factor Store in the model would have resulted in a much-increased SS_{Error}, because SS_{Store} would have been added to SS_{Error}. As a result, factor Brand would have been statistically non-significant. This emphasizes the importance of including all X-factors that contribute to variation.

Further Analysis of Factor Means: The ANOVA and effect tests results tell us that there are statistically significant differences between Brand means and between Store means. They do not tell which of the means are different from each other. To answer this, further analysis, similar to the Compare Means option in Fit Y by X, is needed. Fit Model offers several options. To gain some graphical insight, it is useful to begin with LSMeans plots for the two factors. The pull-down menu of each X-factor produces the desired LSMeans plots (see Figure 11.8 for Brand). We will also be discussing LSMeans Student's t and briefly LSMeans Tukey HSD with their respective options.

Figure 11.8 Pull-Down Menu and LSMeans Plots in Fit Model

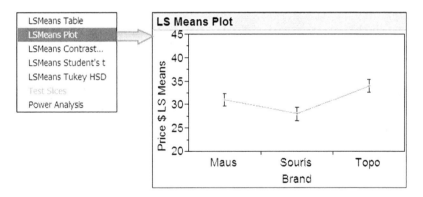

LSMeans Student's t for Brand Level Means: The results for Brand level means are of primary interest. The LSMeans plot on the right in Figure 11.8 shows that Souris is the brand with the lowest mean, while Topo has the highest mean. The LSMeans plot does not show directly whether these means are different with statistical significance. One could use the 95% confidence intervals shown for each mean to infer which means are different with statistical significance. Noticing that the individual 95% confidence intervals of Souris and Topo do not overlap, you could conclude that at least these two means are different with statistical significance. However, a more reliable approach is to use the Crosstab report, the Connecting Letters report, or the Ordered Differences report.

The LSMeans Student's t reports for mean price differences can be selected (Figure 11.9), because the effect tests for Brand p-value is below the $\alpha = 0.05$ threshold. This method gives the shortest confidence intervals for mean differences, but should only be used if the p-value is less than α, and only a few differences are of interest.

First, consider the Crosstab Report. Each cell contains four values:

- The top value is the point estimate of the mean difference between row level and column level means. In the cell representing the intersection between the Maus row and Souris column, the mean difference between Maus and Souris means is +3. The Maus sample mean is 3 dollars higher than the Souris sample mean.

- The second value is the Standard Error of the Mean Difference. This value is used in calculating the confidence intervals. For Maus-Souris the standard error is 0.8165.

- The third and fourth values are the 95% lower and upper confidence limits for the mean difference. The Maus-Souris mean difference confidence interval is from 1.0021 to 4.9979 dollars. This suggests a statistically significant two-sided difference between the Maus and Souris means, since zero is not included.

The Crosstab report for mean differences is a symmetric table. For example, it contains both the Maus-Souris mean difference and the Souris-Maus mean difference. The mean difference and the confidence limits differ by their signs. The standard errors are always positive and are the same irrespective of the order of subtraction in the differences. None of the 95% confidence intervals for mean differences between brands include 0. Thus, all three means are different from each other with statistical significance. The 95% CI for the mean difference between Maus and Souris is from 1.0 to 5.0 (rounded) dollars, the one between Maus and Topo is from –5.0 to –1.0 dollars, and the one between Souris and Topo is from –8.0 to –4.0 dollars.

Figure 11.9 LSMeans Differences Student's t by Brand

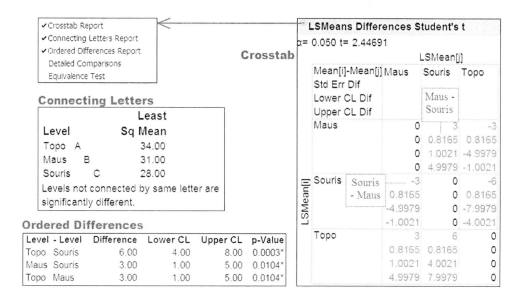

The Connecting Letters report lists factor levels and their factor level means, instead of mean differences. Each mean is labeled with one or more capital letters. Factor level means that share the same capital letters are not different from each other at $\alpha = 0.05$. The Connecting Letters report in Figure 11.9 confirms that all three means are different from each other, since each mean has a different letter.

A useful addition to the analysis is the Ordered Differences report in Figure 11.9, which orders the differences from largest to smallest. For each difference, it gives the 95% confidence interval. These results are a rearrangement of those in the Crosstab report from the largest to the smallest difference. Consistent with the results of the Connecting Letters report, none of the three confidence intervals include 0, and all of the (two-sided) *p*-values for the mean differences are less than 0.05, suggesting that all pairwise differences are statistically significant.

The Crosstabs, Connecting Letters, and Ordered Differences reports are merely different ways to present the same results. The choice depends on the intended purpose of the analysis. There is rarely a reason to include all three versions of LSMeans Differences Student's t in reports or presentations.

LSMeans Differences for Store Level Means: Although Store is a factor of lesser interest, the analysis is presented in this section to practice interpreting JMP output. The LSMeans plot in Figure 11.10 shows the Store means. The means are connected by lines for better visualization of differences. For each mean, a 95% confidence interval is shown. The plot shows that Compmart has the lowest sample mean and that ServMax has the highest one. To answer which pairs of these four means are different from each other with statistical significance, we use the LSMeans Student's t option, noting that *p*-values of the Effect Tests table are below α and the number of comparisons is small.

Figure 11.10 LSMeans Plot of Mean Prices by Store

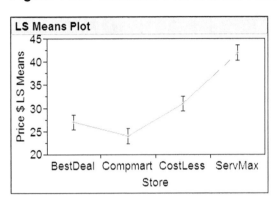

The Crosstab report for LSMeans Differences for Store means is shown in Figure 11.11. This table shows point estimates, standard errors of the point estimate, and 95% confidence limits on all pairwise differences of factor level means. This table compares the differences between row and column level means of the factor Store. The row labeled Compmart contains results on differences between Compmart mean and each of BestDeal, CostLess, and ServMax means. The cell of the column labeled BestDeal shows the following:

1. The mean difference between Compmart and BestDeal is –3.

2. The standard error for the Compmart-BestDeal mean difference is 0.94281.

3. The 95% confidence interval for mean difference Compmart-BestDeal is from –5.307 to –0.693 dollars.

Figure 11.11 LSMeans Differences Student's t by Store

LSMeans Differences Student's t

α= 0.050 t= 2.44691

Crosstab

Mean[i]-Mean[j] Std Err Dif Lower CL Dif Upper CL Dif	BestDeal	LSMean[j] Compmart	CostLess	ServMax
BestDeal	0 0 0 0	3 0.94281 0.69303 5.30697	-4 0.94281 -6.307 -1.693	-15 0.94281 -17.307 -12.693
Compmart	-3 0.94281 -5.307 -0.693	0 0 0 0	-7 0.94281 -9.307 -4.693	-18 0.94281 -20.307 -15.693
CostLess	4 0.94281 1.69303 6.30697	7 0.94281 4.69303 9.30697	0 0 0 0	-11 0.94281 -13.307 -8.693
ServMax	15 0.94281 12.693 17.307	18 0.94281 15.693 20.307	11 0.94281 8.69303 13.307	0 0 0 0

Connecting Letters

Level		Least Sq Mean
ServMax	A	42.00
CostLess	B	31.00
BestDeal	C	27.00
Compmart	D	24.00

Levels not connected by same letter are significantly different.

Ordered Differences

Level	- Level	Difference	Lower CL	Upper CL	p-Value
ServMax	Compmart	18.00	15.69	20.31	<.0001*
ServMax	BestDeal	15.00	12.69	17.31	<.0001*
ServMax	CostLess	11.00	8.69	13.31	<.0001*
CostLess	Compmart	7.00	4.69	9.31	0.0003*
CostLess	BestDeal	4.00	1.69	6.31	0.0054*
BestDeal	Compmart	3.00	0.69	5.31	0.0190*

The symmetry of the Crosstab report for mean differences results in redundant information, because it contains both the Compmart-BestDeal and the BestDeal-Compmart mean difference. The mean difference and the confidence limits differ only by their signs.

The Connecting Letters report in Figure 11.11 shows all four factor level means with a different letter. Therefore each mean is different from all other factor level means at α = 0.05. Similarly, the Ordered Differences report shows all six differences, and none of the intervals include 0. This indicates that all stores charge different average prices. All of the (two-sided) *p*-values are less than 0.05.

Final Interpretation of Results for Wireless Mouse Example: The differences between mean prices of brands are of primary interest. Both the Ordered Differences report and the Connecting Letters report capture these results succinctly. The results show that the mean price differences between all brands are statistically significant at α = 0.05. The most expensive brand is Topo. The cheapest brand is Souris.

It is important to include the factor Store in the model. The mean price differences between stores account for most of the variation. These differences might not be of primary interest. Only by including Store in the model do the Brand differences become apparent and statistically significant. This fact suggests that a price-conscious shopper needs to pay attention to the selection of the store where purchases are made along with the actual brands purchased there.

11.3 Two-Way ANOVA with Equally Replicated Data

Two factors often interact, creating synergistic or antagonistic relationships at certain factor level combinations. The result might be responses that are either larger or smaller than straight additive factor effects provide. In ANOVA, these are modeled by interaction effects. If interaction is anticipated, the data collection should be designed to include replications, because they are used to test the interaction effects. When an equal number (greater than one) of observations for each factor level combination is available, factor level means and the interaction effects due to factor level combinations can be estimated independently of each other. This is why this type of two-way ANOVA is useful.

The data collection design needs to specify the number of levels for each factor and the number *n* of replicate observations per factor level combination. Observations of the different factor level combinations should be obtained in random order as a safeguard against extraneous influences. It would be bad practice to obtain all observations of factor A at level 1, then at level 2, etc.

11.3.1 Model and Sources of Variation

Each observation Y_{tij} can be represented by a model consisting of the effect of the overall mean, the level t effect of factor A, the level i effect of factor B, and the interaction effect of the factor level combination AB_{ti}. Random variation is represented by the error term ε_{tij}. The subscript j refers to the fact that for each factor level combination, there are $n > 1$ observations. The model is:

$$Y_{tij} = \mu + A_t + B_i + AB_{ti} + \varepsilon_{tij}$$

In this model,

- μ is the overall mean
- A_t is the effect of level t of factor A, with t = 1, ..., c
- B_i is the effect of level i of factor B, with i = 1, ..., r
- AB_{ti} is the interaction effect
- ε_{tij} is the residual error term assumed to be Normally distributed as $N(0,\sigma^2)$, j = 1, ..., n

The sum of the effects is zero for each factor and also for the interaction effects. It is further assumed that all observations are statistically independent and the standard deviation σ is constant for all observations. As with unreplicated two-way ANOVA, the model suggests that the mean of level t of factor A is $Mean_{A(t)} = \mu + A_t$. Similarly, the mean of level i of factor B is $Mean_{B(i)} = \mu + B_i$. For the combination of level t of factor A and level i of factor B, the mean is $Mean_{AB(t,i)} = \mu + A_t + B_i + AB_{ti}$.

The variation of each factor is again measured by a Sum of Squares. Each model term has an associated Sum of Squares. The Total Variation of Y is represented by the Total Sum of Squares (SS_{Total}) and is decomposed into two familiar components, the Model Sum of Squares (SS_{Model}) and Error Sum of Squares (SS_{Error}). Model Sum of Squares is the explained variation, while Error Sum of Squares is the unexplained variation. The SS_{Model} is further decomposed into the Sum of Squares due to factor A ($SS_{Factor\,A}$), the Sum of Squares due to factor B ($SS_{Factor\,B}$), and the Sum of Squares due to the interaction effect ($SS_{Interaction\,AB}$).

When each factor level combination is replicated exactly the same number of times, the components are statistically independent. The arrows in Figure 11.12 show the decomposition of SS_{Total} into its four independent parts. The decomposition can be represented by the following equation:

$$SS_{Total} = SS_{Model} + SS_{Error} = SS_{FactorA} + SS_{FactorB} + SS_{InteractionAB} + SS_{Error}$$

Figure 11.12 Decomposition of Total Variation in Two-Way ANOVA with Equal Replication

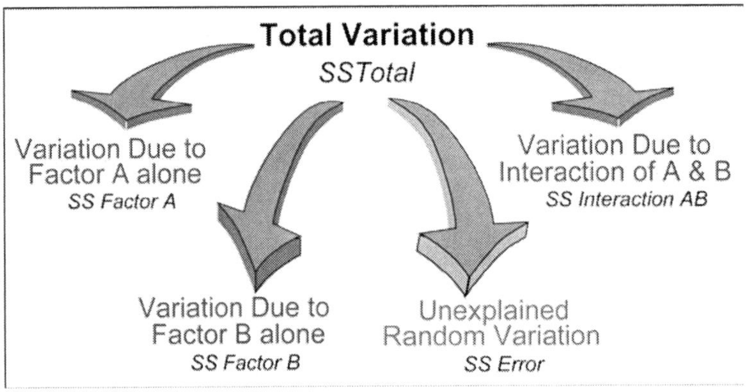

11.3.2 Hypotheses for Each Model Term

The F-Ratio of the ANOVA table tests the overall significance of the model. In the JMP effect tests, each effect is individually tested for significance. For replicated two-way ANOVA, there are three sets of hypotheses, each with its own F-Ratios and *p*-values:

Hypotheses for Factor A: The null hypothesis of the factor A effect test states that the factor A effects are all zero, implying that all factor A level means are equal. The alternative hypothesis states that at least one factor A level effect is not zero. More formally,

$$H_{0(\text{Factor A})}: A_1 = A_2 = ... = A_c \quad \text{versus } H_{A(\text{Factor A})}: \text{at least one } A_t \neq 0$$

Hypotheses for Factor B: The null hypothesis of the factor B effect test states that the factor B effects are all zero, implying that all factor B level means are equal. The alternative hypothesis states that at least one factor B level effect is not zero. More formally,

$$H_{0(\text{Factor B})}: B_1 = B_2 = ... = B_r \quad \text{versus } H_{A(\text{Factor B})}: \text{at least one } B_i \neq 0$$

Hypotheses for Factor Interaction AB: The null hypothesis of the interaction AB effect test states that all interaction effects are zero. The alternative hypothesis states that at least one interaction AB level effect is not zero. More formally,

$$H_{0(\text{Interaction})}: AB_{1i} = AB_{2i} = ... = AB_{ci} = 0 \text{ for all i or } AB_{t1} = AB_{t2} = ... = AB_{tr} = 0 \text{ for all}$$
t versus $H_{A(\text{Interaction})}$: at least one $AB_{ti} \neq 0$

11.3.3 **What Is an Interaction Effect?**

An interaction effect between two factors A and B is present when the magnitude of a factor effect depends on the level of the other factor. The presence of a statistically significant interaction effect between two factors changes the way their effects have to be explained. An interaction requires statements like the following:

1. If factor B is at level 1, then the difference between Y means of level 1 and 2 of factor A is x.

2. If factor B is at level 2, then the difference between Y means of level 1 and 2 of factor A is y, where x ≠ y and so forth.

In dealing with interactions, keep in mind that although interactions have a purely statistical meaning, there might or might not be an underlying physical or theoretical explanation.

Figure 11.13 shows three typical interaction plots of means of two factors, each at two levels. The squares in each plot represent the means of factor level combinations. On the left is a plot showing no interaction. In the center is a means plot showing strong interaction, while on the right is a weaker interaction plot.

Figure 11.13 Three Typical Interaction Plots

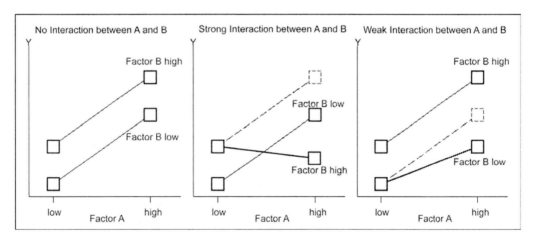

With no interaction, the difference in means between factor B levels high and low does not depend on factor A. The reverse is also true: The difference in means between factor A levels high and low does not depend on factor B. This is shown by the parallel lines between the Y means of factor B high and factor B low.

The interaction graph in the center shows a strong interaction, because the difference in Y means between factor B high and factor B low depends strongly on the level of factor A. When factor A is at low, the Y mean for factor B at a high level is higher than for factor B low. When factor A is high, then the reverse is true. The no interaction situation is shown as a dashed line and square.

The interaction graph on the right shows a weaker interaction than that in the center. The lines are not parallel as in the no interaction situation. The difference in Y means between factor B high and factor B low depends on the level of factor A. The difference is smaller when factor A is low and larger when it is high. The factor B effect thus depends on the level of factor A.

Figure 11.13 shows idealized versions of interaction plots. When interaction plots are constructed from data, sampling variability might move these lines randomly. Thus, data rarely show parallel lines even though no interaction is present. Ultimately, you must rely on the effect tests to determine the significance of the interaction term.

11.3.4 Detail Example: Profitability of Futures Trading Customers of a Financial Institution (Equal Replications)

Situation: A financial institution needs to evaluate the viability of its futures trading operation. Some futures trading customers might be less profitable than others. A small pilot study pulls the records of 12 randomly selected customers, classified according to a factor called Demands on Management with the levels Low, Medium, and High, depending on how much effort management has to spend on each customer. The assumption is that customers with High Demands on Management might be less profitable than Low ones. Customers are also classified according to the factor Customer Actions as either Reactive or Proactive. Reactive customers typically wait for the institution to recommend trading activities. The customers are free to refuse any or all of the recommendations. Proactive customers come to the institution with their own suggestions for trading activities. It is anticipated that Proactive customers are more profitable, because they require less service and consulting time.

Research Question: Do levels of Demands on Management and Customer Actions affect profitability of futures trading accounts? Specifically, how does the profitability of futures customers depend on factor levels of Demands on Management and Customer Actions? Is there a combination of factor levels that is more (or less) profitable than others?

The research question leads to the hypotheses for the main factors and interaction stated in the previous section. The alternative hypotheses specify statistically significant differences in the level means of each main factor and the interaction term. The research question also asks whether certain factor level combinations show a significantly higher mean profitability.

Data Requirements: Figure 11.14 shows $3 \times 2 = 6$ factor level combinations. The tabulation shows two profitability observations for each factor level combination. Profitability is shown in $10000 per customer per year.

- The crosstab approach to presenting these data is very popular in spreadsheets but needs to be converted for JMP analysis. The stacked set-up necessary for JMP is shown in Figure 11.15 as the file called *ProfitFutTradInteract.jmp*.

Figure 11.14 Futures Trading Profitability Data Presented as a Crosstab

Factor: Demands on Management

Factor: Customer Actions	High	Medium	Low
Reactive	19, 11	20, 17	22, 31
Proactive	22, 24	25, 30	52, 60

Y = Profitability per Customer in $10,000

Figure 11.15 Data File *ProfitFutTradeInteract.jmp*

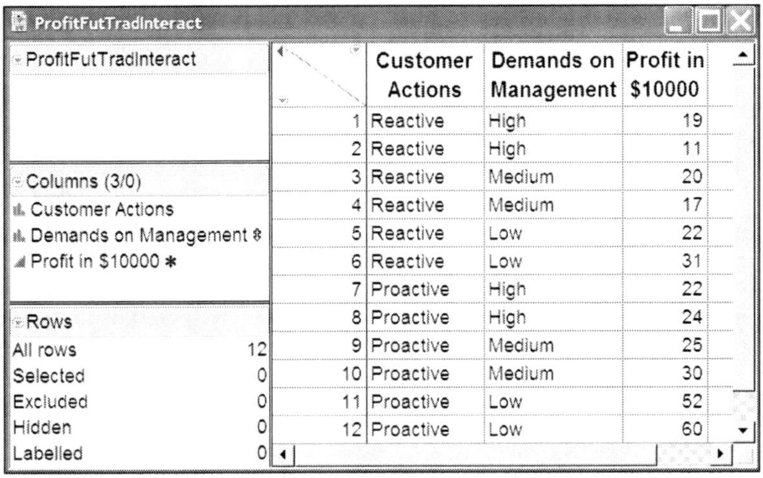

	Customer Actions	Demands on Management	Profit in $10000
1	Reactive	High	19
2	Reactive	High	11
3	Reactive	Medium	20
4	Reactive	Medium	17
5	Reactive	Low	22
6	Reactive	Low	31
7	Proactive	High	22
8	Proactive	High	24
9	Proactive	Medium	25
10	Proactive	Medium	30
11	Proactive	Low	52
12	Proactive	Low	60

ProfitFutTradInteract

- ProfitFutTradInteract

Columns (3/0)
- Customer Actions
- Demands on Management 8
- Profit in $10000 ✱

Rows
All rows — 12
Selected — 0
Excluded — 0
Hidden — 0
Labelled — 0

The data file requires three columns, two for the nominal X-factors, and one for the continuous Y. This example has 12 rows, one row for each observation, because each of the six factor level combinations has two Y observations.

- Column 1 contains Customer Actions with two levels: Reactive and Proactive.
- Column 2 contains Demands on Management with three levels: Low, Medium, and High. (This column is list checked to ensure that levels appear in order Low, Medium, and High, rather than alphabetical.)
- Column 3 contains the Y-variable Profit in $10000.

11.3.5 JMP Analysis of Profitability of Futures Trading Customers

The two main effects and the interaction effect need to be specified in the Model Specification window of the Fit Model platform in JMP. The most convenient way is shown in Figure 11.16.

Figure 11.16 Model Specification for Profitability of Futures Trading

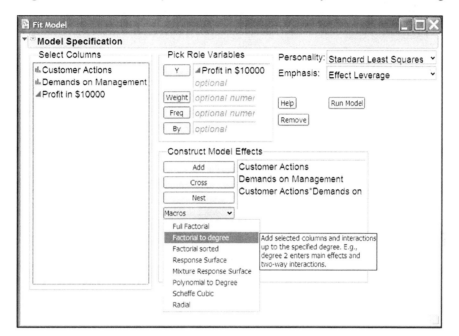

It requires you to do the following:

1. Highlight the two X-factors (**Customer Actions** and **Demands on Management**) in the **Select Columns** list.

2. Use the **Macros** menu and specify **Factorial to degree** = 2, the default value. This will bring the two X-factors and the interaction term into the **Construct Model Effects** section.

3. Specify the continuous Y variable **Profit in $10000**.

4. Click **Run Model** to create the basic output.

The Summary of Fit in Figure 11.17 shows that the overall Mean of Response of all 12 observations is 27.75 (rounded) in $10000. The Root Mean Square Error, the estimated standard deviation around the fitted model, is RMSE = 4.54 (in $10000).

Figure 11.17 Excerpt of Summary of Fit

Summary of Fit	
Root Mean Square Error	4.536886
Mean of Response	27.75
Observations (or Sum Wgts)	12

The overall ANOVA table of Figure 11.18 decomposes the Total Sum of Squares (2264.25) into a Model Sum of Squares (2140.75) and Error Sum of Squares (123.50). The overall F-test is statistically significant at $\alpha = 0.05$, because the *p*-value Prob > F = 0.0010 is smaller than α. At this point you can conclude that at least one of the factors has an effect on profitability.

Figure 11.18 Overall ANOVA Table

Analysis of Variance				
Source	DF	Sum of Squares	Mean Square	F Ratio
Model	5	2140.75	428.150	20.8008
Error	6	123.50	20.583	Prob > F
C. Total	11	2264.25		0.0010*

The Effect Tests table in Figure 11.19 details the decomposition of the Model Sum of Squares by factors including the interaction term. The Customer Actions Sum of Squares is 720.75, Demands on Management Sum of Squares is 1125.50, and Interaction Sum of Squares is 294.50. These three sums of squares add to the $SS_{Model} = 2140.75$.

Figure 11.19 Effect Tests Table of Profitability of Futures Trading by Customer

Effect Tests				
Source	DF	Sum of Squares	F Ratio	Prob > F
Customer Actions	1	720.75	35.0162	0.0010*
Demands on Management	2	1125.50	27.3401	0.0010*
Customer Actions*Demands on Management	2	294.50	7.1538	0.0258*

Figure 11.20 shows that Demands on Management, Customer Actions, and Interaction represent sizable slices of the Total Sum of Squares pie. In this example, the effect tests show both main factors and the interaction effect with statistically significant results, because all their *p*-values are below α. The interaction term is statistically significant with p = 0.0258. This makes the interaction the focus of our analysis, because it looks at the model through the means of factor level combinations. In order to determine which means of factor level combinations are different, use the Interaction Effect plot, Connecting Letters report and possibly the Ordered Differences report. The Crosstab report is also useful in certain applications, but might be very large due to many factor level combinations. Further analysis of main effects is also useful.

Figure 11.20 Decomposition of SS Total in Profitability of Futures Trading Example

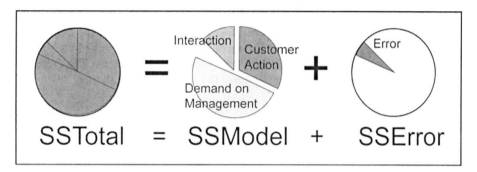

The Interaction Effect plot gives an excellent intuitive understanding of the interaction effects. Without interaction, the Interaction Effect plot should show nearly parallel lines. The Interaction Effect plot for the profitability of futures trading example in the right panel of Figure 11.21 shows non-parallel lines. Slight deviations from parallel are quite acceptable, because the means, represented by squares, are subject to sampling error. However, the p-value $= 0.0258$ of the effect test for interaction indicates that the lines are non-parallel lines with statistical significance at $\alpha = 0.05$.

Figure 11.21 shows that Proactive customers tend to be more profitable than Reactive ones. Also, Low Demands on Management customers tend to be more profitable than High Demands on Management customers. But it also shows that Proactive customers with Low Demand on Management are disproportionately more profitable. However, is this statistically significant? To answer this and other questions, the Connecting Letters report for each factor can be consulted.

The LS Means plots for each model effect can be obtained from the pull-down menu of each variable as shown in Figure 11.8. The LSMeans plot of the interaction term is the interaction plot. Figure 11.21 shows the LSMeans plots for the two Main Effects and the Interaction Effect. The horizontal line is for the overall mean. The factor level means and factor level combination means are at the endpoints of each line.

Figure 11.21 LSMeans Plots for Profitability of Futures Trading Example

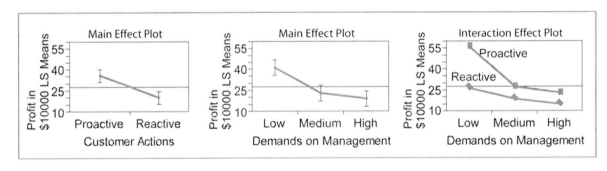

Effect of Factor Customer Actions: Figure 11.22 shows the Connecting Letters report and the Ordered Differences report for the X-factor Customer Actions. Proactive customers have a mean profitability of $355000 (35.55 in $10000 units). Reactive customers have an estimated mean profitability of $200000 (20.00 in $10000 units). This represents a statistically significant difference, because these two levels are marked by different letters A and B. Also, the Ordered Differences report shows a two-sided 95% confidence interval for the mean difference between Proactive and Reactive customers from 9.09 to 21.91 (again in $10000). Since it does not include zero, you can conclude that the difference is statistically significant at $\alpha = 0.05$.

Figure 11.22 Customer Actions: Connecting Letters and Ordered Differences for LSMeans Student's t

LSMeans Differences Student's t				
Level	Least Sq Mean			
Proactive A	35.500000	Connecting Letters Report		
Reactive B	20.000000			
Levels not connected by same letter are significantly different.		Ordered Differences Report		
Level - Level	Difference	Lower CL	Upper CL	
Proactive Reactive	15.50000	9.090627	21.90937	

Effect of Factor Demand on Management: Figure 11.23 shows the Connecting Letters report and the Ordered Differences report for the X-factor Demands on Management. Low Demands on Management customers have an estimated mean profitability of $412500 (41.25 in $10000 units). Medium and High Demands on Management customers show a profitability of $230000 (23.00 in $10000 units) and $190000 (19.00 in $10000 units). There are statistically significant differences between the Low and Medium and Low and High levels, because in the Connecting Letters report the level Low is the only one marked by the letter A. Medium and High are both marked by the letter B, indicating that the difference in mean profitability between Medium and High customers is statistically not significant. The Ordered Differences report shows 95% confidence intervals for the mean difference between the three pairs of mean differences. The two confidence intervals for a difference involving Low do not include zero. So you can conclude that the mean profitability of Low is different from the other two Demands on Management levels at a significance level of $\alpha = 0.05$.

Figure 11.23 Demands on Management: Connecting Letters and Ordered Differences for LSMeans Student's t

Level		Least Sq Mean
Low	A	41.25
Medium	B	23.00
High	B	19.00

Levels not connected by same letter are significantly different.

Level	- Level	Difference	Lower CL	Upper CL
Low	High	22.25	14.40	30.10
Low	Medium	18.25	10.40	26.10
Medium	High	4.00	-3.85	11.85

Interaction Effect: With JMP, it is easy to determine how different legitimate methods to analyze data would modify your conclusion. This can be seen by comparing results from the LSMeans Difference Student's t and Tukey HSD for the interaction term. Shown in Figure 11.24 are the two Connecting Letters reports for the six factor level combinations. Notice that the LSMeans (or Least Sq Means) are not affected by these methods. For example, in both panels, Proactive, Low has an estimated mean profitability of $560000.

For the LSMeans Difference Student's t, Proactive, Low is the only one with the letter A and is significantly different from all other factor level combinations. Four of the remaining five factor level combinations—Proactive, Medium; Reactive, Low; Proactive, High; and Reactive, Medium—are marked with the letter B. The mean profitabilities of these four factor level combinations are not different with statistical significance. Proactive, High; Reactive, Medium; and Reactive, High are marked with the letter C. The mean profitabilities of these three factor level combinations are not different with statistical significance. Thus you can conclude that Proactive, Medium and Reactive, Low are different from Reactive, High with statistical significance.

In this example, the letters B and C create two sets with overlapping factor levels. Proactive, High and Reactive, Medium are both marked with the letters B and C. These two factor levels are not different from a larger set of factor levels. It can be said that Proactive, High and Reactive, Medium are not different from Proactive, Medium; Reactive, Low; and Reactive, High with statistical significance, because they share a letter with these factor level combinations.

Figure 11.24 Interaction Term: Connecting Letters Report—Student's
t versus Tukey HSD

LSMeans Differences Student's t			
α= 0.050 t= 2.44691			
Level			Least Sq Mean
Proactive,Low	A		56.00
Proactive,Medium	B		27.50
Reactive,Low	B		26.50
Proactive,High	B	C	23.00
Reactive,Medium	B	C	18.50
Reactive,High		C	15.00
Levels not connected by same letter are significantly different.			

LSMeans Differences Tukey HSD		
α= 0.050 Q= 3.97999		
Level		Least Sq Mean
Proactive,Low	A	56.00
Proactive,Medium	B	27.50
Reactive,Low	B	26.50
Proactive,High	B	23.00
Reactive,Medium	B	18.50
Reactive,High	B	15.00
Levels not connected by same letter are significantly different.		

In this example, the interaction term has a *p*-value Prob > F = 0.0258. As a result, the LSMeans Differences Student's t can be used. Only a few mean differences are of interest. For example, LSMeans Differences Student's t can be used for the question, Is Proactive, Low more profitable than all the other factor level combinations with statistical significance? However, if all possible pairwise differences are of interest, you might have to rely on LSMeans Differences Tukey HSD.

Six means of factor level combinations allow 15 possible mean differences. An analysis of every possible mean difference is better served with the LSMeans Tukey HSD. It gives wider confidence intervals and because of this, statistically significant differences are more difficult to achieve. The right panel of Figure 11.24 shows that Proactive, Low is still the clear profitability winner, being the only factor level combination with the letter A. All of the remaining five factor level combinations are marked with the letter B, indicating that none of the five mean profitabilities are different from each other at α = 0.05 significance. Notice the difference between Student's t and Tukey HSD results. With Student's t, Proactive, Medium and Reactive, Low are different from Reactive, High, while with Tukey HSD they are not different with statistical significance.

Conclusion: The most compact conclusion from these data can be taken from the Connecting Letters reports. Both Student's t and Tukey HSD suggest that with statistical significance, Proactive, Low is on average more profitable than the other factor level combinations. Customers who are proactive and make low demands on management are more profitable than all the other customers groups.

11.4 Two-Way ANOVA with Unequal Replications

Two-way ANOVA with unequal replications has an unequal numbers of observations for different factor level combinations. Its analysis appears to be similar to two-way ANOVA with equal replications. The model has the same terms: two main factors and one interaction. The steps required to analyze the data and interpret the results are nearly identical. However, there are major differences arising from the unequal number of observations per factor level combination. Two important differences are that the *p*-values of the Effect Tests table are interpreted differently, and that it is absolutely necessary to use LSMeans instead of simple means. A section with a complete case study is available for download.

11.5 Summary

- Two-way ANOVA is useful in explaining the variation of Y when two nominal X factors are available.

- The fixed-effect models used in this chapter always test the equality of means of factor levels or factor level combinations.

- Each factor combination (A_i, B_j) must be observed at least once, and in some cases more than once.

- Analysis is much simplified if an equal number of observations is available for each factor combination (A_i, B_j). If that number is at least two, such data are called equally replicated.

- If more than one observation is available for each factor combination (A_i, B_j), an interaction term can be added to the model. If interaction is anticipated, the data collection should be designed to include replications.

- The effect tests are used to test whether the main factors A or B or the interaction term are statistically significant. Such tests assume that under the null hypotheses the respective level means of the main factors are all equal or that the interaction effects are all zero.

- Do not judge statistical significance from graphs. You need to verify significance by the appropriate *p*-values or from confidence intervals.

- Always start by including both factors in the model using the Fit Model platform in JMP. Avoid starting with analyzing each factor separately (as in Fit Y by X), because it can lead to inadequate conclusions.

- Two-way ANOVA with unequal replications requires different interpretation of the *p*-values of the effect tests. It also requires the use of least squares means in the detailed analysis.

- Differences in means of factor levels are evaluated with the Connecting Letters, Ordered Differences, or Crosstab reports.

11.6 Problems

1. For each of the following situations, identify the appropriate type of analysis of variance that should be conducted—one-way ANOVA, two-way ANOVA without replication, or a two-way ANOVA with replication. Identify the Y-variable and the X-factor(s). Also mention any concerns that you might have about each situation fitting the assumptions of ANOVA.

    a. A corporate ethics officer is comparing the number of sexual harassment complaints made in the past year at different plant locations. Historically, females have filed more complaints than males.

    b. A university must determine whether there are differences in the amount of financial aid given based on veteran status and gender.

    c. A manufacturing facility produces cell phones on two identical production lines, three shifts per day. The quality control manager is reviewing the monthly performance and believes there might be a higher number of defective cell phones produced on one of the manufacturing lines and during the evening and night shifts.

    d. A shoe store is preparing its quarterly order and is reviewing past sales for its most popular lady's dress shoe. The store wants to determine whether there are differences in sales for the four colors stocked: black, navy blue, brown, and beige.

2. Military recruits choosing to specialize in communications are assigned to one of three military bases to receive training. The three bases are located in California, Kansas, and South Carolina. Upon completion of their training, recruits must pass an examination with a minimum score of 80 in order to begin working in communications. Those not achieving a passing score are assigned to another less technical job. The following table shows scores for each of the three bases and the educational level of the recruit. The military is interested in determining whether there are significant differences in the training received at the three bases, as measured by the examination scores.

	High School Diploma	GED	Associate's Degree
California	82	75	84
Kansas	83	80	92
South Carolina	90	85	96

a. Enter the data into a JMP data table in the format needed to carry out a two-way ANOVA.

b. Determine whether at least one of the bases is on average different from the others. Use the Fit Model platform to obtain an analysis of variance table. Write out the appropriate hypotheses. Find the associated F-Ratio and *p*-value. What is your conclusion?

c. From the analysis of variance options for the Military Base factor, select LSMeans Student's t. If at least one of the mean scores differs, determine specifically which means differ.

d. Give 95% confidence intervals for the pairwise differences in mean score.

e. Why was it important to take education level into account in this analysis?

3. A county uses four assessors to determine the values of residential properties for tax purposes. To see whether the assessors differ in their valuations, five houses, each representing a particular style, were selected. Each assessor was asked to determine the market value of these houses. The data are contained in *HouseAssessments.jmp*.

   a. Write down the relevant statistical hypotheses that need to be tested.

   b. Is there any statistically significant difference in the mean house values between assessors? If so, which assessors differ? Explain how you arrived at your answer.

   c. Since differences between assessors were of primary importance, could you have used one-way ANOVA? If your answer is no, explain why not. If your answer is yes, run the one-way ANOVA and comment on the results.

   d. Could you have included an interaction term in this analysis? If your answer is yes, run the model in JMP and comment on the results. If your answer is no, explain why an interaction term could not have been included.

   e. (Optional) Based on your results, suggest actions that could be taken to achieve more consistency in assessments.

4. A market research firm conducts an extensive survey of customer satisfaction of bank branches writing different types of loans. Eighteen customers are asked to fill out a questionnaire in which a satisfaction score ranging from 0 (absolutely

dissatisfied) to 100 (completely satisfied) is recorded. Bank branches were selected according to business volumes of small, medium and high. Customers were stratified according to type of loan, either secured or personal. Customers who had both loan types were excluded from the survey. The basic question to be addressed is whether size of branch or type of loan affects the customer satisfaction score. The results are as follows.

ID	Branch Size	Loan Type	Satisfaction	ID	Branch Size	Loan Type	Satisfaction
1	Small	Secured	83	10	Small	Personal	76
2	Small	Secured	81	11	Small	Personal	71
3	Small	Secured	78	12	Small	Personal	69
4	Medium	Secured	77	13	Medium	Personal	73
5	Medium	Secured	72	14	Medium	Personal	70
6	Medium	Secured	76	15	Medium	Personal	67
7	High	Secured	67	16	High	Personal	69
8	High	Secured	67	17	High	Personal	67
9	High	Secured	64	18	High	Personal	65

a. Enter the data into a JMP data table in the format needed to carry out a two-way ANOVA.

b. Based on the problem statement given, write out the appropriate hypotheses that should be tested.

c. Determine those factors that significantly influence customer satisfaction. Use the Fit Model platform to obtain the analysis of variance table. Find the F-Ratios and p-values associated with the hypotheses to be tested. What is your conclusion?

d. From the Fit Model platform, obtain the interaction plot. What is the meaning of interaction? Describe the nature of the interaction in this case.

e. List the Branch Size, Loan Type combinations that are significantly different.

5. Com.com, a new and growing organization, is evaluating three possible vendors for a repair contract for laptops. They would like to choose a vendor with the speediest service. Com.com's IT manager prepares seven repair tasks. Each task is completed by each vendor. The time from calling the vendor with the problem to reinstatement of the laptop to standard use is given in hours. The results of the evaluations are as follows:

Task	Vendor		
	A1Cheap	ComeLate	DoQuick
A	39	47	48
B	50	48	55
C	68	74	70
D	54	58	52
E	44	45	54
F	22	28	38
G	46	42	57

The data are contained in a file called *ServiceVen.jmp*

    a. What advantages does Fit Model analysis using two X-factors (Vendor and Task) offer over the Fit Y by X platform using only Vendor as the single X-factor? Does it lead to the same or different recommendation of a vendor?

    b. What does the Effect Test for Task measure and does this affect your conclusion about vendors?

    c. Why is there no interaction term in this ANOVA?

    d. In Fit Model, what do the LS Means plots for Task and Vendor show? How do these plots relate to the Least Squares Means in the tables?

    e. Give your summary conclusions from this analysis.

6. A study needs to determine whether bed occupancy in a hospital depends on either the care level or ownership. The Y-variable is bed occupancy expressed in percentage of total beds occupied in a month. The X-factors are care level and ownership of hospitals. Hospitals are classified into one of three care levels: tertiary is the highest care level, and primary is the lowest. Hospitals can be owned by the public (typically a local, state, or federal government), a not-for-profit corporation, or an investor-owned stock corporation. There are three levels for each of the two X-factors, and so there are three times three or nine factor level combinations. The data are composed of three columns and 27 rows. This indicates that each of the nine factor combinations has been sampled three times. Data for a two-way ANOVA with replications are given in *HospitalOccupancy.jmp*.

    a. Specify a two-way ANOVA model with an interaction term and run the standard ANOVA analysis.

    b. What are the key outputs from the ANOVA table? List them and explain them.

    c. What is the purpose of the Effect Tests table? Why is it needed in addition to the overall ANOVA table?

d. What does an LSMeans plot show in general?

e. What does the LSMeans plot for Care Level show?

f. What does the LSMeans plot of Ownership show in this specific application?

g. What does the LSMeans plot for the interaction term show?

h. Examine and interpret the results of the LSMeans Differences Student's t using both the Connecting Lines report and the Ordered Differences report.

i. Remove the interaction term from the Fit Model dialog box and rerun the analysis of variance. Write down the corresponding model. Compare the Mean Squared Errors from the two models (with and without interaction). What are the consequences of omitting the interaction term from the model?

j. State your final conclusions using plain English but quoting relevant evidence. List the necessary assumptions that make these valid conclusions.

7. A manufacturer of consumer durables wants to test the effectiveness of different advertising campaigns on sales (in 1000 units per test market). The company decides to run three different advertising programs labeled A, B, and C in 12 different test markets. Each ad program is repeated four times. The marketing VP wants to combine the ad research with the presence or absence of additional promotional discounts. The sales data (in units of 10000) for the 12 test markets with the factors AdProgram and Promotion are given in *AdvertEffect.jmp*.

a. State the hypothesis of the ANOVA and the Effect Tests tables. Interpret and explain how they help you in answering questions about advertising and promotion effects on sales.

b. Based on LS Means plots, what can you say about advertising and promotion effects on sales, of course subject to verification of statistical significance with *p*-values?

c. Interpret the interaction plot between Promotion and AdProgram.

d. Use LS Means Student's t to select significant differences between means corresponding to the interaction between AdProgram and Promotion. Check the confidence intervals as to whether they contain 0. Briefly summarize their meaning.

8. Stroke patients enter a rehabilitation hospital with the goal of regaining as much function as possible. This is accomplished through a variety of therapies. It is desirable to discharge patients to outpatient status as quickly as possible. The file *StrokeTherapy.jmp* contains data on length of stay and the amount of physical and occupational therapy received.

a. Analyze the data to determine whether the levels of occupational and physical therapy are related to length of stay.

  b. Recommend the therapy combination that results in the shortest length of stay.

9. The Westville Meat Processing Plant has added a new product line that has required changes in the speeds at which the assembly lines are run. With rising health care costs, there is concern that an increase in line speed might lead to increases in accidents and workers' compensation costs. The number of injuries over the last year for the three shifts (day, evening, and night) and three line speeds (slow, medium, and fast) can be found in the data file *WestvilleIndustrialInjuries.jmp*.

  a. Perform the appropriate analysis of variance to address the issue.

  b. Briefly summarize your findings. Include a statistical summary of the data, conclusions, and supporting statistical evidence, and the assumptions needed to make your findings valid.

11.7 Case Study: Fish Catch near Oil Rig

Business Problem

Three oil rig designs are to be evaluated for their friendliness towards fish. To make this evaluation, a small net is cast around each of three designs at depths of 30 feet and 50 feet. The size of the catch is taken as a measure of environmental friendliness.

Tasks

You are asked to evaluate the data and make a recommendation as to which if any design is superior with respect to friendliness towards fish using the data below. The catch in pounds of fish is recorded as follows:

ID	Depth in ft.	Design	Catch	ID	Depth in ft.	Design	Catch
1	30	A	1219	0	50	A	1064
2	30	A	1147	1	50	A	1162
3	30	A	1300	2	50	A	1101
4	30	B	1047	3	50	B	741
5	30	B	1088	4	50	B	805
6	30	B	1125	5	50	B	772
7	30	C	1095	6	50	C	774
8	30	C	1160	7	50	C	892
9	30	C	1154	8	50	C	838

Summarize your analysis and recommendation in the form of a technical report.

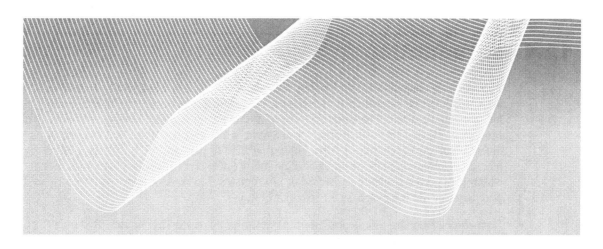

Chapter 12

Proportions

12.1 Introduction 316

12.2 Proportions from a Single Sample 317

    12.2.1 Introduction to Proportions 317

    12.2.2 Detail Example: A Survey to Evaluate Brand Preference 318

    12.2.3 Sample Size Considerations with Single Proportions 323

12.3 Chi-Square Test for Equality of k Proportions 327

    12.3.1 Introduction to Chi-Square 327

    12.3.2 Detail Example: Default Rates on a Specific Type of Loan (CBA Loan) by Credit Rating 327

    12.3.3 The Pearson Chi-Square Statistic 331

12.4 Summary 334

12.5 Problems 335

12.6 Case Study: Incomplete Rebate Submissions 337

12.1 Introduction

The next two chapters deal with categorical data. As discussed in Chapter 1, JMP uses the terms *nominal* or *ordinal* for such data. Categorical data occur regularly in many business applications. The simplest form of nominal data is called binary, with only two categories. Binary variables occur, for example, when parts are classified as good or defective or applicants for a job are either qualified or unqualified. Similarly, polling and marketing research questionnaires contain many questions with binary variables. For example, a person is either registered or not registered to vote, or an interviewee either purchased or did not purchase a particular brand.

Data can be recorded as categorical outcomes directly or can first be recorded on a continuous scale and then converted to categories. For example, a lending institution classifies applicants for a particular loan type with a credit score at or above 720 as qualified and those with a score below 720 as unqualified. The continuous variable Credit Score is changed into a binary variable for use in making loan decisions.

Categorical variables can have more than two categories. An engineered part might have the correct size, or might be too small or too large. A business might have several potential suppliers for a service or a component. Each supplier represents a distinct category of the nominal variable Supplier. Ordinal variables occur when the outcome categories can be ranked in some way. Likert scales are often treated as ordinal. For example, the response categories Very Dissatisfied, Dissatisfied, Neutral, Satisfied, and Very Satisfied can be ranked by degree of satisfaction.

This chapter deals with methods for proportions of binary outcomes from a single sample for which either the exact binomial distribution model or an approximate Normal distribution model is used. For comparisons across several samples, an approximate Chi-Square distribution model is used. Other methods are available, but are beyond the scope of this book.

12.2 Proportions from a Single Sample

12.2.1 Introduction to Proportions

An important application of a single proportion is polling. Here are some other examples:

- Customer Relations asks new customers whether they have a Favorable or Unfavorable opinion of a product or service.

- A human resources department asks employees whether or not they support a change to the employee benefit package.

- Manufacturing keeps track of the proportion of defective units produced on a given day.

- Marketing Research tests a new product concept by asking potential customers whether they would or would not buy such a product.

Assumptions: The model for a single proportion can be used whenever the outcome Y satisfies three assumptions: (1) Y is binary, taking on only one of two values. (2) The probabilities of the outcomes are constant. (3) Outcomes in a sample are probabilistically independent. Fortunately, many business processes satisfy these assumptions.

A simple example of a process that satisfies all three assumptions is the flip of a coin. Assume that a coin has a probability of heads P[heads] = 0.5 and a probability of tails P[tails] = 0.5. These are binary outcomes with only heads or tails possible. The probabilities do not change from coin flip to coin flip. Therefore, the probability of heads or tails is constant. Lastly, the outcome of one trial in no way affects the probability of outcomes at the next trial and so the outcomes are independent. Even observing 10 consecutive heads in a row does not change the probability that the eleventh outcome is also heads. The model of a coin toss can be extended to business applications.

Examples of binary outcomes satisfying the first assumption are (Yes, No), (Good, Defective), (Approve, Do not approve), (Favorable, Other), or simply (0, 1). Even non-binary outcomes are often transformed into binary outcomes. In marketing polls, respondents are asked to classify their sentiment towards a brand as Favorable, Unfavorable, or No Opinion. These three possible response categories can be transformed into binary responses. One is the Favorable rating. It represents the proportion of respondents who selected Favorable. The complement is the Other rating, representing the proportion of respondents who selected either Unfavorable or No Opinion.

The second assumption is that the probability of an outcome remains constant. In the manufacture of many products, you need to ask whether or not the probability of a defective product remains constant during a work shift, during a day, or even during a week. In a marketing poll, the Favorable rating of a brand often changes rapidly as special events and advertising have their effect. This is why it is important that such polls are taken in a narrow time window and that it is understood that any conclusions are strictly speaking valid only for the specified time window.

The third assumption is that the outcomes are independent. Statistically, you can often meet this assumption by taking a random sample. However, random samples do not guarantee independent observations, because not all processes have independent outcomes. In sampling a river for a pollutant, samples that are close in time at a fixed location might not give independent samples. Non-scientific polls, such as non-random interviews of respondents at a particular shopping mall, violate this assumption as well.

12.2.2 Detail Example: A Survey to Evaluate Brand Preference

Situation: In marketing research, respondents are asked their view of a brand of computer accessory. A typical result of such a survey would be a statement that Brand A has a Favorable rating of 55% ± 6%. Where does such a statement come from? How is it calculated? Consider the question, What is your opinion of Brand A? Choose one of the following responses: (a) Favorable, (b) Unfavorable, (c) No Opinion.

The Favorable rating is the proportion (or percentage) of respondents answering with Favorable. As mentioned, these three answers can be dichotomized by combining two of the categories into one. For a Favorable rating, response (a) represents the Favorable category, whereas responses (b) and (c) are combined to form a complementary category. For a random respondent, the probability of the respondent's answer being Favorable can be represented as P[Favorable rating for respondent i] = p, where p represents the unknown proportion of people in the population that would give a Favorable rating to Brand A.

Problem: A corporation has become concerned that the brand image of its product A has suffered from recent bad publicity about reliability problems. The corporation needs to estimate whether or not deterioration in the Favorable ratings due to bad publicity has occurred. Before the reliability problems became public, Favorable ratings hovered around 55% Favorable and 20% Unfavorable, with 25% of respondents having No Opinion.

This problem statement leads to a hypothesis test of H_0: $p = 0.55$ versus H_A: $p < 0.55$. As stated in Chapter 8, we write H_0 as a simple hypothesis in which the parameter is specified in an equality. Our reason for using the equality is that the significance level α and resulting p-values are calculated from this parameter value. A different way to write the null hypothesis is as the complement of the alternative hypothesis. This would result in a null hypothesis H_0: $p \geq 0.55$.

Data Requirements: The poll is to be conducted using a list of well over 10,000 customers as the frame. Most surveys contain more than one question. However, in this example, only the Favorable rating is at issue. Only one column—Response Image of A—is needed, and its values are Favorable, Unfavorable, and No Opinion. For the test to determine whether or not the Favorable rating decreased, we use the formula in Figure 12.2 to generate a second column, Image A favorable, containing only two values, Favorable and Other, where Other is substituted for Unfavorable and No Opinion. The column containing the Respondent ID numbers is not needed in this analysis. An excerpt of the data file *ImagePoll.jmp* is shown in Figure 12.1. This file of the raw data contains one row for each respondent. Altogether, there are 320 rows.

Figure 12.1 Excerpt of *ImagePoll.jmp* Data Table

ImagePoll			Respondent ID	Response Image of A	Image A favorable
		1	197	Favorable	Favorable
Columns (3/0)		2	389	No Opinion	Other
Respondent ID		3	415	Unfavorabl	Other
Response Image of A		4	486	No Opinion	Other
Image A favorable		5	569	Favorable	Favorable
		6	656	Unfavorabl	Other
Rows		7	668	Favorable	Favorable
All rows	320	8	672	No Opinion	Other
Selected	0	9	682	Favorable	Favorable
Excluded	0	10	684	No Opinion	Other
Hidden	0	11	704	No Opinion	Other
Labelled	0	12	817	Favorable	Favorable

The column Response Image of A is transformed into the column Image A favorable by the formula in Figure 12.2.

Figure 12.2 Formula to Create the Image A favorable Column

If	*Response Image of A* == "Favorable" ⇒ "Favorable"
	else ⇒ "Other"

Similar statistical results can be obtained from summarized data. In this case, the outcome levels are listed in one column and the frequencies in a second column, as shown in Figure 12.3. The frequencies in *ImagePollSummary.jmp* with 163 Favorable and 157 Other responses add up to 320. The data in *ImagePoll.jmp* and in *ImagePollSummary.jmp* will yield the same results.

Figure 12.3 Image Poll Data in Summary Form

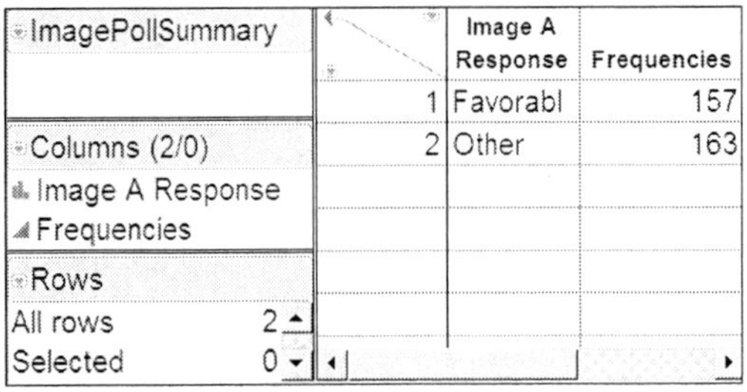

ImagePollSummary		Image A Response	Frequencies
		1 Favorabl	157
Columns (2/0)		2 Other	163
Image A Response			
Frequencies			
Rows			
All rows	2		
Selected	0		

Anticipated Results: As the problem statement suggests, the image might have suffered somewhat from the adverse publicity.

Analysis of Data with Binary Outcomes in JMP: With the raw data in *ImagePoll.jmp* and the Distribution platform, specify a single nominal variable (Image A favorable) as shown in Figure 12.4.

Figure 12.4 Variable Specification for Raw Data

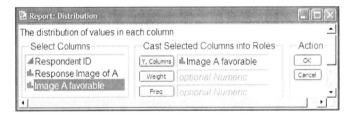

The same results can be produced with *ImagePollSummary.jmp* as shown in Figure 12.3. The only difference is the additional specification of the Frequencies column in the **Freq** role as shown in Figure 12.5.

Figure 12.5 Variable Specification for Summary Data

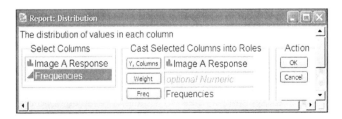

Clicking **OK** produces basic results for a nominal variable including a histogram of the nominal categories and, in the **Frequencies** panel, the Counts and the estimated proportions labeled Prob. The basic results are shown in Figure 12.6.

Figure 12.6 Histogram, Counts, and Estimated Proportions

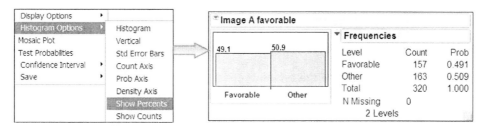

The histogram has been enhanced by adding the percentages of responses in each of the categories. By clicking the red triangle in the variable name, you can select from the contextual menu a variety of histogram options as shown on the left of Figure 12.6. In this example, percentages were chosen, because percentages were used in the problem statement.

Polling results are often presented as confidence intervals analogous to those for a single mean from a Normal distribution discussed in an earlier chapter. As with confidence intervals for the mean, those for single proportions are selected from the **Confidence Interval** submenu by clicking the red triangle. Choose **Confidence Interval** with the desired confidence level as shown on the left in Figure 12.7. Both the confidence level and the sample size influence the width of the confidence interval. Narrow confidence intervals can be achieved by a relatively large sample size or a lower confidence level. Large sample sizes require more sampling resources. Lower confidence levels come at the price that the resulting confidence interval has a higher probability of being wrong. JMP presents the counts in each category, the estimated proportions Prob and confidence intervals for each category. The 95% confidence interval for Favorable is from 0.436 to 0.545, and the confidence interval for Other is from 0.455 to 0.564. The intervals for the two complementary categories have the same width, $(0.545 - 0.436) = (0.564 - 0.455) = 0.109$.

Figure 12.7 Estimated Proportions and 95% Confidence Intervals

Display Options	▸	
Histogram Options	▸	
Mosaic Plot		
Test Probabilities		
Confidence Interval	▸	0.90
Save	▸	0.95
		0.99
		Other...

Confidence Intervals					
Level	Count	Prob	Lower CI	Upper CI	1-Alpha
Favorable	157	0.491	0.436	0.545	0.950
Other	163	0.509	0.455	0.564	
Total	320				
Note: Computed using score confidence intervals.					

Traditionally, the tests and confidence intervals for proportions are based on the Normal approximation to the binomial distribution. In Figure 12.7, the note below the table of confidence intervals states that JMP calculates the intervals with the score method. This is a superior method that yields slightly different confidence intervals than the traditional ones.

Is the Favorable rating below the historical 55%? A left-sided hypothesis test is the way to answer this question. The null hypothesis says that no decrease in the Favorable proportion occurred. JMP offers a **Test Probabilities** option (Figure 12.6) that is analogous to the **Test Mean** option for the single mean of a Normal distribution.

The **Test Probabilities** option opens a window to specify the null hypothesis, as shown in Figure 12.8. The estimated proportions are shown as **Estim Prob**. Enter one value in **Hypoth Prob**. In this problem, the hypothesized value is 0.55 for the Favorable rating. Select the type of alternative hypothesis by clicking the button labeled **probability less than hypothesized value**. The item in parentheses indicates that JMP uses the exact one-sided binomial test.

Figure 12.8 Specifying Null Hypothesis and Alternative Hypothesis and Left-Sided Results

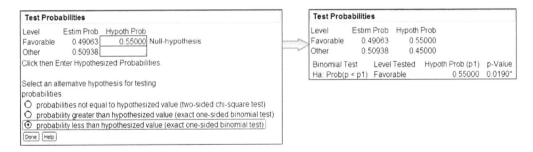

Clicking **Done** produces the results on the right in Figure 12.8. With a left-sided alternative hypothesis, the *p*-value = 0.0190 is below α = 0.05. Reject the null hypothesis that the proportion of Favorable is 0.55 (or higher). Conclude that the Favorable brand image has fallen significantly from the earlier 0.55. It requires a judgment call to characterize the drop from 55% to 49% as a major or minor decrease, especially in view of the 95% confidence interval for the Favorable ranging from 43.6% to 54.5% (Figure 12.7).

12.2.3 Sample Size Considerations with Single Proportions

One of the most important questions before many statistical studies is, How large a sample is necessary to obtain reasonable results? Sample size calculations are important, because they determine the resources required. Smaller samples require fewer resources, but samples that are too small will not produce estimates with the desired precision. Results will be inconclusive. Samples that are too large can be a waste of resources. Sample size selection considers several factors that are likely to produce meaningful results. JMP's Sample Size platform reduces sample size calculations to specifying a few parameters.

Example: Market Research for "Welcome On-Board" Program: Retail banks have a great turnover of retail customers. New customers open accounts, begin to do business with the bank for a while, and then leave to do business with another bank. In order to reduce the turnover of customers, a bank has initiated a "Welcome On-Board" program for new customers. For a period of six months, this program is intended to show new customers the advantages of doing business with this bank, to raise the bank's image with them, and ultimately to create customer loyalty. Data from the past indicate that after six months, 37% of customers who did not undergo the "Welcome On-Board" program rated the bank favorably. To simplify the problem, the favorability rating is a Yes response to the question, "Would you recommend this bank to your friends?"

Bank management thinks that an improvement of 6% (from 37% to 43%) should be discovered with a high probability 0.90. If there is no change due to the program, then that should be discovered with probability 0.95. How large a sample of new customers is needed? In statistical language, we can rephrase this problem in terms of a hypothesis test. The null hypothesis is H_0: $p = 0.37$ and associated with it is the one-sided error probability $\alpha = 0.05$ (taken as $1 - 0.95$ from above). The alternative hypothesis is H_A: $p > 0.37$. The problem also states that when $p = 0.43$, then the sample should be able to discover it with probability 0.90. (The power of the test is supposed to be 0.90 when $p = 0.43$). This is sufficient to translate into the JMP parameters.

Sample Size for One Proportion in JMP 8: JMP 8 changed calculations for one-proportion sample sizes versus previous versions. The Sample Size and Power platform can be accessed on the DOE menu as shown in Figure 12.9. Clicking the **One Sample Proportion** button produces the windows shown on the right. Default specifications are Alpha = 0.05 and Proportion = 0.1. The **Null-Proportion**, **Sample Size** and **Power** fields are blank. A toggle menu enables sample sizes for two-sided or one-sided alternative hypotheses.

Figure 12.9 Sample Size for One Sample Proportion

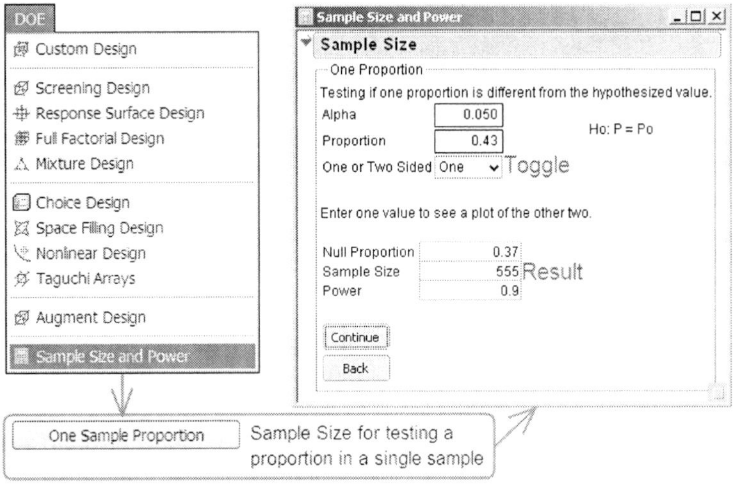

JMP offers five fields for parameters. With four fields properly specified, the fifth will be calculated. In order to calculate the sample size, we leave the **Sample Size** field blank. The remaining four fields are specified as follows:

- **Null Proportion** should be 0.37, because H_0: p = 0.37.
- **Alpha** should be $\alpha = 0.05$, because the problem specifies a one-sided $\alpha = 1 - 0.95 = 0.05$.
- **Proportion** should be 0.43. Although the one-sided alternative H_A: p > 0.37 is non-specific, the problem selects 0.43 as a specific alternative that should be discovered with probability 0.90 when it is true.
- **Power** should be 0.90, because **Power** represents the probability of rejecting H_0 when a specific alternative hypothesis is true.
- Toggle **One or Two Sided** to **One**.

Thus, there are two pairs of parameters: **Null Proportion** is associated with **Alpha** and **Proportion** with **Power**. JMP will calculate the required sample size as soon as you click **Continue**. The result is given rounded to whole numbers. In this example, 555 respondents are needed to obtain data that meet the specified criteria.

Sample Size and Population Size: One curious fact about sample size calculations for one sample proportion is that there is no requirement to specify population size. There is a simple reason for this. For moderately large populations, say with a size of at least 10,000 elements, the absolute sample size is more important than the sample size relative to the population size. One consequence of this is that a poll regarding the approval rating of a presidential candidate requires the same sample size as a poll for a senatorial candidate in any of the 50 states. In fact, even a mayoral race of a city with, say, 30,000 registered voters requires the same sample size. Certain non-statistical, i.e., organizational, aspects might be easier in a small-town poll, but the absolute sample size is the same.

Figure 12.10 gives sample sizes for different 95% sampling margin of errors for a Null Proportions = 0.5. The sampling margin of error (MoE) is the half-width of a confidence interval. The sampling margin of error is often expressed as ±% instead of as a proportion. For example, a 95% sampling margin of error of ± 3% (0.03) requires a sample size of 1058, while one of ± 1% (0.01) requires 9619. This latter result is one reason why there are few polls of that size.

Figure 12.10 Sample Size versus Margin of Error

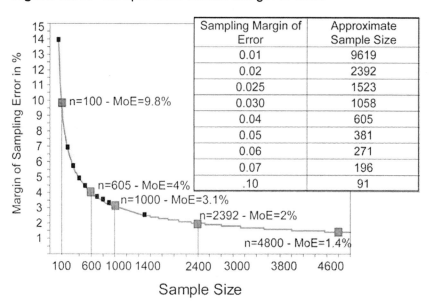

Sampling Margin of Error	Approximate Sample Size
0.01	9619
0.02	2392
0.025	1523
0.030	1058
0.04	605
0.05	381
0.06	271
0.07	196
.10	91

The graph of sample size versus margin of error is not a straight line, because the sampling margin of error involves the standard deviation divided by the square root of the sample size. Thus in order to double the precision of a poll from ±4 % to ±2%, a nearly a fourfold increase in the sample size from 605 to 2392 is required.

Example of Political Poll: A newspaper commissions a poll to estimate the Favorable rating of a candidate for national office. The target population is all registered voters in the 50 states. The Favorable rating is expected to be around 50%. The actual Favorable rating is to be estimated with a sampling margin of error no larger than ± 3%. How large a sample is needed for the poll?

This confidence interval problem is rephrased as a hypothesis test with H_0: p = Null-Proportion = 0.50 versus H_A: p ≠ 0.50. In terms of the JMP parameters, proceed as follows:

- **Null Proportion** should be 0.50, because H_0: p = 0.50. How do pollsters guess Baseline or Null Proportions? In polling, the worst-case scenario requiring the largest sample size for a specific sampling margin of error is when the Null Proportion is 0.50. Sometimes experience or the result of a smaller pilot poll help in specifying better guesses.

- **Alpha** should be 0.05 with a two-sided alternative hypothesis, because the common confidence level for polls is a two-sided 95%, leading to a two-sided $\alpha = 0.05$.

- **Proportion** may be 0.53 or 0.47. (In this example, there is no difference which value is entered, because when Null Proportion is 0.5, the binomial distribution is symmetric and yields identical sample sizes. When Null Proportion ≠ 0.5, confidence intervals by the score method are not symmetric resulting in different sample sizes.) The difference |Null Proportion - Proportion| = 0.03 is the half-width of the desired 95% confidence interval or the sampling margin of error. The smaller the difference, the larger the required sample size, everything else being equal.

- In sample size calculations, the **Power** is often left unspecified, implicitly choosing a *Power of 1–β = 0.5*. Recall that β is the probability of accepting the null hypothesis when the alternative is true. JMP requires that a value be entered in the **Power** field of the Sample Size calculator.

The required sample size for a poll with a sampling margin of error ±3% is 1058, assuming the Null Proportion is 0.5.

12.3 Chi-Square Test for Equality of k Proportions

12.3.1 Introduction to Chi-Square

Testing for the equality of proportions from several categories or groups occurs frequently with business data. For example:

- A casino is interested whether the roulette wheels from different manufacturers produce the same proportion of zeros.

- A bank is interested whether the default rate depends on the type of loan (or other characteristics such as region or type of business).

- A manufacturer of power generation equipment needs to know whether three production lines for making turbine blades have the same proportion of defectives.

- A creator of software is looking for differences in acceptance of its main line software product across several operating systems.

- An insurance provider needs to know differences in relative frequency of malpractice claims by medical specialties.

The number of sampling units considered in this type of analysis is often quite large, often larger than 100. Consequently, data are often summarized in tables. An individual observation is classified according to two variables: a qualitative (nominal, ordinal, categorical) X-variable with k categories, and another binary (dichotomous, nominal) Y-variable.

12.3.2 Detail Example: Default Rates on a Specific Type of Loan (CBA Loan) by Credit Rating

Situation: A bank wants to explore default rates of a loan type called a CBA loan. CBA loans are intended for small businesses with annual revenues from $500,000 to $1,000,000. Each customer can have only a single CBA loan at any one time. Customers are grouped by a credit rating that is based on credit scores. Credit scores consider a customer's payment history, the capacity to repay the loan, the length of the credit history, the number of times a customer applied for a loan within the last 18 months, and a customer's current mix of credit. The qualifying credit score levels are summarized in a credit rating. Customers with a score from 640 to 679 receive a good credit rating, with a credit score from 680 to 729 being a very good credit rating. Those with a credit score of 730 and above are rated excellent.

Problem: The bank needs to know whether default rates of the three customer groups are different. This leads to a hypothesis test of H_0: $p_{good} = p_{very\ good} = p_{excellent}$ versus H_A: at least one of the proportions is different.

Data Requirements: Data on the status of such loans are available. The X-variable is the credit rating with three categories: "good", "very good", and "excellent". This X-variable has a List Check property defined so that the categories are properly ordered. The Y-variable is the status of the loan with the two outcome categories "default" or "perform".

The left panel of Figure 12.11 shows a portion of the raw data file with 1442 rows, one per customer. The variables are shown in two nominal columns. Column 1 is Credit Rating with three categories. Column 2 is Loan Status with two categories. There are six different combinations of outcome categories: (good, default), (good, perform), (very good, default), (very good, perform), (excellent, default), and (excellent, perform). Such data are often analyzed in a summarized form as shown on the right in Figure 12.11. This summary data table has only six rows labeled by the first two columns. The third column is the Frequency column and represents the count of respondents for each (Credit Rating, Loan Status) combination. The six frequencies sum to 1442.

Figure 12.11 Comparison of Raw Data and Summary Data Files of CBA Loans

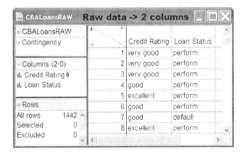

Anticipated Results: It is expected that the proportion of default loans is smaller in customer groups with higher credit ratings.

Analysis of CBA Loans in JMP: The Fit Y by X platform is used for testing the equality of several proportions. The binary variable (Loan Status) is the **Y, Response**, while the other nominal variable (Credit Rating) is the **X, Factor**. The role for each variable affects only the presentation of results. Depending on whether raw data or summary data are used, two slightly different variable specifications are needed in the Fit Y by X variable specification window (see Figure 12.12 and Figure 12.13).

Figure 12.12 Variable Specification for Raw Data

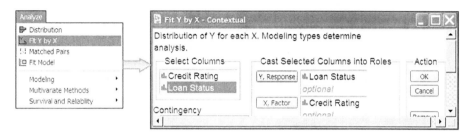

To analyze *CBALoanRAW.jmp*, Loan Status is the binary **Y, Response** and Credit Rating is the **X, Factor** with three categories (see Figure 12.12). For the summarized data in *CBALoans.jmp*, the **Y, Response** and **X, Factor** are specified with Frequency in the role as **Freq** as in Figure 12.13.

Figure 12.13 Variable Specification for Summarized Data

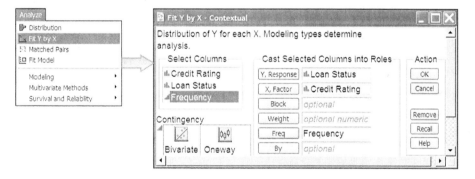

Clicking **OK** produces the standard output for the Fit Y by X platform when two nominal variables are specified. The most visible output is the Mosaic plot shown in Figure 12.14. The plot shows the proportions of performing and default loans for each of the three credit rating categories. In this example, the proportion of default in each category is very low. This makes it difficult to distinguish the magnitude of differences in proportions from the Mosaic plot. You can tell that the "excellent" category has the lowest proportion of default cases. But, is it significantly lower? That will be determined with the Pearson Chi-Square Test, the results of which are shown in Figure 12.16.

Figure 12.14 Mosaic Plot for CBA Loans

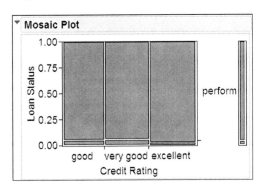

The basic results also include the Contingency Table shown in Figure 12.15. The contingency table is a cross-classification of the frequencies by Loan Status and Credit Rating. The row and column totals are not labeled but can be found on the margin. By default, the six central cells contain four values. The first entry represents the frequencies, here labeled "Count", of each combination. The other three default values are Total %, Column %, and Row %.

In this example, the row percentages are of primary interest, because they represent the percentage of "default" and "perform" by Credit Rating category. Row % is the percent of observations in a cell relative to the row total. With this in mind, Figure 12.15 has been modified to show only Count and Row %. (Total % and Column % are eliminated by deselecting them from the context menu. Observe the change in check marks on the left in Figure 12.15). For the combination (good, default) the cell frequency is 24, the row total is 460, and the Row % is 100(24/460) = 5.22.

Figure 12.15 Modified Contingency Table for CBA Loans

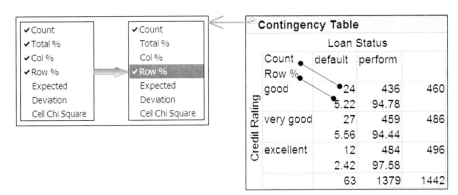

The problem statement asks to compare the default rates for each category of credit rating. This requires a test for whether the percentages of default are equal for the three credit rating categories. The contingency table of Figure 12.15 lists the two Loan Status categories in the top row of the contingency table. The row percentages are those of the three categories of Credit Rating: The default loans show that good loans defaulted at 5.22 %, very good loans at 5.56 %, and excellent loans at only 2.42 %.

One of the basic results of contingency table analysis is called "Test". There are two tests listed, the Likelihood Ratio test and the traditional Pearson (Chi-Square) test. In most practical applications, these two tests result in *p*-values that are very close to each other. The question is, Are the proportions of default different in the three Credit Rating categories? Here we use the Pearson Chi-Square statistic to test the null hypothesis that all proportions are equal. The alternative hypothesis stipulates that at least one proportion differs. Figure 12.16 shows the value of the Pearson Chi-Square test statistic at 6.943 with a *p*-value of 0.0311. This *p*-value indicates that the default proportions differ significantly using $\alpha = 0.05$.

Figure 12.16 Pearson Test with *p*-Value

Test	ChiSquare	Prob>ChiSq
Likelihood Ratio	7.612	0.0222*
Pearson	6.943	0.0311*

Reject the null hypothesis that all three proportions are the same. The Chi-Square statistic does not tell which proportions differ. The discussion of Figure 12.15 indicates that the proportion default in the "excellent" category is lower than the other two, while the proportions of default in the "good" and "very good" categories are nearly equal. This suggests that the "excellent" category is different from the other two.

12.3.3 The Pearson Chi-Square Statistic

The previous section used the Pearson Chi-Square statistic to calculate the *p*-value. The null hypothesis for testing the equality of k proportions is H_0: $p_1 = p_2 = \ldots = p_k$. Here p_i stands for the proportion of one of the two outcome categories of Y in category i of the X-variable. In the CBA loans example, Loan Status is the Y-variable with two outcome categories for loans—"default" or "perform". The X-variable is Credit Rating with three outcome categories—"good", "very good", and "excellent". The proportion p_i of default loans in credit rating category i is p_i, with i ranging from 1 to 3. The default proportion of loans with good credit ratings is p_1, that with very good ratings is p_2, and that with excellent credit ratings is p_3. The null hypothesis states that the proportion of defaults is the same across all credit rating categories. The alternative hypothesis states that at least one rating category has a different proportion of defaults. The Pearson Chi-Square statistic tests whether the null hypothesis can be supported by the data.

To calculate the Pearson Chi-Square statistic, begin with the contingency table of observed frequencies. Observed frequencies were shown as Count in the JMP contingency table and are the first entry in each cell of Figure 12.15. Next, calculate the cell frequencies expected if the null hypothesis were indeed true. In a world without variation and assuming H_0 is true, the proportions of default loans would be the same, irrespective of the credit rating category.

How can you use the data to estimate these two proportions? Assuming H_0 is true, there is no difference in the proportion of default and performing loans between the three credit ratings. The data can be aggregated into one overall group and used to estimate the overall proportion of default and performing loans. The overall frequency of default loans in the data is $24 + 27 + 12 = 63$. The total number of loans in the sample is 1442. The overall proportion of default loans is $63/1442 = 0.0437$ or 4.37 %. Likewise, the overall proportion of performing loans is estimated from $(436 + 459 + 484)/1442 = 0.9563$. Note that $0.0437 + 0.9563 = 1$.

The expected default and performing loan frequencies, assuming H_0 is true, are calculated from these overall proportions and the total loans in each credit rating category. There are 460 loans with a good credit rating. You would expect 4.37 % of these to be in default. Thus, the expected default frequency for good loans is $0.0437*460 = 20.0971$. Likewise, for the 486 loans with a very good credit rating you would expect $0.0437*486 = 21.233$ loans to default. For the 496 loans with excellent credit rating the expected frequency of default loans is $0.0437*496 = 21.6699$. The expected number of performing loans is similarly calculated. For example, the 460 loans with a good rating $0.9563*460 = 439.903$ are expected to perform. Notice that $439.903 + 20.097 = 460$. The expected frequencies represent the fitted model under the null hypothesis. They are shown as the second entry in each cell of Figure 12.17.

As mentioned before, the Pearson Chi-Square statistic looks at how well the data match the fitted model. For each cell in the contingency table, the difference between the observed frequencies and the expected frequencies is recorded as the Deviation. The Deviation is shown as the third entry in each cell of the contingency table in Figure 12.17.

Figure 12.17 Entries to Understand the Pearson Chi-Square Statistic

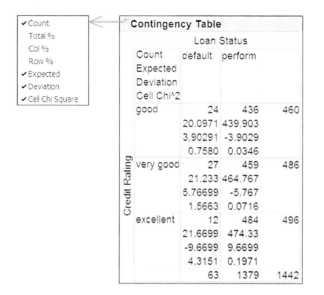

The cell deviations are squared and standardized by dividing them by the expected cell frequency. For each cell, calculate

$$\text{Cell Chi}\wedge 2 = \frac{(\text{Observed - Expected})^2}{\text{Expected}} = \frac{\text{Deviation}^2}{\text{Expected}}$$

For the cell representing the combination of good credit rating and default, this calculation is $(3.90291)^2/20.0971 = 0.7580$. The Cell Chi^2 is the fourth entry in each cell of the contingency table in Figure 12.17. The menu selection for this table is on the left in Figure 12.17.

The Pearson Chi-Square statistic is the sum of all Cell Chi^2. In the CBA loan example, this is Pearson Chi-Square = 0.7580 + 1.5663 + 4.3151 + 0.0346 + 0.0716 + 0.1971 = 6.943. Under H_0 and certain other assumptions, the Pearson statistic follows a Chi-Square distribution with $(k-1) = 2$ degrees of freedom where, in this example, k is the number of credit rating categories. JMP calculates the *p*-value from this Chi-Square distribution.

Whenever H_0 is true, you would expect the deviations between the observed frequencies and expected frequencies to be small. Small deviations result in small Cell-Chi^2 and these in turn result in small Pearson Chi-Square statistics with high *p*-values. The null hypothesis cannot be rejected whenever the Pearson statistic is smaller than the critical value from the Chi-Square distribution or the p-value is greater than the α significance level.

In the CBA loan example, the Pearson Chi-Square statistic = 6.943 has a p-value = 0.0311. It could also be compared to the critical value of the Chi-Square distribution with $(3 - 1) = 2$ degrees of freedom. For $\alpha = 0.05$, select the $(1 - \alpha)$ quantile, i.e. choose Chi[0.95,2] = 5.99 (see Chapter 6). Since 6.943 is greater than 5.99, the null hypothesis is rejected. The default rates by group are different for at least one of the three credit rating categories.

12.4 Summary

- Single proportions are analyzed for situations such as determining the favorability rating of a political candidate, the acceptance rate of a new product, or the proportion of defectives in a production process.

- Proportions are based on binary data. If the data are not in binary form, they need to be transformed into binary form for this type of analysis.

- Continuous scales can be transformed into binary scales by assigning an outcome at or below a threshold to one category and an outcome above that threshold to the second category.

- The analysis of proportions presented in this chapter assumes that the outcome probabilities remain constant throughout the sample. It also assumes independent observations.

- The analysis for a single proportion is similar to one-sample methods of continuous data.

- JMP calculates confidence intervals for proportions by the score method. Unlike traditional approximate confidence intervals, score intervals are usually non-symmetric, i.e., (UCL – Estimate) ≠ (Estimate – LCL).

- The sample size required for a desired statistical precision is an important consideration in polls.

- The parameters needed to estimate the required sample size for a single proportion are the Null Proportion with its associated α and the tested Proportion with its associated Power. You also need to know whether one-sided or two-sided conclusions are desired.

- The Pearson Chi-Square statistic is used to calculate the p-values for tests of equality of several proportions.

12.5 Problems

1. For each of the following situations, determine whether the assumptions for a test of hypothesis or a confidence interval for a proportion are satisfied. Indicate which assumptions are satisfied and which are not.

   a. An environmental activist attending a political fundraising event conducts an informal poll to estimate the proportion that will support upcoming environmental legislation.

   b. An automobile dealership asks clients to rate the service they obtained on a scale of 1 to 5, with 1 being very dissatisfied and 5 being very satisfied.

   c. An industrial engineer is responsible for a manufacturing process that historically has produced 0.5% defectives. In order to ensure that the process has not changed, he randomly samples and inspects 10 items periodically.

2. Cancelled appointments at a medical practice result in lost revenue and other inefficiencies. The practice is instituting a new policy where a $20 fee will be assessed for cancellations made within 24 hours of the scheduled appointment. Historically 8% of appointments were cancelled within 24 hours. One month after the policy was initiated, the office manager randomly sampled 230 patients and recorded whether the cancellation fee was incurred in the file *MedicalAppointments.jmp*. Has the new policy been effective at reducing cancellations?

3. A mail order company offers free shipping on orders over $100. In a random sample of 640 recent orders, 272 received free shipping. Compute a 95% confidence interval on the proportion of orders that receive free shipping. Write a brief interpretation of the confidence interval.

4. A small business that makes custom-designed furniture believes that some deliveries take longer than expected due to factors such as incorrect or incomplete directions to the delivery location or problems assembling the furniture. Such deviations from the schedule cause other deliveries to be late and require the company to pay overtime to the drivers. The file *FurnitureDeliveries.jmp* contains data from deliveries made during the last month. Scheduled and actual delivery times are recorded to the nearest 5 minutes.

   a. Describe the data.

   b. Find a 95% confidence interval for the proportion of deliveries that exceed the scheduled delivery times.

   c. Give an interpretation of this confidence interval.

   d. Does the mean number of minutes a delivery is late exceed 5 minutes?

5. A non-profit organization has sent out 1200 solicitations by mail requesting donations. In the past, they have received contributions from 5% of those

contacted in this manner. Of the 1200 mailings, 25 were returned as undeliverable and 77 contributions were received. Was this mailing consistent with historical response levels?

6. In a process that produces light emitting diodes (LEDs), 80% will be of sufficient quality for industrial lighting applications. The quality engineer has found a new vendor who produces a key raw material with higher purity than the raw material currently being used. The engineer would like to test the new material to see if an increase in the yield of industrial grade LEDs will result. He believes an increase of 4% in yield will be sufficient to justify the new vendor's higher cost product. He would like to choose a sample size that will detect a 4% increase in yield with 90% probability. If there is no change in yield, this should be discovered with 95% probability. Find the sample size that meets the quality engineer's requirements.

7. A city mayor has called a press conference to announce his candidacy for statewide office in the next election.

a. The mayor's campaign manager wishes to conduct a poll the day following the announcement to assess the mayor's favorability rating. She wishes to assume a worst-case scenario for the sample size; therefore, a Null Proportion of 0.5 should be used. How many voters should be selected to obtain an estimate of favorability within ±2%?

b. Suppose that there is only time to poll approximately 1500 voters. What sampling margin of error can be obtained with this sample size?

8. A company takes orders through the Web, telephone, or mail. There is a perception that error rates differ for the various ordering channels. Specifically, some items received by customers are not what they ordered. There are a number of possible causes for these errors, such as difficulty in reading handwriting and inaccurate electronic data. The following table summarizes recent experience in terms of number of orders.

	Correct Orders	Incorrect Orders
Web	695	19
Telephone	502	58
Mail	165	9

a. Enter the data into JMP so that a Chi-Square test can be conducted.

b. Give graphical and numerical summaries of the data.

c. Set up and conduct a Chi-Square test to determine whether the three ordering methods share a common error proportion.

9. A clinical trial is evaluating three new formulations for an allergy medication. The principal investigator wishes to test whether the proportions of side effects (e.g., drowsiness, sore throat) are the same for each of the three formulations. The file *SideEffects.jmp* contains the data reported.

   a. Give a statistical summary of the data.

   b. Set up and conduct a test of hypothesis to determine whether the proportions of reported side effects are the same for each of the three formulations.

   c. If there is a difference between the formulations, use the cell Chi-Square values to determine which formulations are different.

10. Customer service at a department store processes returned merchandise. Different procedures are followed, depending on whether the purchase was made with cash, by debit card, by credit card, or by store charge account. A random sample of recent purchases was obtained. The data are contained in the file *MerchandiseReturns.jmp*. Do the four forms of payments have the same return rate?

12.6 Case Study: Incomplete Rebate Submissions

Thirty percent of mail-in rebate submissions are incomplete. In such cases, the requestor will be sent a letter informing them of the missing item (for example, UPC code, proof-of-purchase) and will be allowed 30 days to resubmit the missing items in order to receive the rebate. Recently, consumers have been offered an online rebate submission option. The rebate processor would like to randomly sample online submissions to see whether there has been a significant drop in incomplete submissions. The submissions are performed in a step-by-step fashion, but without the internal checking for errors and omissions that will become available at a later date. There are limited resources available to conduct this evaluation.

Prepare a table showing the sample sizes for all possible combinations of confidence level (90% and 95%), reduction in incomplete submissions (4% and 8%), and power (0.80 and 0.90). There will be a total of eight different combinations. Write a short paper presenting your table. Describe how sample size changes as confidence level, power, and reduction in incomplete submissions change. Recommend a confidence level, power, and percentage reduction that you think are appropriate for this situation and explain your rationale.

Chapter 13

Tests for Independence

13.1 Statistical Independence of Two Nominal Variables   340

    13.1.1 Introduction   340

    13.1.2 Review of Conditional Probability and Independent Events   341

    13.1.3 Statistical Hypotheses of Independence of Two Nominal Variables   345

    13.1.4 Example 1: Executive Transfers   346

13.2 Stratification in Cross-Classified Data   352

    13.2.1 What Is Stratification of Cross-Classified Data?   352

    13.2.2 Example 1: Consumer Preference of Two Cola Brands   353

    13.2.3 Example 2: On-Time Performance of Package Delivery Companies   357

    13.2.4 Example 3: Mortality after a High-Risk Procedure in Two Hospitals   362

13.3 Summary   364

13.4 Problems   364

13.5 Case Study: Financial Management Customer Satisfaction Survey 366

13.6 References 367

13.1 Statistical Independence of Two Nominal Variables

13.1.1 Introduction

Data on nominal variables are summarized as frequencies for each category. With two nominal variables, the joint frequencies of the categories can be crosstabulated and analyzed for the presence of a relationship or association. Many business problems involve such associations. In marketing research, associations between consumer behavior and demographic, socio-economic, and similar characteristics are sought. In production, you might be interested in product or production characteristics by shift, production line, or product line. The following are some specific questions for a relationship between two nominal variables:

- Is a customer's satisfaction with a car (high, medium, or low) affected by the after-sale service effort of the car maker (minimal, adequate, strong, very strong)?

- Does consumer preference (high, medium, or low) depend on the type of cola (Cherry, Vanilla, or Traditional)?

- Do customer complaints (yes or no) vary by type of product (electronic, computers and accessories, audio equipment)?

- Does gaming software have the same product acceptance (high, medium, or low) in different regions of the country (Northeast, South, Midwest, West)?

- Is the probability of a defect type in engine parts (cracking, dimensional, surface) associated with the production lines (A, B, C) at which those parts are manufactured?

This chapter looks at the independence of two nominal variables. Independence between X and Y is most precisely defined in terms of probability concepts. Independence is a case in which the outcome of Y is probabilistically not affected by the outcome category of X. When two nominal variables are not independent, they are said to be associated or related. X and Y are related if the outcome category of Y depends on the outcome category of X. Association is a term that is used in connection with measuring the degree of dependence or relatedness between Y and X.

For example, high customer satisfaction with a car brand would be expected when the car maker strongly emphasizes after-sale service. Similarly, expect low customer satisfaction with a car brand where the maker neglects after-sale service. Of course, there are other factors that influence customer satisfaction, so the single nominal variable "Level of after-sale service" might not explain customer satisfaction completely. However, you can say that a relationship between Y and X exists when the level of X affects the outcome of Y in some way. A customer owning a car from a manufacturer with a high level of after-sale service should be more satisfied, with a higher probability, than a customer owning a car from a maker with a poor after-sale service. Lastly, the strength of association between level of after-sale service and customer satisfaction could be measured.

The approach to testing statistical independence between two nominal (categorical, qualitative) variables is very similar to that of testing the equality of proportions. The only difference is that instead of Y being a binary outcome variable, Y can have more, i.e., r, outcome categories. Nevertheless, many important applications can be stated in terms of two nominal variables, each with binary outcomes. Suppose a company produces a product in two locations called Atown and Becity. The company wishes to find out whether the proportion of customer complaints differs by where the product is made. Location is a nominal variable with two categories, Atown and Becity, The outcome variable is a nominal variable with possible outcomes Complaint or No Complaint. In this context, two questions can be answered by the same statistical method: (1) Does the proportion of customer complaints differ by production location? (2) Does the proportion of customer complaints depend on production location?

For statistical analysis, these two questions are the same. If the proportions of customer complaints differ by production location, customer complaints are said to depend on production location. Testing statistical independence between two random variables requires an understanding of the terms "independence" and "dependence" as used in probability.

13.1.2 Review of Conditional Probability and Independent Events

The conditional probability of an event B given an event A is

$$P[B|A] = \frac{P[A \cap B]}{P[A]}$$

where $P[A \cap B]$ is the probability that both events A and B occur jointly. Similarly, the conditional probability of an event A given an event B is

$$P[A|B] = \frac{P[A \cap B]}{P[B]}$$

From these two definitions, the multiplication law of probabilities is derived: $P[A \cap B] = P[A] \cdot P[B|A] = P[B] \cdot P[A|B]$. Two events A and B are probabilistically (stochastically) independent if $P[B|A] = P[B]$ or $P[A|B] = P[A]$. The resulting multiplication rule for probabilities of independent events $P[A \cap B] = P[A] \cdot P[B]$ defines the independence of the two events A and B.

Example 1 of Independent Joint Probabilities: Consider the simple experiment of tossing two coins. The only possible outcomes for each coin are Heads or Tails. Each outcome is equally probable and is assigned the probability 0.5. Thus for any coin P[heads] = 0.5 and P[tails] = 0.5.

The entire experiment can be represented with a cross-classification table of probabilities as shown in Table 13.1. The marginal probabilities p_{H1}, p_{T1}, p_{H2}, and p_{T2} represent the probabilities of an outcome of a single coin toss and are all equal to 0.5. The probabilities on each margin sum to 1. The four outcome combinations of flipping two coins are (heads, heads), (heads, tails), (tails, heads), and (tails, tails). In tossing two coins the probability of obtaining (heads, heads) is simply the product of obtaining one head with each coin, i.e., $p_{H1} \cdot p_{H2} = (.5)(.5) = .25 = p_{H1,H2}$. The probabilities of these joint outcomes of a toss of two coins are called joint probabilities. In this example, the joint probabilities in Table 13.1 can be represented as the product of the two marginal probabilities. Therefore, the joint probabilities of the toss of two coins are considered statistically independent. The four joint probabilities are (.5)(.5) = 0.25 and sum to 1.

Table 13.1 Independent Probabilities of the Toss of Two Coins

	Heads on first toss	Tails on first toss	Marginal probability of second toss
Heads on second toss	(.5)(.5) = 0.25 = $p_{H1,H2}$	(.5)(.5) = 0.25 = $p_{T1,H2}$	0.5 = p_{H2}
Tails on second toss	(.5)(.5) = 0.25 = $p_{H1,T2}$	(.5)(.5) = 0.25 = $p_{T1,T2}$	0.5 = p_{T2}
Marginal probability of first toss	0.5 = p_{H1}	0.5 = p_{T1}	1

Example 2 of Independent Joint Probabilities: A company has found that the Family Status of its executives is as follows: 40% have two-career marriages, 30% have one-career marriages, and 30% are unattached. The firm found that executives, when offered a transfer to another location, made the following decisions: 45% of executives refused a transfer to another location, 30% accepted a transfer but with reservations concerning the location and the timing of the transfer, while 25% accepted outright without reservations. Assuming that family status and transfer decision are two independent events, a table of joint probabilities can be constructed from the marginal probabilities of rows and columns. In Table 13.2, the number of rows r = 3, the number of columns c = 3, and percentages have been expressed as probabilities.

Table 13.2 Independent Probabilities of Executive Transfer Example

		Transfer Decision			
		Refused (C1)	**Accepted with reservations (C2)**	**Accepted without reservations (C3)**	**Marginal probability of rows**
Family Status	Executives with two-career family (R1)	$(.4)(.45) = 0.18$ $= p_{C1,R1}$	$(.4)(.3) = 0.12 =$ $p_{C2,R1}$	$(.4)(.25) = 0.10$ $= p_{C3,R1}$	$0.40 = p_{R1}$
	Executives with one-career family (R2)	$(.3)(.45) =$ 0.135	$(.3)(.3) = 0.09$	$(.3)(.25) = 0.075$	$0.30 = p_{R2}$
	Unattached executive(R3)	$(.3)(.45) =$ 0.135	$(.3)(.3) = 0.09$	$(.3)(.25) = 0.075$	$0.30 = p_{R3}$
	Marginal probability of columns	$0.45 = p_{C1}$	$0.30 = p_{C2}$	$0.25 = p_{C3}$	1.00

Example of Dependent Joint Probabilities: Taking sample observations one by one and not replacing them before the next observation is drawn is called sampling without replacement in a finite population. Unlike with flipping a coin or similar experiments, the joint probabilities of the sample outcomes are non-independent, because the population changes from one draw to the next. As a simple example, consider drawing two cards, one by one, from a deck of 52 cards. Assume this is a regular deck with exactly four aces and 48 non-aces.

The probability of obtaining an ace on the first draw is P[Ace on first trial] = $p_{Ace\ 1}$ = 4/52. The unconditional or marginal probability of an ace on the second trial is also $p_{Ace\ 2}$ = 4/52. However, the probability of obtaining an ace on the first and second drawing of a card is

$$\frac{4}{52} \cdot \frac{3}{51} = \frac{12}{2652}$$

and is not equal to the marginal product

$$(p_{Ace\,1} \cdot p_{Ace2}) = \frac{4}{52} \cdot \frac{4}{52} = \frac{16}{2704}$$

(see Table 13.3). The only two possible outcomes for each card are "ace" or "non-ace". The probabilities are not independent, because the number of cards in the deck decreases by 1 with each card withdrawn.

Table 13.3 Dependent Probabilities of Selecting Two Cards from a Deck of 52

	Ace on first trial	Non-ace on first trial	Marginal probability of second trial
Ace on second trial	$\frac{4}{52} \cdot \frac{4}{52} \neq \frac{4}{52} \cdot \frac{3}{51} = p_{A1A2}$	$\frac{48}{52} \cdot \frac{4}{52} \neq \frac{48}{52} \cdot \frac{4}{51} = p_{N1A2}$	$\frac{4}{52} = p_{Ace\,2}$
Non-ace on second trial	$\frac{4}{52} \cdot \frac{48}{52} \neq \frac{4}{52} \cdot \frac{48}{51} = p_{A1N2}$	$\frac{48}{52} \cdot \frac{48}{52} \neq \frac{48}{52} \cdot \frac{47}{51} = p_{N1N2}$	$\frac{48}{52} = p_{No\,Ace\,2}$
Marginal probability of first trial	$\frac{4}{52} = p_{Ace\,1}$	$\frac{48}{52} = p_{No\,Ace\,1}$	1

The marginal probability of drawing an ace on the second trial is the sum of the two joint probabilities $p_{Ace1Ace2}$ and $p_{NoAce1Ace2}$. It is

$$\left(\frac{4}{52} \cdot \frac{3}{51} + \frac{48}{52} \cdot \frac{4}{51} \right) = \frac{4}{52} = p_{Ace\,2}$$

The other marginal probabilities are calculated in a similar manner. Since the joint probabilities cannot be represented as the product of the marginal probabilities, the two outcomes of ace on first trial and ace on second trial are dependent. After one joint probability is determined to be non-independent, the entire distribution is considered non-independent.

13.1.3 Statistical Hypotheses of Independence of Two Nominal Variables

Two nominal variables X and Y are said to be independent whenever the cell or joint probability distribution can be written as the product of the two marginal distributions. This definition of independence is useful in formulating the null hypothesis for the test of independence. Also, two nominal variables X and Y are statistically independent if the conditional distribution of one variable (e.g., Transfer Decision) is identical for each level of the other variable (e.g., Family Status). This definition of independence is related to testing the equality of several proportions discussed in the previous chapter.

Statistical Hypotheses and Chi-Square Statistic for Testing Independence

The null hypothesis states that the cell or joint probabilities can be written as the product of the corresponding row and column marginal probabilities. Using the notation of Table 13.2, the null hypothesis is H_0: $p_{C_i R_j} = p_{C_i} \, p_{R_j}$. The alternative hypothesis is H_A: $p_{C_i R_j} \neq p_{C_i} \, p_{R_j}$ for at least one cell (i, j).

Under the null hypothesis the expected cell frequencies for the i-th row and j-th column cell are

$$\text{Expected frequency of cell (i, j)} = \frac{\text{Marginal total of row i} \times \text{Marginal total of column j}}{\text{Overall total}}$$

The differences between observed and expected frequencies are called *deviations*. Large deviations can be taken as evidence against the null hypothesis and make you less inclined to accept the null hypothesis that X and Y are independent. Small deviations are taken as supporting the null hypothesis. A statistic is needed that summarizes the deviations in a standardized way. This Chi-Square statistic is the sum of the Cell Chi-Square (Cell Chi^2 in JMP). For each combination of category i of X and category j of Y, the deviation is calculated as the difference between observed frequency and expected frequency. The Cell Chi-Square value is:

$$\text{Cell Chi - Square of cell (i, j)} = \frac{\left(\text{Deviation of cell (i, j)}\right)^2}{\text{Expected of cell (i, j)}}$$

for each cell of the cross-classification. The values of Cell-Chi-Square of cell (i, j) are summed to yield the Pearson Chi-Square:

$$\text{Pearson Chi - Square} = \sum_{\text{all cells}} \frac{\left(\text{Deviation of cell (i, j)}\right)^2}{\text{Expected of cell (i, j)}}$$

This test statistic is used to calculate the *p*-value under the null hypothesis.

The larger the deviations between observed and expected cell frequencies (divided by expected cell frequencies for standardization), the larger the Pearson Chi-Square statistic. Cell Chi-Square values can be used to identify combinations of categories that contribute to a rejection of the null hypothesis, if indeed it is rejected.

Assuming the null hypothesis is true, the Pearson Chi-Square statistic follows a Chi-Square distribution with $(r - 1)(c - 1)$ degrees of freedom, where r is the number of rows and c is the number of columns in the cross-classification. To complete the test, compare the calculated *p*-value with the α significance level. If $p < \alpha$, reject H_0. If $p \geq \alpha$, do not reject H_0.

13.1.4 Example 1: Executive Transfers

Situation: The human resources department of a large international business is charged with investigating the reasons why a fairly large number of its executives do not accept offers to transfer to new positions in different locations.

Problem: Recently, a large number of executives have refused or asked to delay transfers to other locations, even though transfers are usually in connection with a promotion. It is claimed that family status has an important role in the transfer decision by executives.

Data Requirements: The two variables that are of particular interest are the transfer decision and family status. Transfer decisions of executives are assigned to one of three groups:

- outright refusal
- acceptance of transfer but with reservations as to timing and location
- outright acceptance of transfer as offered

Family status is categorized as follows:

- two-career families
- one-career families
- unattached

Both variables are nominal with three categories. Data on 285 executives are available. For each executive, only the most recent transfer offer was considered. The cross-classified data table is shown in Table 13.4.

In JMP, the contingency table shown in Table 13.4 requires a data table (*ExecutiveTransfer.jmp*) with three columns, one column each for X and Y, and one column for the counts representing the cell frequencies. Each cell in the table, not including the marginal frequencies, requires a separate row. For r rows and c columns there are r × c cells.

Table 13.4 Contingency Table of Transfer Decisions versus Family Status

		Transfer Decision			
		Refused	**Accepted with reservations**	**Accepted without reservations**	**Marginal row totals**
Family Status	**Executives with two-career family**	65	29	21	115
	Executives with one- career family	31	30	21	82
	Unattached executive	26	29	33	88
	Marginal column totals	122	88	75	285

The JMP data table of Figure 13.1 shows the three columns with r × c = 9 rows. Column 1 contains the nominal variable Family Status with three outcome categories (Two-career, One-career, and Unattached). Column 2 contains the nominal variable Transfer Decision, also with three outcome categories (Refused, AcceptWITHres, AcceptWITHOUTres). Column 3 contains the counts for the combinations of outcome categories. Column 3 must be numerical. The Frequency column sums to 285.

Figure 13.1 JMP Data Table for Data from Contingency Table

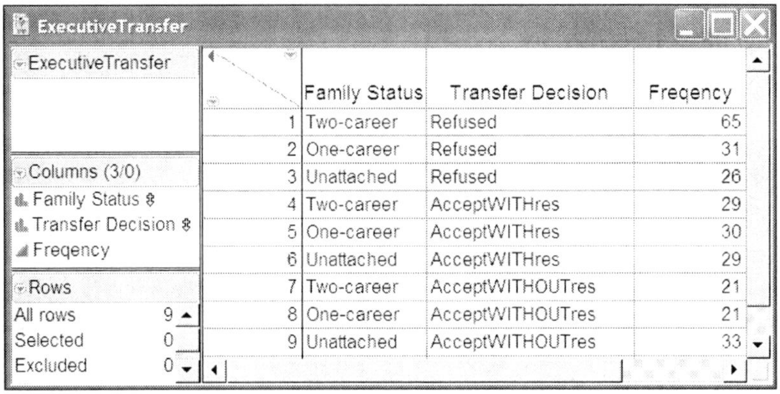

	Family Status	Transfer Decision	Freqency
1	Two-career	Refused	65
2	One-career	Refused	31
3	Unattached	Refused	26
4	Two-career	AcceptWITHres	29
5	One-career	AcceptWITHres	30
6	Unattached	AcceptWITHres	29
7	Two-career	AcceptWITHOUTres	21
8	One-career	AcceptWITHOUTres	21
9	Unattached	AcceptWITHOUTres	33

Family Status and Transfer Decision both show the List Check symbol. List Check is used to rearrange the values of character columns in a different order from the default of alphabetical. For example, the outcomes of Family Status have been arranged in the desired order: Two-career, One-career, and Unattached. The alphabetical default order would have been One-career, Two-career, and Unattached.

Analysis of Executive Transfer in JMP

Using the Fit Y by X platform, specify the variables as shown in Figure 13.2. The role of the two nominal variables is interchangeable. Nevertheless, it is a good habit to select the variable of interest as the **Y, Response** and the other variable as the **X, Factor**. Enter the variable Frequency in the field labeled **Freq**. Clicking **OK** produces the basic output consisting of a Mosaic plot, a contingency table with four default entries in each cell, as well as the test statistics for independence with associated *p*-values.

Figure 13.2 Variable Specification Window for Chi-Square Test of Independence

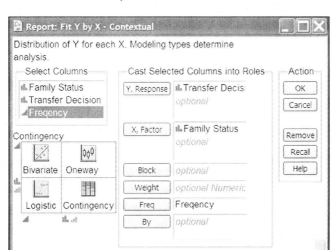

The Mosaic plot, shown in the left panel of Figure 13.3, gives a visual summary of the data. The narrow vertical mosaic on the right represents the overall proportions for the Y-variable Transfer Decision. The main Mosaic plot contains nine rectangles arranged in three columns. The horizontal axis labels each column according to its Family Status. The widths of the three vertical columns in the main Mosaic plot represent the relative frequency of each category of Family Status. In this sample, Two-career executives are more frequently represented than either One-career or Unattached executives. The heights of the rectangles of each mosaic represent the relative frequencies of each cell in that column. As per the legend on the right axis, the bottom rectangle in each column represents Refused. The rectangle in the middle is for AcceptWITHres. The topmost rectangle is for AcceptWITHOUTres.

The Mosaic plot of the data in the left panel of Figure 13.3 shows that the proportion of executives who refused transfers is largest in the Two-career category, less in the One-career category and even less in the Unattached category. The topmost rectangles, representing AcceptWITHOUTres, show the opposite. The smallest rectangle by height is for Two-careers and the largest is for Unattached. The tiles of the narrow strip to the right of the row labels represent the average proportions across the cells in each row.

Figure 13.3 Mosaic Plot for Executive Transfers Compared with Hypothetical Independence

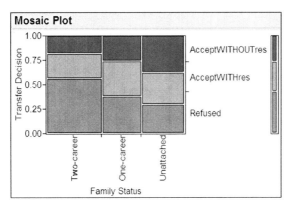 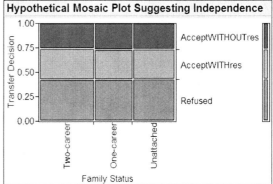

The Mosaic plot suggests that the data do **not** support the assumption of independence between Family Status and Transfer Decision. The visual impression needs to be confirmed with the *p*-value of the Pearson Chi-Square test before any definite conclusions can be drawn.

If the variables were independent, as assumed under the null hypothesis, the proportion of Refused, AcceptWITHres, and AcceptWITHOUTres should show rectangles of approximately similar height. The right panel of Figure 13.3 shows a Mosaic plot for hypothetical data that strongly suggests independence of Transfer Decision and Family Status. The height of each rectangle representing a Transfer Decision is about the same for all three types of Family Status.

Figure 13.4 Contingency Table of 285 Executive Transfer Decisions by Family Status

Contingency Table

Count	Refused	AcceptWITHres	AcceptWITHOUTres	
Total %				
Two-career	65	29	21	115
	22.81	10.18	7.37	40.35
One-career	31	30	21	82
	10.88	10.53	7.37	28.77
Unattached	26	29	33	88
	9.12	10.18	11.58	30.88
	122	88	75	285
	42.81	30.88	26.32	

(Family Status on vertical axis; Transfer Decision across top)

In the contingency table provided by the Fit Y by X platform, the variable specified as **Y, Response** appears across the top, while the **X, Factor** is now on the vertical axis. In Figure 13.4, **Row %** and **Col %** have been deselected from the menu obtained by clicking the red triangle. Count and Total % remain. Count represents the observed frequencies. Total % represents the observed frequencies (counts) of each cell as a percent of the total number of observations. For example, 65 observations fall into the (Refused, Two-career) combination representing 22.81 percent of the total of 285 observations.

The contingency table contains the numbers needed to calculate the Pearson Chi-Square test statistic. Additional options for display within the contingency table include expected frequencies, deviations of observed and expected frequencies, and the Cell Chi Squares. However, the key number to confirm the rejection of the null hypothesis is the *p*-value of the Pearson Chi-Square statistic.

In addition to the Pearson test, JMP also displays the Likelihood Ratio test. Both tests use the Chi-Square distribution with $(r-1)(c-1)$ degrees of freedom. The test statistics and *p*-values of the two methods usually are in close agreement. Figure 13.5 shows the test statistics in the column labeled ChiSquare and gives the corresponding *p*-values in the column labeled Prob>ChiSq. For Pearson $p = 0.001 < \alpha = 0.05$, so reject H_0. The row labeled Likelihood Ratio gives almost identical results.

Figure 13.5 Test Statistics and *p*-Values for the Chi-Square Test

Tests		
Test	ChiSquare	Prob>ChiSq
Likelihood Ratio	18.242	0.0011*
Pearson	18.363	0.0010*

Conclude with statistical significance that Family Status and Transfer Decision are not independent. Under certain assumptions, you might interpret this result as suggesting that Family Status affects the outcome of Transfer Decision, suggesting a cause-and-effect relationship. However, the statistical model for independence does not claim, nor imply, a cause-and-effect relationship. Nevertheless, this empirical evidence is often used together with non-statistical models and arguments to support cause-and-effect interpretations.

13.2 Stratification in Cross-Classified Data

13.2.1 What Is Stratification of Cross-Classified Data?

This chapter treats relationships between two nominal variables. However, with cross-classified data, as with all other data, *lurking variables* can hide, alter, or even reverse a relationship between two variables. Lurking variables are variables that mask a relationship. They need to be introduced into the analysis; otherwise, you might arrive at incorrect conclusions.

The simplest way to analyze data with nominal lurking variables is to make separate tables, one table for each category of the lurking variable. Each of these tables is then analyzed separately using Chi-Square analysis. The results of the individual tables are combined for a summary interpretation. (More advanced tools are available in JMP, but are beyond the scope of this book.) The lurking variable used in constructing the different tables is called the stratification variable, because it is used to form non-overlapping subgroups of the data. Non-overlapping subgroups of populations are called strata. A single subgroup is called a stratum.

The following three examples show three aspects of how stratification might change conclusions. All of these examples involve 2 × 2 tables.

The first example concerns consumer preference of two brands of cola. The stratification variable Income Level shows that one of the subgroups has a brand preference pattern that is contrary to the pattern of the overall table.

The second example is about on-time performance of two package delivery companies. This example shows a situation in which the stratification variable representing the delivery method reveals a counter-intuitive result. The overall table shows no significant relationship between on-time performance and package delivery company. When the stratification variable is applied, the same company is best in both regular and express mail, a fact that was masked by the unequal sampling of packages. In the package delivery example, the stratification variable changes the conclusions from non-significant to significant.

In the third example of hospital mortality after a high-risk procedure, the inclusion of the stratification variable Health of Patient changes the conclusions from significant to non-significant. Differences that are apparent in the overall table are not due to the original variables, but in these examples can be explained by the stratification variable.

13.2.2 Example 1: Consumer Preference of Two Cola Brands

This example is designed to show how a stratification variable can reveal a pattern of association that is hidden in the overall table.

Situation: A soft drink company wants to revise its marketing development strategy. The marketing research department is to explore the brand positioning of its cola against that of the major competitor. At issue are the strength of consumers' generic and brand-specific likes and dislikes of colas. People have a generic like or dislike when they either like or dislike cola. People have a brand-specific like or dislike when they like cola, but only a specific brand.

Problem: The company has been losing market share for the past 18 months. Every percentage loss in market share results in a considerable loss in profits and requires large expenditures if the company wants to regain that market share. The company needs a better understanding of consumers' cola preferences.

Table 13.5 Contingency Table of Observed Frequencies

		Cola B		
	Count	NO	YES	Total
Cola A	no	84	32	116
	yes	48	122	170
	Total	132	154	286

Data Requirements: A random sample of 286 consumers is asked to record their purchasing habits of two brands of cola, Cola A and Cola B. Consumers are asked whether or not they purchased brand A or brand B cola on their most recent visit to the supermarket, irrespective of quantity purchased. The resulting data are summarized in Table 13.5.

Consumers might have purchased one of the brands, both, or neither. Table 13.5 indicates that 84 of the 286 consumers did not purchase cola. One hundred and twenty two purchased both colas, while 48 purchased Cola A exclusively and 32 Cola B exclusively. These results need to be entered into a JMP data table. A single row is needed for each of the four outcome categories. Two of the three columns identify the brand of cola, and their rows indicate whether or not (yes or no) that brand has been purchased. The third column, called Freq ALL respondents, specifies the frequencies with which these four outcomes occurred. Figure 13.6 shows the resulting JMP data table.

Figure 13.6 Data File *ColaPreference.jmp*

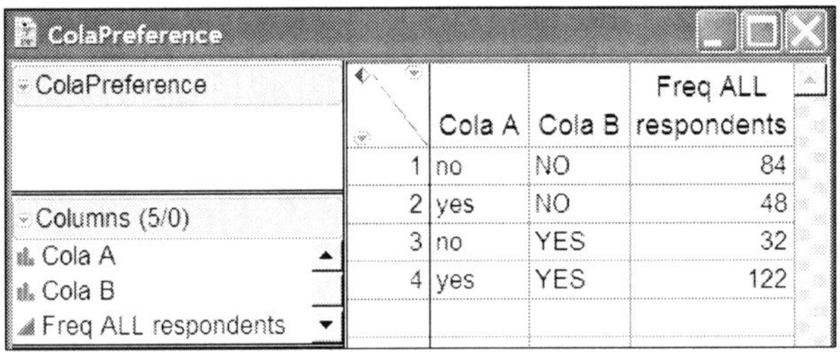

Analysis of Consumer Preference of Two Cola Brands in JMP

In order to analyze the association between preferences for Cola A and Cola B use the Fit Y by X platform. Frequencies need to be specified in the variable specification window (Figure 13.7).

Figure 13.7 Variable Specification Window for Fit Y by X

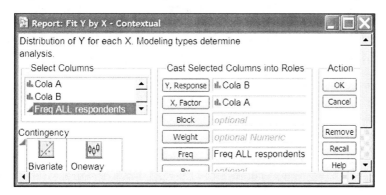

The default results of the Fit Y by X platform for two nominal variables are the Mosaic plot, the contingency table and the table of tests. The Mosaic plot on the left in Figure 13.8 shows that the largest two rectangles are for the (yes, YES) and the (no, NO) outcomes, suggesting a mostly generic like or dislike of colas, because respondents either bought both brands or neither brand.

Figure 13.8 Mosaic Plot and Contingency Table of Preferences for Cola A and Cola B

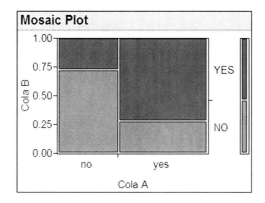

The contingency table on the right in Figure 13.8 confirms that 42.66 percent of respondents fall into the (yes, YES) outcomes and 29.37 percent fall into the (no, NO) outcomes. This is over 72 percent of all respondents, implying that consumers either buy or do not buy cola. This indicates that the majority of respondents either dislike or like Cola, irrespective of brand. The results show only a moderate brand preference for either Cola A or Cola B. Therefore, the generic preference seems to be confirmed.

The Pearson Chi-Square test in Figure 13.9 indicates that this relationship is highly significant. The Pearson Chi-Square test statistic = 54.15 with p < 0.0001. You can reject the null hypothesis that there is independence between Cola A and Cola B purchases. Both the Pearson and Likelihood Chi-Square test statistics approximate their p-values. Figure 13.9 also gives p-values for Fisher's Exact Test, used only for 2 × 2 tables and especially useful for small sample sizes. The p-value of Fisher's Exact Test is exact. Furthermore, there are separate p-values for left-sided, right-sided, and two-sided tests. Both the Pearson and the Likelihood tests only test for two-sided alternative hypotheses. In this case, you could conclude that the proportion of respondents answering that they bought Cola B is higher when they also responded that they bought Cola A, than if they did not buy Cola A.

Figure 13.9 Tests of Independence for Cola Preferences

Tests		
Test	ChiSquare	Prob>ChiSq
Likelihood Ratio	55.783	<.0001*
Pearson	54.150	<.0001*
Fisher's Exact Test	Prob	Alternative Hypothesis
Left	1.0000	Prob(Cola B=YES) is greater for Cola A=no than yes
Right	<.0001*	Prob(Cola B=YES) is greater for Cola A=yes than no
2-Tail	<.0001*	Prob(Cola B=YES) is different across Cola A

Analysis with Stratification: Respondents reported their income as high or low as a third variable. Suppose the 286 respondents are separated in two tables, one for the 76 high-income respondents and a second for 210 low-income respondents, as shown in Table 13.6. High-income results are shown in the left column, low-income results in the right column.

Table 13.6 Cola Preferences by Income Level

Contingency Table - High I.			
		Cola B	
Count Expected	NO	YES	
no	4	30	34
	19.6842	14.3158	
yes	40	2	42
	24.3158	17.6842	
	44	32	76

(Cola A on vertical axis)

Contingency Table - Low I.			
		Cola B	
Count Expected	NO	YES	
no	80	2	82
	34.3619	47.6381	
yes	8	120	128
	53.6381	74.3619	
	88	122	210

(Cola A on vertical axis)

Tests - High Income

Test	ChiSquare	Prob>ChiSq
Likelihood Ratio	62.744	<.0001*
Pearson	53.708	<.0001*

Tests - Low Income

Test	ChiSquare	Prob>ChiSq
Likelihood Ratio	206.937	<.0001*
Pearson	171.177	<.0001*

In high income group, strong brand preference for either Cola A or Cola B.

In low-income group, strong like or dislike of cola, but no brand preference.

Both tables show significant association. However, the high-income group does not fit the overall pattern. Purchases in this group are concentrated in the (no, YES) and the (yes, NO) categories. This indicates a strong preference for either Cola A or Cola B, i.e., a strong brand preference. Low-income respondents follow the pattern observed in the overall table, preferring either cola soft drinks or not, but with less strong feelings about the brand of the soft drink. The pattern observed with high-income respondents was hidden in the overall table, because of the relatively smaller sample size.

13.2.3 Example 2: On-Time Performance of Package Delivery Companies

This example is designed to show how stratification, judging from the overall table, can lead to counter-intuitive results in the stratified tables. It is an example in which a non-significant association in the overall table hides significant associations in the stratified tables.

Situation: A mail order firm needs to send packages on a regular basis. The firm wants to guarantee its customers swift delivery times. It has used the MessEx package delivery service exclusively. Recently it has been approached by Bagone, another package delivery service. Bagone promises to match MessEx's performance at a slightly lower cost. The mail order firm is primarily concerned about on-time delivery, although reduced cost would be a welcome bonus. On-time delivery to the mail order firm simply means that a package arrives on or before the promised date and time. A package is considered late if it is not at the customer's premises at the promised time, whatever that time is. Two delivery methods are available. Regular delivery is cheaper but takes longer. Express delivery is available at extra charge.

Problem: The firm needs to evaluate on-time performance of the two competing package delivery services. Marketing wants to choose the service with the higher proportion of packages delivered on time. The question is whether on-time delivery is independent of package company.

Data Requirements: The file *PackageServiceStrat.jmp* in Figure 13.10 shows eight different categories defined by the combination of categories in Package Company, On Time Performance, and Delivery columns. The numbers in column 4, Frequency, sum to 700, indicating that 700 packages were mailed. Figure 13.11 shows that 350 packages were mailed using Bagone and 350 using MessEx. Column 2, On Time Performance, records that 321 packages arrived late and 379 arrived on time. It can be similarly verified that 350 packages were sent by Regular mail and 350 by Express mail.

Figure 13.10 Data of Package Services Performance Including Stratification Frequencies

	Package Company	On Time Performance	Delivery	Frequency
1	Bagone	on time	Regular	21
2	Bagone	late	Regular	49
3	MessEx	on time	Regular	138
4	MessEx	late	Regular	142
5	Bagone	on time	Express	166
6	Bagone	late	Express	114
7	MessEx	on time	Express	54
8	MessEx	late	Express	16

PackageServiceStrat

Columns (4/0)
- Package Company
- On Time Performance
- Delivery
- Frequency

Rows
All rows 8

Analysis of On Time Performance and Package Companies in JMP

If On Time Performance is independent of the Package Company, then the proportion of on-time delivery should be comparable for the two services. The variable specification for this overall analysis is On Time Performance as the **Y, Response**, Package Company as the **X, Factor** and Frequency in the **Freq** field.

Figure 13.11 Mosaic Plot and Contingency Table for Combined Frequencies of Package Delivery

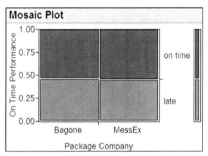

Contingency Table			
	On Time Performance		
Count Expected Cell Chi^2	late	on time	
Bagone	163 160.5 0.0389	187 189.5 0.0330	350
MessEx	158 160.5 0.0389	192 189.5 0.0330	350
	321	379	700

Test	ChiSquare	Prob>ChiSq
Likelihood Ratio	0.144	0.7045
Pearson	0.144	0.7045

Fisher's

Exact Test	Prob	Alternative Hypothesis
Left	0.6755	Prob(On Time Performance=on time) is greater for Package Company=Bagone than MessEx
Right	0.3808	Prob(On Time Performance=on time) is greater for Package Company=MessEx than Bagone
2-Tail	0.7616	Prob(On Time Performance=on time) is different across Package Company

The two "late" rectangles of the Mosaic plot on the left in Figure 13.11 are approximately the same height, supporting the assumption of independence. Both Package Companies have approximately the same on-time and late proportions. With Bagone, the portion of late packages is $163/350 = 0.47$, and with MessEx it is $158/350 = 0.45$. This graphical impression is confirmed by the small differences between Count and expected frequencies, as well as the small Cell Chi^2 in the contingency table on the right in Figure 13.11.

The Pearson Chi-Square statistic $= 0.144$ with an approximate p-value $= 0.7045$, but since this is a 2×2 contingency table, the exact p-value p $= 0.7616$ can be calculated and is preferable. It suggests that Package Company and On Time Performance are independent. The data indicate that there is no significant difference in the proportion of packages late or on time between Bagone and MessEx.

Stratified Analysis by Regular and Express Mail Packages

The 700 packages were sent by two different delivery methods—350 packages with Regular mail and 350 with Express mail. Separating the frequencies into two tables, one for Regular and one for Express mail, allows a more detailed analysis. These two tables might show a hidden relationship. On Time Performance might well depend on the delivery method. Delivery serves as the stratification variable. The separate analysis by Regular and Express mail is performed by entering the stratification variable Delivery as the **By** variable. This produces a separate analysis for each category of the stratification variable.

Figure 13.12 Variable Selection for Stratified Analysis By Delivery

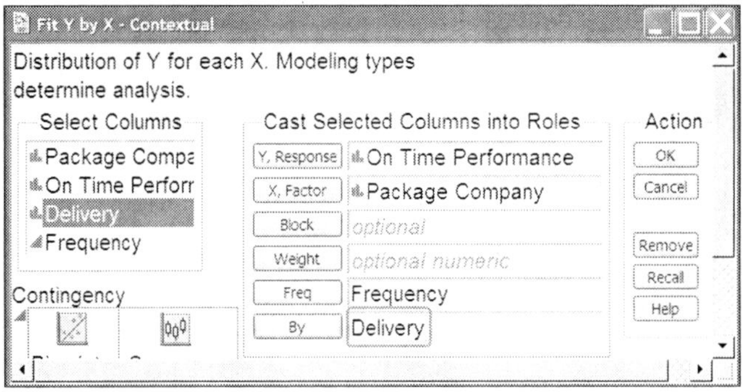

The results are shown side by side in Table 13.7. In each cell of the contingency table for the 350 Regular mail packages, the observed and expected frequencies differ by 10.8 in absolute value. In the contingency table for the 350 Express mail packages, the difference between observed and expected frequencies is 10 in absolute value.

Regular Mail Packages: The results for the Regular mail packages are given in the left column. The Pearson Chi-Square statistic = 8.402 with a *p*-value = 0.0037 shows that there is a significant difference between Package Company in the proportion of late Regular mail packages. The *p*-value confirms that On Time Performance of Regular mail is associated with Package Company. The proportion of late packages is 0.70 for Bagone, but only 0.51 for MessEx. For Regular mail packages, MessEx is significantly more often on time than Bagone.

Express Mail Packages: The results for Express mail packages are given in the right column. The Pearson Chi-Square statistic = 7.649 with p = 0.0057 shows that there is again a significant difference between Package Companies in the proportion of on time Express mail packages. On Time Performance of Express mail packages is also significantly associated with Package Company. The proportion of late packages is 0.41 for Bagone, but only 0.23 for MessEx. MessEx is again significantly more often on time than Bagone.

How can MessEx be the better package service for both Regular and Express mail packages, when overall there was no difference between the package companies? The two Mosaic plots give the answer. The column widths, showing the sample sizes by package company, exhibit different patterns. Overall, 350 packages were sent by each carrier. However, for Regular mail only 70 (20%) of the packages were Bagone's and 280 (80%) were MessEx's. It is just the reverse for Express mail packages. Thus the unequal allocation of packages to package company has hidden the fact that MessEx is on time with a significantly higher proportion of packages than Bagone. The overall table should be scrapped, because it suggests a misleading conclusion. Stratification leads to the correct conclusion. MessEx is proportionately more often on time.

Table 13.7 Comparison of Regular and Express Mail Package Delivery Status

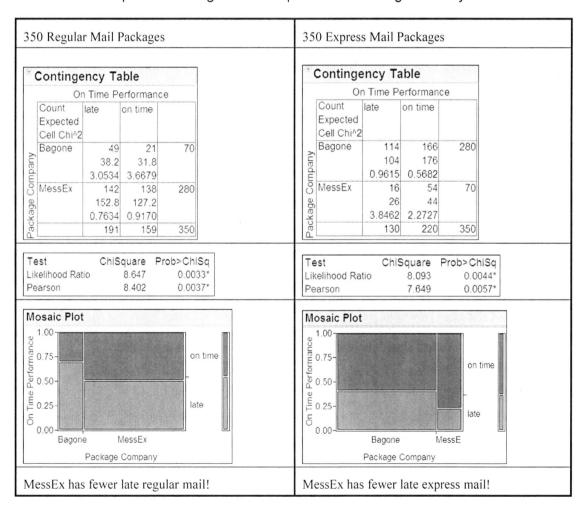

13.2.4 Example 3: Mortality after a High-Risk Procedure in Two Hospitals

This example is designed to show that a stratification variable might reduce a significant association in the overall table to a non-significant association in the stratified table. In such a case, you might conclude that observed differences are mostly due to the stratification variable.

Situation: In healthcare, it is common to compare mortality outcomes of different healthcare providers. This example contains a comparison of the mortality after a high-risk procedure between two hospitals. One consideration in the success or failure of a procedure is the health of the patient on which the procedure is performed. Patients in poor health often have poorer outcomes following therapeutic intervention. Thus, a comparison between two hospitals needs to include a measure of the health of patients that these hospitals serve.

Problem: Compare the mortality following a high risk-procedure between two hospitals.

Data Requirements: The data in this case consist of the hospital name and treatment outcome. Hospital A is to be compared to Hospital B. For each case, a patient's death or survival is recorded. The frequencies of the four combinations of Hospital and Treatment Outcome are also required. Lastly, using the health of patients as a stratification variable, the frequencies will be given separately for patients in good and in poor health. These data allow three different analyses – all patients, patients in good health, and patients in poor health. The complete data are in *Mortality_in_Hospital.jmp*.

Analysis of Mortality after High Risk Procedure in JMP

All Patients: Consider the contingency table on the left in Figure 13.13 with mortality of *all* patients in two hospitals after a procedure. Is Treatment Outcome from the procedure associated with Hospital? Is the mortality the same or is it different in Hospitals A and B? The Pearson Chi-Square statistic = 5.282 with 1 degree of freedom has a p-value = .0216. (The exact two-sided p-value is 0.0247.) The results are significant and Hospital B appears to have significantly higher mortality. The table and the Pearson statistic suggest with statistical significance that Mortality and Hospital are associated.

Figure 13.13 All Patients and Those in Good and Poor Health Condition by Outcome and Hospital

Contingency Table - All Patients

Treatment Outcome

Count Row %	Died	Survived	
A	120	1880	2000
	6.00	94.00	
B	40	960	1000
	4.00	96.00	
	160	2840	3000

Test	ChiSquare	Prob>ChiSq
Likelihood Ratio	5.534	0.0186*
Pearson	5.282	0.0216*

Contingency Table - Good Health

Treatment Outcome

Count Row %	Died	Survived	
A	15	485	500
	3.00	97.00	
B	10	490	500
	2.00	98.00	
	25	975	1000

Test	ChiSquare	Prob>ChiSq
Likelihood Ratio	1.032	0.3096
Pearson	1.026	0.3112

Contingency Table - Poor Health

Treatment Outcome

Count Row %	Died	Survived	
A	105	1395	1500
	7.00	93.00	
B	30	470	500
	6.00	94.00	
	135	1865	2000

Test	ChiSquare	Prob>ChiSq
Likelihood Ratio	0.610	0.4347
Pearson	0.596	0.4402

Patients Separated by Good and Poor Health: Next separate the observations of All Patients with Patient Health as the stratification variable by entering it in the **By** field at variable selection. Construct separate tables for patients in Good health and in Poor health. The center table in Figure 13.13 shows the contingency table for patients in Good health. The Pearson Chi-Square statistic = 1.026 with a p-value = 0.311. (The exact two-sided p-value is 0.4184.) For patients in Good health, there is no significant difference in treatment outcome between the two hospitals. Hospital and Treatment Outcome are independent variables.

The right table in Figure 13.13 shows the contingency table for patients in Poor health. These patients are estimated to have a higher probability (135/2000 = 0.0675) of dying after surgery than patients in Good condition (25/1000 = 0.025). The Pearson Chi-Square statistic = 0.596 with a p-value = 0.4402. (The exact two-sided p-value is 0.4728.) For patients in Poor health, there is no significant difference in treatment outcome between the two hospitals. Hospital and Treatment Outcome are independent variables.

Using patient health as a stratification variable has turned the significant difference in treatment outcome of the overall table into two non-significant differences between the two hospitals. The result suggests very strongly that the significant difference observed in the overall table is due to a different mix of patients in good and poor health between the hospitals. Both hospitals treated 500 patients in good health. However, Hospital A treated 1500 patients in poor health, while Hospital B only treated 500 patients in poor health. This sampling difference causes a significant association in the overall table, but disappears in the stratified tables. The major variable that accounts for differences in treatment outcomes seems to be patient health.

13.3 Summary

- The methods described in this chapter are for the analysis of two nominal variables.

- Statistical methods regarding the independence of two nominal variables look at the existence of a relationship or an association between the two variables.

- Two independent variables are also said to be unrelated or not associated. It could also be said that two dependent variables are related or associated.

- A lack of statistical independence is often used as evidence to support conclusions about cause and effect relationships. However, statistics does not prove the existence of cause and effect relationships. Other explanations and evidence are required.

- Cross-classification tables of the joint frequencies of the outcomes of two variables in a sample are also called contingency tables and are often analyzed using the Chi-Square statistic.

- Stratification of a contingency table is performed using a third categorical variable. The third variable is used to split the original contingency table into layered, or stratified, subtables.

- Stratification offers a more refined analysis that might or might not lead to clearer results.

13.4 Problems

1. A corporation is preparing its employee benefits package for the coming year. In addition to selecting health coverage, employees can choose to purchase long-term disability insurance. Employees pay for this insurance through payroll deduction. There are three different employee classes: salaried, non-union hourly, and union hourly. The following table shows the pertinent data for the long-term disability insurance program.

	Did not elect long-term disability	Elected long-term disability
Salaried	265	108
Non-union Hourly	311	148
Union Hourly	628	342

    a. Enter the data into JMP so that a test for independence can be conducted.

    b. Give graphical and numerical summaries of the data.

    c. Interpret the result of the Chi-Square test to determine whether election of long-term disability insurance is dependent on employee class.

2. A restaurant offers takeout meals (in addition to sit-down service) for lunch and dinner at their three locations (Riverside, Westdale, and Northview) in the same metropolitan area. A recent review has raised concern regarding Riverside's performance for dinner takeouts (which yield a higher return than lunch). The data can be found in *takeout_meals.jmp*.

    a. Describe the data.

    b. Is there any evidence that the meal (lunch or dinner) is dependent on location? Explain how you arrived at your answer.

    c. Suggest a lurking variable that might be useful in this analysis. Explain why you selected this variable.

3. A sports medicine clinic is compiling data on overuse injuries (e.g., tendonitis, stress fracture) affecting the lower extremities. By obtaining a better understanding of the relationship between sports and joint injury (e.g., foot, ankle), the clinic can provide preventative services such as pre-season clinics and sport-specific training programs. The clinic needs to decide how specialized the clinics should be. This will determine the number and type of different clinics to be offered. For example, they might offer a clinic that focuses on a specific sport, a clinic that focus on a specific joint, or a combination of both. The data are contained in *sports_injuries.jmp*. Analyze the data and determine whether joint injury is dependent on sport. Give your recommendation for what clinics should be offered. Summarize your findings in a memo using appropriate graphs, statistics, and statistical tests.

4. The April 18, 2001 issue of the *Journal of the American Medical Association* contains the article "Prior alcohol consumption and mortality following acute myocardial infarction" (K. J. Mukamal, et al.) which published data from a study of alcohol consumption in patients with acute myocardial infarction (heart attack). Patients were grouped into non-drinkers of alcohol, drinkers of less than seven drinks per week, and drinkers of seven or more drinks per week prior to myocardial infarction. The patients were also classified as to whether they suffered congestive heart failure during the three years after myocardial infarction.

Drinks/week	Congestive Heart Failure	
	No	Yes
None	750	146
< 7	590	106
> 7	292	29

Analyze the data to determine whether there is dependence between congestive heart failure and alcohol consumption. Write a paragraph summarizing the results. Can cause and effect be established from this data?

5. An appliance center delivers and installs refrigerators, freezers, and ranges. Recent experience shows that of 564 refrigerator deliveries, 34 resulted in damage to the appliance. Similarly, of 233 freezer deliveries, 9 resulted in damage, and of 55 range deliveries, 6 resulted in damage. With a damage rate for range deliveries considerably higher compared to the other appliances, the manager is planning a special training session focused on range delivery and installation. The training session will last two hours and will be mandatory for all 35 delivery personnel. Do you think this training is necessary? Justify how you arrived at your conclusion.

6. A company developed a new flavor of barbecue sauce and has conducted taste tests in each of the four regions where the sauce will be sold. The results are contained in the file *bbq_sauce.jmp*.

   a. Set up and conduct a test of hypothesis using $\alpha = 0.05$ to determine whether preference for the new flavor sauce depends on region.

   b. Examine the Cell Chi-Square values. What do these values suggest about differences between regions?

   c. The market researchers also recorded the gender of each of the respondents. Reanalyze the data, taking gender into account.

   d. Discuss how this analysis can be applied to future market research and advertising campaigns for the barbecue sauce.

13.5 Case Study: Financial Management Customer Satisfaction Survey

Business Problem

A small firm provides financial management services to individuals and small businesses. Each year the firm sends a customer satisfaction survey to a randomly selected sample of clients. The survey can be completed in about 10 minutes and asks clients to rate the firm on two important characteristics: quality of service provided and responsiveness. The survey responses are contained in the file *FinancialAdvisorSurvey.jmp*.

Task

Analyze the survey data and prepare a report, not to exceed two pages, that gives a statistical summary of the data and identifies dependencies between the variables. Based on the results, identify actions the firm could take to improve customer satisfaction.

13.6 References

Mukamal, K. J., M. Maclure, J. E. Muller, J. B. Sherwood, and M. A. Mittleman. "Prior alcohol consumption and mortality following acute myocardial infarction." *Journal of the American Medical Association*. 2001 Apr 18; 285(15):1965–70.

Chapter 14

Simple Regression Analysis

14.1 Introduction   370

    14.1.1 General Simple Regression Problem and Use   372

    14.1.2 Data Requirements   372

    14.1.3 Basic Results   373

14.2 Detail Example: Yield in a Chemical Reactor   374

    14.2.1 Situation, Research Question, Data Requirements, Anticipated Results   374

    14.2.2 Outputs Related to Problem Statement   376

14.3 JMP Analysis of the Yield in a Chemical Reactor Example   377

    14.3.1 Simple Regression Results with the Fit Y by X Platform   377

    14.3.2 Fit Y by X Results   378

    14.3.3 Summary of Relevant Output   380

14.4 Interpretation of Basic Regression Outputs   381

    14.4.1 Estimates of Simple Regression Equation   384

    14.4.2 t-Ratios to Test Significance of b_0 and b_1 385

    14.4.3 Root Mean Square Error (RMSE)   386

    14.4.4 Additional Simple Regression Results   387

14.5 How Good Is the Regression Line? 393
    14.5.1 Scatterplot 393
    14.5.2 Significance of Slope Coefficient 393
    14.5.3 RMSE Relative to Std Dev[Mean] 393
    14.5.4 RSquare: Coefficient of Determination R^2 393
    14.5.5 Plots to Verify Simple Regression Assumptions 395
    14.5.6 Residual Plots 396
14.6 Important Considerations 397
    14.6.1 Slope Estimates b_1 Are Sensitive to Outliers 397
    14.6.2 Hazards of Extrapolation Beyond the Range of the Data 399
    14.6.3 Confidence Intervals for b_0 and b_1 401
14.7 Summary 401
14.8 Problems 402
14.9 Case Study: Lost Time Occupational Injuries 405

14.1 Introduction

Simple regression analysis is a very useful tool for analyzing two continuous variables. It serves to explain the variation in the Y-variable in terms of the X-variable with linear or curvilinear relationships. Numerous applications in business lend themselves to analysis by simple regression analysis. Here are some examples:

- In order to evaluate the effectiveness of advertising expenditure, a company needs to determine the relationship between sales (Y) and advertising expenditures (X).

- A builder needs to know the relationship between housing starts (Y) and interest rates (X).

- Retailers need to understand the relationship between expenditures on clothing (Y) and family income (X).

- A business strategist needs to quantify the relationship between cost of a product (Y) and number of units produced (X) for determining future price levels.

- A chemical engineer needs to understand the relationship between yield of a chemical reaction (Y) and the amount of energy used in the reactor (X).

Simple regression fits a straight line between two continuous variables, X and Y. Figure 14.1 illustrates three uses of simple regression:

1. obtaining a reasonable summary of the data

2. quantifying the average change in Y per unit change in X

3. predicting Y_0 at a specific value X_0

Figure 14.1 Three Uses of Simple Linear Regression

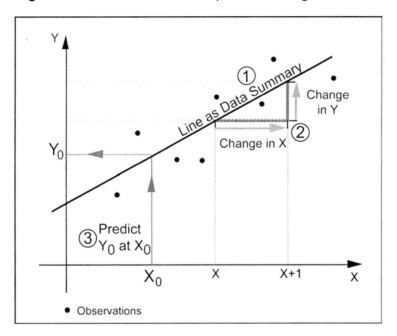

The dots represent eight observations. The line itself summarizes the data either graphically or numerically. The slope of the line is an estimate of the average change in Y as the X-variable changes by one unit. Regression results can also be used to predict average Y-values or individual future Y-values from specified X-values (X_0).

In many applications, straight-line simple regression will provide a good approximation for the relationship between Y and X. Simple regression analysis also allows curvilinear relationships either by a change of scale (e.g., taking the logarithms or inverses) in one or both variables or by adding higher order terms (such as quadratic terms) to the model. The Y-variable is often referred to as the *dependent variable*, and the X-variable is called the *independent variable*. Other names for the Y- and X-variables express different views of the role of each variable, as shown in Table 14.1.

Table 14.1 Names for Y- and X-Variables

Variable	Names
Y	Dependent, Explained, Response, Predicted, Performance, Output
X	Independent, Explanatory, Control, Predictor, Input

14.1.1 General Simple Regression Problem and Use

Simple regression analysis is used to estimate an empirical relationship between two continuous variables X and Y. Simple regression models are used to get more precise estimates of the mean of Y or a predicted value of a future Y outcome at a particular value of X. A linear statistical relationship between Y and X is applicable when the mean of Y tends to vary linearly with X. Fitting a straight line to (X, Y) data is useful when the slope of the fitted line is significantly different from 0. We then say that the linear relationship is statistically significant.

An empirical relationship makes sense mostly when there is an underlying (non-statistical) theory that explains or postulates a relationship between X and Y. The regression relationship does not imply a cause and effect relationship between X and Y.

14.1.2 Data Requirements

Observations need to be in pairs of (X, Y). An observation that lacks either the X or the Y measurement is incomplete. Most software packages will not include incomplete observations in the calculations unless special tools are applied. The (X, Y) observations are points in a two-dimensional scatterplot. The spreadsheet format requires two columns, one each for Y and X. Rows represent individual observations. Simple regression analysis assumes that the X-variable is not subject to random variation. For that reason, it is different from correlation analysis (see the next chapter), where both variables have an equal role and are treated as Y-variables. The difference between one-way ANOVA and simple regression analysis is that the X-variable in ANOVA is nominal or ordinal, while in regression analysis it is continuous.

14.1.3 Basic Results

The following basic results are required in almost any application:

- slope and intercept of the fitted regression equation
- test of the slope to be significantly different from zero using the *p*-value of the associated t-Ratio
- evaluation of the goodness of fit of the data to the simple regression model from the Root Mean Square Error, the RSquare, and data plots
- Assessment of the effect of outlier observations if present

Table 14.2 summarizes the basic outputs from simple regression analysis and gives brief summary interpretations of each.

Table 14.2 Basic Outputs of Simple Linear Regression with Definitions and General Interpretations

Basic Outputs	Definition	Interpretation or Purpose
Fitted model equation	Straight-line $y = b_0 + b_1 X$	Compare this model to the base line model of the overall mean $\overline{Y}$ without X.
Slope b_1	Estimated average change in Y per unit change in X	Expresses the linear empirical relationship between Y and X.
Intercept b_0	Estimated average of Y at X = 0	Interpret the intercept if X = 0 is within or near data range; otherwise use it as fitting constant.
Yhat at $X_0 = \hat{Y}_{X_0}$	$\hat{Y}_{X_0} = b_0 + b_1 X_0$	Estimated average value of Y at X_0. Also referred to as the Predicted Y or simply Yhat.
RMSE	Estimated standard deviation around a fitted model (line)	Standard deviation estimate of Y_i around the fitted line is different from the standard deviation around the overall mean $\overline{Y}$.

(continued)

Table 14.2 (*continued*)

Basic Outputs	Definition	Interpretation or Purpose
RSquare = R^2	Proportion of the Total Sum of Squares explained by the model	Values of R^2 are between 0 and 1. R^2 indicates how well the equation fits the data.
Confidence interval	Interval to contain the mean of Y at X_0 with stated confidence	Indicates the precision with which the mean of Y at X_0 can be estimated using available data.
Prediction interval	Interval to contain a single future outcome of Y at X_0 with stated confidence	Indicates the precision with which a future outcome of Y at X_0 can be predicted using available data. Prediction intervals are wider than confidence intervals.

14.2 Detail Example: Yield in a Chemical Reactor

Before we describe the problem, recall that a statistical model looks for empirical relationships and does not answer questions about cause and effect. Regression analysis establishes a *statistical* rather than a *causal* relationship. It does not require a perfect fit. A scatter of data around the line is expected.

14.2.1 Situation, Research Question, Data Requirements, and Anticipated Results

Situation: A producer of a plastic material needs to establish a relationship between the amount of agitation of the ingredients in the reactor and the yield of the chemical. The plastic is produced in distinct batches. The ingredients are added to a reactor and then mixed and heated. The mixing and heating is called agitation. Both consume costly energy. The general expectation is that yield increases with increased agitation. If the relationship is steep, additional agitation might result in profitable yield. If there is no or only a small relationship, then agitation beyond the minimum is a waste of a costly input.

In this example, the following variables are used to measure agitation and yield:

- Agitation is measured by the variable HP and represents horsepower as a composite variable of the combined effort used in agitation.
- Yield represents the percentage of useful product in the final total output of a batch.

Research Question: Management needs to establish the relationship between agitation and yield. Two questions need to be addressed:

- Is there a statistically significant linear relationship between HP and Yield?
- Does the linear relationship hold over the entire practical range of agitation?

Data Requirements: As mentioned before, simple regression requires both a continuous X- and Y-variable. The data consist of pairs of observations (HP, Yield) = (X_i, Y_i) from eight batches: (10.0, 83.9), (14.1, 88.4), (12.0, 84.2), (15.6, 89.3), (9.4, 80.1), (13.2, 87.7), (10.3, 82.6), (14.5, 90.2). Figure 14.2 shows the data file *HorsePowerYield.jmp* with eight experimental runs. Column 1 is a continuous variable labeled HP for horsepower. Column 2 is a continuous variable labeled Yield given as a percentage.

Figure 14.2 JMP Data Table for Horsepower versus Yield Example

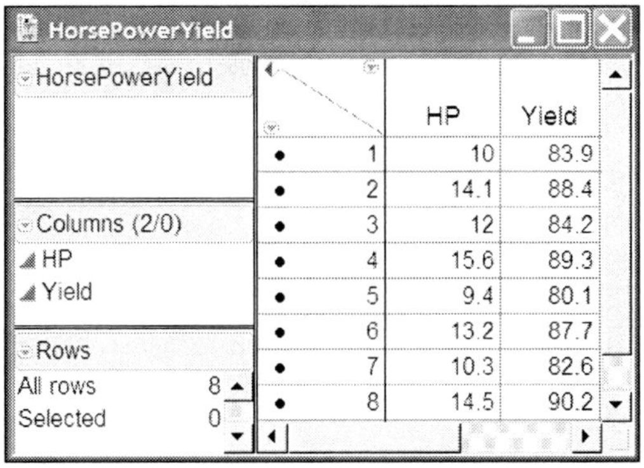

	HP	Yield
1	10	83.9
2	14.1	88.4
3	12	84.2
4	15.6	89.3
5	9.4	80.1
6	13.2	87.7
7	10.3	82.6
8	14.5	90.2

Anticipated Results: In Figure 14.3, the scatterplot of Yield versus Horsepower indicates a linear relationship with a positive (increasing) slope within the data range. The linear relationship is not perfect, but is expected to be statistically significant.

Figure 14.3 Scatterplot of Horsepower versus Yield

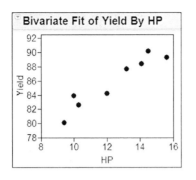

14.2.2 Outputs Related to Problem Statement

In explaining the data, the competing models are between either using the mean of Y without regard to HP or using the linear relationship between Yield and HP. Using only the mean of Yield is the simplest model, and we call it the *base model*. The straight-line fit is the *alternative model* and requires two parameters: an intercept and a slope. Figure 14.4 shows the base model in the left panel and the alternative model in the right panel.

Which model gives a better explanation of the data? The "winning" model should be as simple as possible, but adequate to explain the data for the purposes of the problem statement. The regression model should be significantly better than the base model. Simple regression requires us to test whether the slope is significantly different from 0. The graphs suggest that the straight-line fit is a better explanation of the data, a conclusion that is supported by a significance test for the slope of the line (see the next section).

Figure 14.4 Base and Alternative Models

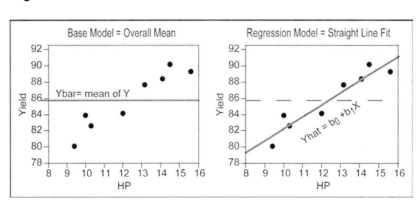

In practice, it is possible that neither model provides a good fit. In other words, neither the overall mean of Y nor the straight line provides adequate explanations of the data. In that case, you might have to resort to other models, for example by adding different X-variables to the current model, or by transforming the current variables in the model.

14.3 JMP Analysis of the Yield in a Chemical Reactor Example

Simple regression can be analyzed with either the specialized Fit Y by X platform or the more general Fit Model platform. Fit Y by X emphasizes graphical output and easily fits simple, curvilinear, and polynomial relationships. Fit Model requires more preparatory work. Both platforms use least squares to estimate all necessary model statistics.

14.3.1 Simple Regression Results with the Fit Y by X Platform

Always make sure that both Y and X are continuous variables in the JMP data table. Proceed as follows:

Step 1: Select **Fit Y by X** from the **Analyze** menu as shown in Figure 14.5.

Step 2: Place the Y- and X-variables into their respective roles. Yield is **Y, Response** and HP is **X, Factor**.

Step 3: Click **OK.** The scatterplot appears as shown previously in Figure 14.3.

Step 4: Select **Fit Line** from the menu by clicking on the red triangle of the scatterplot.

Figure 14.5 Fit Y by X Platform to Fit Straight Line

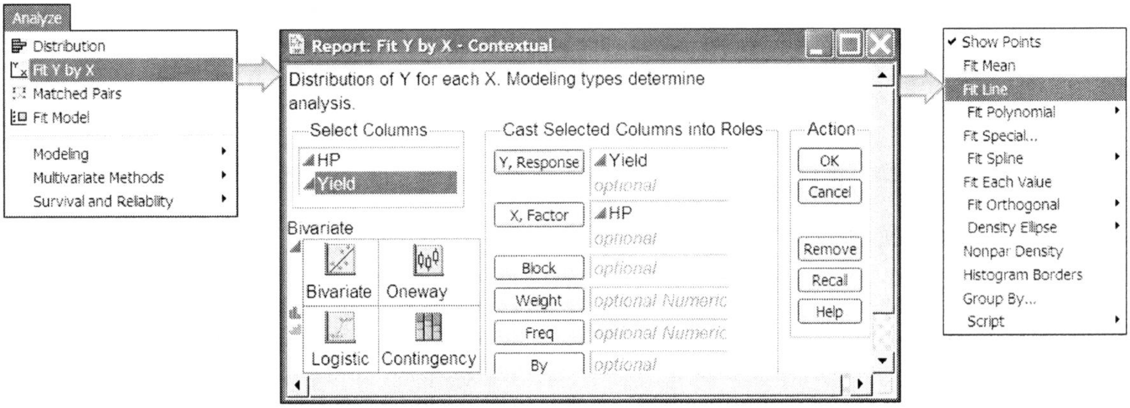

Use the scatterplot to verify that all the data observations appear as expected and are not abnormal or unexpected. Abnormalities could be the result of data input errors or might suggest that the anticipated model is inadequate. In Figure 14.3, the scatterplot suggests a linear or straight-line relationship without outliers. The **Fit Mean** selection will add the base model and draw a horizontal line at Ybar. **Fit Line** adds a straight line to the scatterplot in Figure 14.3.

Fit Y by X offers other models for fitting a single Y and X. Some models are quite difficult mathematically, but are useful for specific applications in business. The following are among the simplest models:

- Polynomial models include quadratic and cubic terms, and can be very useful for fitting many curvilinear relationships (see the next chapter).

- Fit Special lets the user choose logarithmic, exponential, square root, and inverse transformations (see the next chapter).

- Fit Spline allows data smoothing by fitting many connected smaller pieces.

14.3.2 **Fit Y by X Results**

This section focuses on standard output of simple regression analysis. The most useful outputs of the Fit Mean table (Figure 14.6) are the standard deviation of Y around the overall mean (Std Dev [RMSE]), which will be compared to the RMSE of the fitted line.

Figure 14.6 Results for the Fit Mean Base Model

Fit Mean	
Mean	85.8
Std Dev [RMSE]	3.603173
Std Error	1.273914
SSE	90.88

In Figure 14.6, the Mean is the overall mean, which is Ybar = 85.8, calculated as the simple average of all (eight) Y-values Std Dev [RMSE] = 3.603 is the standard deviation of the Y around the overall mean of Y. Std Dev [RMSE] = $\sqrt{90.88/7}$ = 3.603 is the root of the SS_{Total} = 90.88 (see the ANOVA Table in Figure 14.10) divided by its degrees of freedom = (n − 1) = 7. The Std Error is the standard deviation divided by the square root of n = 8.

Figure 14.7 shows the Summary of Fit from the Fit Line output. The overall mean = 85.8 is the same as in Fit Mean. The Root Mean Square Error is RMSE = 1.228, considerably smaller than StdDev[RMSE] = 3.603 of Fit Mean. This drastic reduction in RMSE suggests that the line explains much of the variation in Y. The RSquare or R^2 = 0.90 measures the proportion of the variability explained by the model. RSquare is often used as a measure of fit in regression and related techniques. Since RSquare is always between 0 and 1, a value of 0.90 is relatively high.

Figure 14.7 Results for More Complex Model (Straight-Line Model)

Summary of Fit	
RSquare	0.900365
RSquare Adj	0.883759
Root Mean Square Error	1.22847
Mean of Response	85.8
Observations (or Sum Wgts)	8

Although many lines could be fitted to a scatter of points, the most common method is the least squares method. The fitted least squares equation has an intercept b_0 = 67.503802 and slope b_1 = 1.4769887. Fit Y by X displays this equation as in Figure 14.8.

Figure 14.8 Straight-Line Regression Formula for HP versus Yield Example

Yield = 67.503802 + 1.4769887 HP

The Parameter Estimates table (Figure 14.9) shows the estimated intercept b_0 and slope b_1 (labeled HP) in the Estimate column with related statistics in other columns. The Std Errors of the parameters measure how precise the parameter estimates are. The t-Ratios = Estimate/(Std Error) are used to test the null hypothesis that a parameter (slope, intercept) equals 0, i.e., H_0: $\beta_i = 0$ versus H_A: $\beta_i \neq 0$. The Prob > |t| is the associated two-sided p-value. The p-value of HP is 0.0003, and so we can reject H_0 that the slope is zero. The slope is significantly different from 0. The regression equation is statistically significant! (The p-value of the intercept is <.0001. We reject H_0 that the intercept is zero. More on that later.)

Figure 14.9 Straight-Line Parameter Estimates for HP versus Yield Example

Parameter Estimates				
Term	Estimate	Std Error	t Ratio	Prob>|t|
Intercept	67.503802	2.522419	26.76	<.0001*
HP	1.4769887	0.200585	7.36	0.0003*

Figure 14.10 contains the Analysis of Variance (ANOVA) table. It shows the degrees of freedom associated with the Model and the Error term. The degrees of freedom of Error are those of the RMSE. In simple regression, they are $n - 2 = 8 - 2 = 6$. The single degree of freedom for Model is for fitting a single slope. The ANOVA table also gives the breakdown of the Sum of Squares: $SS_{Total} = SS_{Model} + SS_{Error}$, i.e., $90.88 = 81.825 + 9.055$ from which the RSquare is calculated. The overall F-Ratio = 54.2198 is used to test the overall significance of the model relative to the base model. In simple regression, the F-Ratio and the t-Ratio of the slope are related.

Figure 14.10 Analysis of Variance Table for HP versus Yield Example

Analysis of Variance				
Source	DF	Sum of Squares	Mean Square	F Ratio
Model	1	81.825174	81.8252	54.2198
Error	6	9.054826	1.5091	Prob > F
C. Total	7	90.880000		0.0003*

14.3.3 Summary of Relevant Output

A summary of important output from the Yield versus HP simple regression analysis is shown in Table 14.3.

Table 14.3 Output Summary for HP versus Yield Example with Comments

Output	Value	Comments
Mean Yield	85.8%	Overall mean Yield for all observed values of HP.
Model Equation	Yhat = 67.50 + 1.477·X	Yhat is the estimate of the mean Y for a specific HP value. The slope b_1 = 1.477 is significantly different from 0 with p = 0.0003. The intercept b_0 = 67.50 is the estimated average of Y at X = 0. X = 0 is outside the range of X-values, so b_0 is treated as a fitting constant.
RMSE	1.228	The standard deviation around the fitted line is RMSE = 1.228 and is smaller than the standard deviation (3.603) around average Yield.
RSquare	0.900	R^2 indicates that simple regression explains 90% of the total sum of squares (around average Yield).
Scatterplot	Figures 14.3 and 14.4	Shows the closeness of observations to average Yield and to the fitted line. It helps identify outlier values.

14.4 Interpretation of Basic Regression Outputs

The simple regression model is a straight-line equation with a random error term needed to account for the fact that not all observations fall exactly on the straight line. The model equation is

$$Y_i = \beta_0 + \beta_1 \cdot X_i + \varepsilon_i$$

Figure 14.11 shows the straight-line equation represented by $\mu_{Y|Xi} = \beta_0 + \beta_1 \cdot X_i$ and the distribution of errors ε_i at selected X-values.

Figure 14.11 Simple Regression Model with Error Distributions

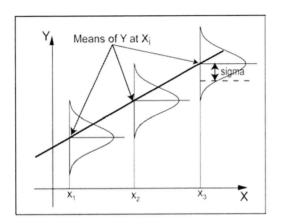

The straight line represents the means of Y for given values of X. The bell-shaped curve around each mean shows the error distribution around the mean. The usual assumption is that the error distribution is Normal with constant standard deviation σ for all X.

In estimating the model parameters, the Greek letters of the model are replaced by Roman letters to in dicate that the least squares estimates are based on data. Statisticians often indicate the estimate of a model parameter by placing a hat (^) over the Greek letter. In regression this is the "hat" in Yhat, except that the ^ is over a Roman letter, because Y is not a parameter. Table 14.4 summarizes the true parameters and their estimates.

Table 14.4 Review of Terms of Simple Regression Model

True parameters	Statistical estimates	Terminology
β_0	b_0	Intercept
β_1	b_1	Slope
ε_i	$e_i = Y_i - Yhat_i$	Residual of i-th observation
σ	RMSE	Standard deviation around fitted line
$\mu_{Y|X0} = \beta_0 + \beta_1 \cdot X_0$	Yhat at X_0 $\hat{Y}_{X0} = b_0 + b_1 \cdot X_0$	Average of Y at X_0 or Predicted Y at X_0

The least squares criterion fits a model to data by choosing parameters (slope, intercept) that minimize the squared vertical deviations of the raw data from the fitted model. This involves a set of linear equations whose solution is best left to statistical software. Figure 14.12 shows the vertical deviations from which the sum of the squared deviations is derived. (This will be the SS_{Error} or SSE in future discussions.)

Figure 14.12 Least Squares Minimizes Vertical Deviations

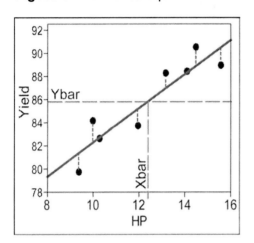

A line fitted by least squares has several characteristics that are of interest when interpreting output derived from this method:

1. The fitted line always goes through the "center of the data," i.e., the averages of X and Y (Xbar,Ybar) = $(\bar{x}, \bar{y})$.

2. The fitting method penalizes large deviations from the fitted line, especially at the extremes of the X-values. This leads at times to counter-intuitive results.

3. Least squares fitting does not require data to be normally distributed. Confidence and prediction intervals are however sensitive to data from non-Normal distributions.

4. The least squares method yields estimated error terms (residuals) that average to zero.

The simple regression model needs the estimates b_0, b_1, and RMSE, the residual standard deviation or the standard deviation around the line. A number of quantities are used in answering a variety of specific questions. These quantities include the following:

- estimate Yhat of the average of Y at a specific X
- coefficient of determination R^2 or RSquare
- confidence intervals for the true average of Y at a specific X
- prediction intervals for a future value of Y at a specific X

14.4.1 **Estimates of Simple Regression Equation**

This section discusses the interpretation of the slope b_1 and the intercept b_0.

Slope Estimate b_1: *The slope coefficient b_1 is the estimated average change in Y per unit change in X.* It is used to test whether the regression equation is a significantly better fit than the base model of the overall mean of Y. Figure 14.13 shows that as the value of the predictor variable increases by one unit from X to X + 1, the estimated average of Y changes from $Yhat_X = b_0 + b_1 \cdot X$ to $Yhat_{X+1} = b_0 + b_1 \cdot (X + 1)$. The difference $Yhat_{X+1} - Yhat_X$ is the estimated slope b_1.

Example—Yield versus Horsepower: In the Yield versus HP example, $b_1 = 1.47699$ estimates that for each additional unit of horsepower applied, the mean yield increases by 1.47699. A positive slope is consistent with process knowledge that more agitation results in a higher yield. The measurement units of the slope b_1 are (Units of Y) divided by (Units of X). The slope coefficient describes the rate of change in the mean of Y as X increases by one unit. In the example, the units are Yield/Horsepower.

Figure 14.13 Slope and Intercept in Simple Regression Model

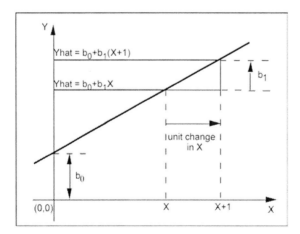

Intercept Estimate b_0: *The intercept coefficient b_0 is the estimated average value of Y when X = 0* and might or might not have a relevant interpretation, depending on whether a value of X = 0 makes sense in the problem context. When a value of X = 0 does not make sense, the intercept of the simple regression equation is of little or no interest. The intercept is included in the equation and is treated as a *fitting constant* to complete the straight line, but has no other function or interpretation.

Example—Yield versus Horsepower: The value $b_0 = 67.5038$ has little meaning here, because it represents the mean of Y when no agitation (HP = 0) is applied to the system. In this application, the yield would not result in a useful outcome without the benefits of some agitation. Also, the range of observed HP values is in the range $9.4 \leq X \leq 15.6$. Thus, b_0 represents an extrapolation far beyond the range of the data and should be left without interpretation.

14.4.2 t-Ratios to Test Significance of b_0 and b_1

For significance tests of the intercept and the slope coefficient, most statistical software packages assume default null hypotheses that both the slope and the intercept are zero. The t-Ratios of b_0 and b_1 test whether the intercept and slope are significantly different from 0. The associated p-value for the slope coefficient b_1 is used to decide whether the simple regression model is significantly better than the base model of a single overall mean of Y. The standard null hypothesis that the slope $\beta_1 = 0$ implies that there is no linear relationship between X and Y. If the slope is significantly different from zero, then assume that a linear equation is better than using the overall mean.

The standard null and alternative hypotheses for both parameters are:

$H_0: \beta_i = 0$ versus $H_A: \beta_i \neq 0$

When i = 0 the hypothesis applies to the intercept, and when i = 1 to the slope. The *p*-values of the respective t-Ratios are used to test these hypotheses. Because the null hypothesis states that the parameter is equal to 0, the t-Ratio of the Parameter Estimates table is

$$t_i = \frac{b_i - 0}{s(b_i)} = \frac{b_i}{s(b_i)}$$

where $s(b_i)$ is the standard error of the parameter estimate. For simple regression, the formula for the standard error of the slope is

$$s(b_1) = \frac{\text{RMSE}}{\sqrt{\sum_{i=1}^{n}(X_i - \overline{X})^2}}$$

The formula shows that $s(b_1)$ depends on the following:

- RMSE (smaller RMSEs result in smaller Std Errors)
- the choice of the X values (spread out X-values result in smaller Std Errors)
- the sample size (larger numbers of observations are associated with smaller Std Errors, because the sum under square root contains more positive terms)

The degrees of freedom (df) of the RMSE (or the SS_{Error}), t-Ratios, and the standard errors $s(b_i)$ are $(n-2)$. The t-Ratio associated with b_1 and the F-Ratio of the ANOVA table test the same hypothesis $H_0: \beta_1 = 0$ versus $H_A: \beta_1 \neq 0$, resulting in the same *p*-values. The null hypothesis for the intercept states that the estimated average value of Y is also 0 when $X = 0$. As with the interpretation of b_0, this null hypothesis is of interest only when $X = 0$ makes sense and is not a dangerous extrapolation.

Example—Yield versus Horsepower: The t-Ratios $t_i = b_i / s(b_i)$ with $n-2$ degrees of freedom are calculated from the ratio of the Estimate and the Std Error column and test whether or not each coefficient is significantly different from zero. In the example, $t_0 = 67.5038/2.522 = 26.8$ and $t_1 = 1.47699/0.2006 = 7.36$. The low *p*-values ($<0.0001$ for b_0 and 0.0003 for b_1) indicate that both slope and intercept are significantly different from 0. The fitted regression line is a significantly better model to explain the data than the simple average of Yield.

14.4.3 Root Mean Square Error (RMSE)

The RMSE is the standard deviation around the fitted model, i.e., the regression line. The RMSE estimates the parameter σ, the standard deviation of the model error term ε_i with $(n-2)$ degrees of freedom. The RMSE is the square root of the Least Squares Sum of Squared Deviations of the Y_i from the estimated mean values Yhat divided by the error degrees of freedom, i.e.,

$$\text{RMSE} = \sqrt{\sum_{\text{all } i}(Y_i - \hat{Y}_i)^2 / (n-2)}$$

Example—Yield versus Horsepower: To evaluate how good the regression equation is, compare this RMSE $=1.228$ from the Summary of Fit table with the standard deviation of the Ys around the overall mean labeled Std Dev [RMSE] $= 3.603$ from the Fit Mean table. The regression RMSE $= 1.228$ is considerably smaller than the Std Dev [RMSE] $=$

3.603. This shows that the regression model (through its slope parameter) has explained much of the variability in Y. The variability around the fitted line is much smaller than the variability around the (zero slope) overall mean.

14.4.4 Additional Simple Regression Results

Additional results from the options menu include predicted values, confidence intervals, and prediction intervals.

Yhat: Estimated Average or Predicted Value of Y at X_0

$\hat{Y}_{X_0} b_0 + b_1 \cdot X_0$ stands for two related quantities with a subtle difference in interpretation: (1) estimated mean of Y at X_0, and (2) predicted value of Y at X_0. The first is the estimated mean of all Y that might be observed at X_0. The second is the predicted value of a single future observation of Y observed when $X = X_0$. This distinction is important in deciding between confidence and prediction intervals.

Obtaining Yhat in JMP

In the Fit Y by X platform, predicted values are accessible from the Linear Fit pull-down menu on the left below the scatterplot. Figure 14.14 shows the menu from Linear Fit. With the selection of **Save Predicteds**, the Yhats (Predicted Yield) will appear as a new column. Predicted Ys can also be calculated for X-values that are not part of the original data. You need to add rows to the JMP table with values only for X, leaving Y blank.

Figure 14.14 Fit Y by X—Save Predicteds to Data Table

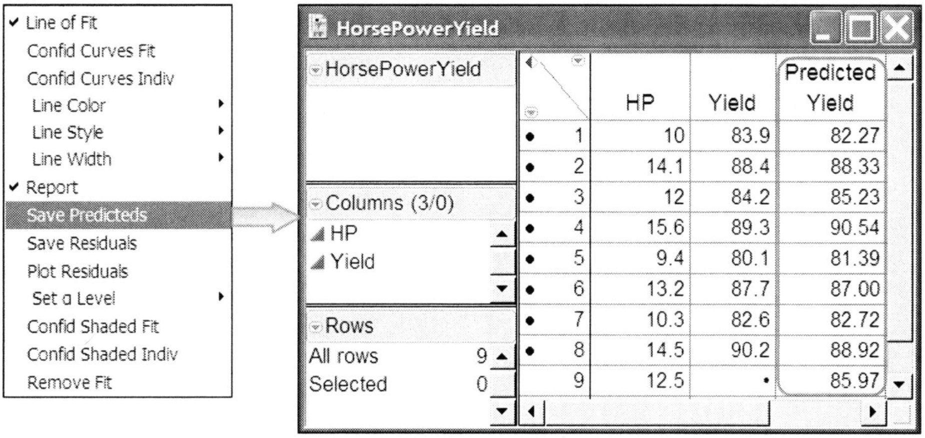

When the regression equation is used for prediction within the range of observed X-values, it is called *interpolation*. Making such predictions is considered safe from a statistical point of view. When the equation is used for predictions outside the observed range of X-values, it is called *extrapolation*. Extrapolation is risky as it assumes that the functional relationship observed within the X-range continues to apply outside the X-range. In practice, extrapolation may be necessary, but should always be done with extreme caution and complete documentation.

Example—Yield versus Horsepower: An estimate of Yield for a horsepower of 12.5 is needed by the process engineer. To obtain this, an additional row (Row 9) with X = 12.5 and no Y value is added to the data table. The prediction formula in the data table calculates Predicted Y = 85.97 at X = 12.5 as an interpolated value.

Confidence Intervals for the Mean of Y at X_0

Ninety-five-percent confidence intervals for the mean of Y at a specific X-value X_0 are useful when the average of Y at a specific X-value is of interest. In JMP, confidence intervals can be obtained in two ways. In Fit Y by X, graphs of the confidence intervals are drawn into the scatterplot window. In Fit Model, the confidence limit values can be saved to the data table.

Obtaining Confidence Intervals in JMP using Fit Y by X

Select **Confid Curves Fit** from the Linear Fit pull-down menu. This will draw the confidence curves on the scatterplot as shown in Figure 14.15. Approximate values for the confidence limits can be obtained using the JMP Crosshairs tool. Regression confidence curves are not straight lines. They are narrowest (in a vertical direction) at Xbar and widen as one moves away from the average point $(\overline{X}, \overline{Y})$. This indicates that the estimates of the average of Y are most precise at the center $(\overline{X}, \overline{Y})$ of the data and get increasingly less precise as you move away from the center, a good reason to be very cautious when estimating outside the range of the data.

Figure 14.15 Straight-Line Fit with 95% Confidence Curves

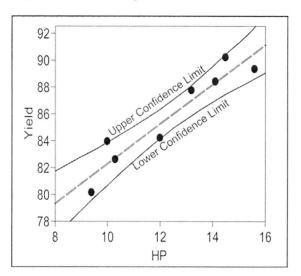

Confidence intervals (CI) for the mean require the t-distribution quantile and the standard error of the mean of Y at X_0. The formula for this standard error shows why there is a curve instead of a straight line for connected CI.

Figure 14.16 Standard Error for Mean Confidence Interval at X_0

$$s(\text{Mean of Y at } X_0) = \underset{1}{\text{RMSE}} \sqrt{\frac{1}{\underset{2}{n}} + \frac{\overset{3}{(X_0 - \overline{X})^2}}{\sum_{i=1}^{n}\left((X_i - \overline{X})^2\right)}}$$

The equation for the standard error of the mean of Y at X_0 has four terms:

1. RMSE: A smaller RMSE yields a narrower confidence interval. The RMSE can be influenced by fitting a better model, but is often a representation of the inherent variability of the process being modeled.

2. Sample size n: A large sample size yields a narrower confidence interval. The sample size n can be increased by taking more observations.

3. Position of X_0 versus Xbar: The closer X_0 is to Xbar, the smaller the numerator in the second term under the square root sign with the result of a narrower confidence interval, and vice versa.

4. A large sum of squares in the denominator of the second term under the square root sign results in narrower confidence intervals. This sum of squares can be increased by taking observations at the extremes of the range of X.

Obtaining Confidence Intervals in JMP Using Fit Model

To save the confidence limits in the JMP data table, use the Fit Model platform by specifying the variables as shown in Figure 14.17.

Figure 14.17 Variable Selection in Fit Model

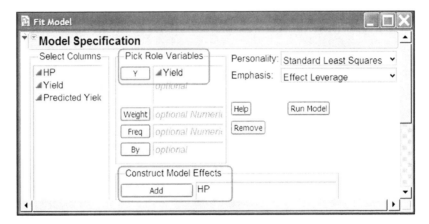

Click the red triangle and select **Save Columns → Mean Confidence Interval** as shown in Figure 14.18. This action saves the limits of the 95% upper and lower confidence interval in two separate columns in the data table. The file shows the original data (HP, Yield) and the Predicted Yield (Yhat). In the columns labeled Lower and Upper 95% Mean Yield, it shows the 95% confidence limits for the mean of Y at each X. For example, when HP = 10, the 95% confidence interval for the mean yield is from 80.69 to 83.86.

Figure 14.18 Save Limits of 95% Confidence Interval to Data Table

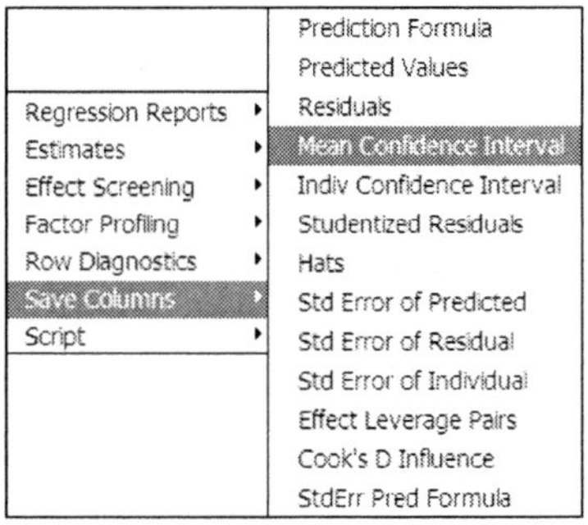

			HP	Yield	Predicted Yield	Lower 95% Mean Yield	Upper 95% Mean Yield	Lower 95% Indiv Yield	Upper 95% Indiv Yield
	Prediction Formula		10	83.9	82.27	80.69	83.86	78.88	85.67
	Predicted Values		14.1	88.4	88.33	86.97	89.68	85.03	91.63
Regression Reports ▸	Residuals		12	84.2	85.23	84.15	86.31	82.03	88.42
Estimates ▸	Mean Confidence Interval		15.6	89.3	90.54	88.64	92.45	86.99	94.10
Effect Screening ▸	Indiv Confidence Interval		9.4	80.1	81.39	79.58	83.20	77.88	84.90
Factor Profiling ▸	Studentized Residuals		13.2	87.7	87.00	85.86	88.14	83.79	90.21
Row Diagnostics ▸	Hats		10.3	82.6	82.72	81.24	84.19	79.37	86.07
Save Columns ▸	Std Error of Predicted		14.5	90.2	88.92	87.44	90.40	85.57	92.27
Script ▸	Std Error of Residual		12.5	•	85.97	84.90	87.03	82.78	89.15
	Std Error of Individual								
	Effect Leverage Pairs								
	Cook's D Influence								
	StdErr Pred Formula								

Figure 14.18 also shows 95% prediction limits on a future observation of Y (labeled Lower and Upper 95% Indiv Yield) that will be discussed next.

Prediction Intervals for a Single Future Y at X_0

Prediction intervals are used to bracket a single future observation of Y at a specified X_0. Prediction intervals are wider than confidence intervals, because they contain two sources of variability:

1. Variability because the true slope and intercept are not known: This variability is also present in confidence intervals and in theory can be reduced by sampling additional observations (X,Y). Its components are the same as in Figure 14.16 and are also marked ①, ②, ③, and ④ in Figure 14.19.

2. Variability of future outcomes of Y at X_0: This source of variability (marked by ⑤ in Figure 14.19) is not present in confidence intervals. The constant value 1 for the uncertainty in observing the future cannot be reduced by additional sampling. This value is typically the largest value of the three terms under the square root sign. Only a better model with a smaller RMSE can improve that part of the prediction.

Figure 14.19 Standard Error of Individual Prediction at X_0

$$s(\text{Individual of Y at } X_0) = \text{RMSE} \underbrace{\sqrt{\underbrace{\frac{1}{\underbrace{n}_{2}}}_{1} + \underbrace{\frac{\overbrace{(X_0 - \overline{X})^2}^{3}}{\underbrace{\sum_{i=1}^{n}\left((X_i - \overline{X})^2\right)}_{4}}}_{} + \underbrace{1}_{5}}}$$

The formula for the standard error for a single future outcome of Y at X_0 is thus only slightly different from that for the confidence interval.

Confidence intervals and prediction intervals can be obtained in JMP using either the Fit Y by X or the Fit Model platform. To obtain prediction intervals in JMP in the Fit Y by X platform, select Confid Curves Indiv from the Linear Fit pull-down context menu. This will draw the confidence curves on the scatterplot as shown in Figure 14.20. The Fit Model platform allows prediction limits to be saved to the data table. As in Figure 14.18, click on the Save Columns pull-down window and select Indiv Confidence Intervals. This will add two columns to the data table, one containing the lower 95% prediction limit, the other the upper 95% prediction limit. Numerical results are stored in the data table and can be printed separately. As noted earlier, prediction intervals are wider than confidence intervals of the same confidence level.

For example, the 95% prediction interval for a single future Yield at HP = 10 is from 78.88 to 85.67. Notice that the prediction interval is considerably wider than the confidence interval at the same X (80.69 to 83.86).

Figure 14.20 Straight-Line Fit with 95% Prediction Intervals

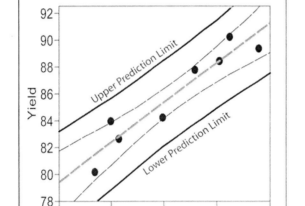

14.5 How Good Is the Regression Line?

So far, we have discussed several methods to evaluate the fitted model. These included the scatterplot, the significance level of the slope, and the RMSE relative to the standard deviation around the overall mean of Y, called Std Dev[Mean] in the Fit Mean option. In this section, we discuss the very popular coefficient of determination, R^2, and show how to use residual plots to verify some assumptions of simple regression analysis.

14.5.1 Scatterplot

A scatterplot provides a visual measure of the goodness of the regression line. The closer the observations fit the line, the better the fit. Scatterplots are also useful for identifying alternative models, such as a curved line, or in identifying possible outliers.

14.5.2 Significance of Slope Coefficient

The *p*-value associated with the t-Ratio measures the significance of the slope under H_0. However, you can have a significant slope with a poorly fitting equation. It is quite possible to have a rather poorly fitting line and a very low *p*-value, because the number of observations is very large or because of the presence of some outliers. A higher *p*-value could be because of too few observations or the fact that the X-values do not provide a sufficient range relative to the variability along the Y-axis.

14.5.3 RMSE Relative to Std Dev[Mean]

The size of the RMSE relative to Std Dev[Mean] (the estimated standard deviation of Y around the sample mean of Y) is a relative measure of how much better the line fits. It does not indicate whether better models exist.

14.5.4 RSquare: Coefficient of Determination R^2

R^2 measures the proportion of the Total Sum of Squares that is explained by fitting the simple regression model. A high R^2 indicates that a large portion of the baseline variation in Y has been accounted for by fitting the X-variable. In the ANOVA table, use the basic identity that $SS_{Total} = SS_{Model} + SS_{Error}$, and observe that a relatively smaller SS_{Error} will result in a relatively larger SS_{Model}, because

$$R^2 = \frac{SS_{Model}}{SS_{Total}} = \frac{SS_{Total} - SS_{Error}}{SS_{Total}} \quad \text{with } 0 \le R^2 \le 1$$

In the example, $R^2 = 81.8252/90.88 = 0.9003$ indicates a very good fit. You could see the good fit from the graph of observations with the fitted line. RSquare is a unitless number and allows comparisons across different models.

$R^2 = 0$ indicates no linear relationship. $R^2 = 1$ indicates a perfect linear fit in which all observations are on a straight line. Figure 14.21 shows examples with positive and negative slope, but no variability around the lines, resulting in each case in $R^2 = +1$.

Figure 14.21 Perfect Positive and Negative Linear Relationships with RSquare = 1

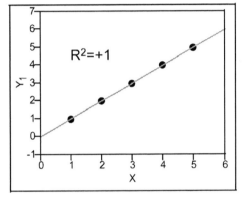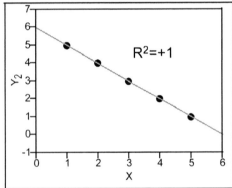

Relying too much on R^2 as a measure of fit is not advised, because it sensitive to the choice of the X-values and to possible outliers. Figure 14.22 shows two plots with the same straight-line fit and ostensibly with the same variability around the line. In the left plot there are no observations for $X = 1$ and $X = 5$. The straight-line fit results in $R^2 = 0.5$. The right plot has the same observations at $X = 2, 3,$ and 4, and additional observations at $X = 1$ and $X = 5$. The additional observations do not alter the fitted line, nor do they change variability around the line. With the additional observations at the extremes of this X-range, the fitted line has $R^2 = 0.75$. This suggests the importance of sampling from a wide range of X-values.

Figure 14.22 RSquare Changes with Observations Added at Extremes of the X-Range

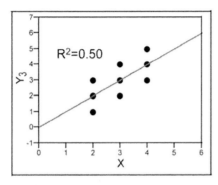
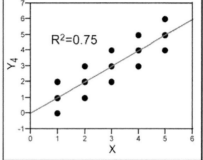

Even one outlying observation can considerably increase the value of R^2, even though virtually no other relationship is discernible from the data. Figure 14.23 shows that six observations in the left plot result in $R^2 = 0$. Adding one more observation at the extreme of the (X, Y) range raises the R^2 to 0.68 and could lead to different conclusions about the fit.

Figure 14.23 Outlier Influence on R^2 and Regression Line

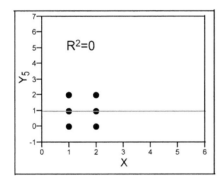
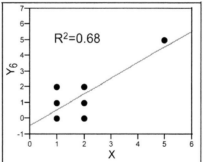

14.5.5 Plots to Verify Simple Regression Assumptions

The standard assumptions of simple linear regression analysis include the following:

1. There is constant variability of residuals around the regression model throughout the range of X-values. The plot of residuals versus predicted Y will check this assumption.

2. Normal distribution of residuals around the regression model is especially important for prediction intervals. The Normal plot of residuals will check this assumption.

Predicted Yield and the residuals are related by the equation that residuals are the difference between observed values and corresponding fitted values, i.e., Residual = (Observed – Predicted Yield). Figure 14.24 shows observed Yield, the Predicted Yield (Yhat), and the Residual Yield.

Examining residuals in plots might reveal several important aspects of the fit. Look for the existence of large residuals or outliers. Also, check whether the residuals are within the expected magnitude defined by the problem, or whether residual patterns support the assumptions of randomness, equal variance, and normality. Some of those aspects are explored in the following sections.

Figure 14.24 Residuals of Horsepower versus Yield Example

HP	Yield	Predicted Yield	Residual Yield
10	83.9	82.27	1.63
14.1	88.4	88.33	0.07
12	84.2	85.23	-1.03
15.6	89.3	90.54	-1.24
9.4	80.1	81.39	-1.29
13.2	87.7	87.00	0.70
10.3	82.6	82.72	-0.12
14.5	90.2	88.92	1.28

The following characteristics are shared by all least squares regression fits:

- In least squares regression (with intercept), the sum of the residuals is always zero.
- The sum of the predicted values equals the sum of the observed Y values.

The magnitudes of residuals in Figure 14.24 are all small. Residual values range from – 1.67 to 1.28. For this application, this is considered a fairly good fit. In other applications, a closer fit might be required.

14.5.6 Residual Plots

Residual Plot versus Predicted: A plot of residuals versus Predicted Ys can be used to find possible relationships such as whether residuals increase with increasing predicted values. The left panel of Figure 14.25 shows a residual by predicted plot for the chemical reactor example. The data points appear random, suggesting that the model assumptions are adequately met.

Normal Quantile Plot of Residuals: Determine whether the residuals follow a Normal distribution with the Normal Quantile plot from the Distribution platform. Look for a straight-line relationship between residuals and Normal scores. The residuals of the Normal Quantile plot of the right panel in Figure 14.25 all fall on or near a straight line, suggesting normality of the residuals.

Figure 14.25 Residuals Plots versus Predicted Y and Normal Quantile

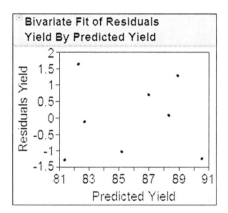 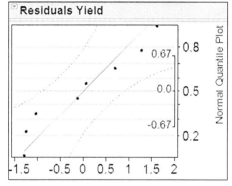

14.6 Important Considerations

14.6.1 Slope Estimates b_1 Are Sensitive to Outliers

When data include outliers (observations that do not fit the pattern or model as the remaining observations do), the least squares method might produce counterintuitive results. Examine a simple example of five observations (Figure 14.26).

Figure 14.26 Data for Outlier Effect Example

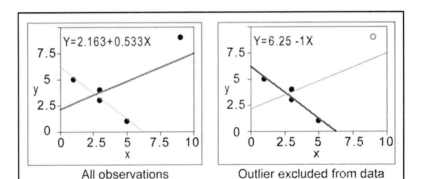

Figure 14.27 shows that depending on whether observation (9, 9) is included in the fit, the slope estimate can be positive or negative. Slopes that change signs depending on the existence of outliers are very problematic.

Figure 14.27 Effect of Outliers On Slope Estimate

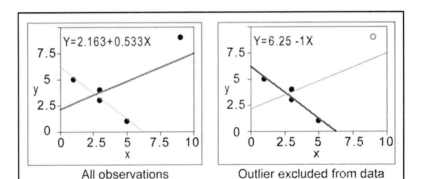

Outliers can be legitimate data points. When they represent values that are precisely and accurately measured following the data collection protocol, they should be included in the data. You should try to find a better model that reconciles the outliers as part of the overall pattern. When outliers are bad values, because of errors in the data collection, or circumstances that are not part of the model assumptions, it is permissible to eliminate outlier observations and reanalyze the data. In reports, such problematic data should be identified and reasons for their elimination clearly stated.

In JMP, the analysis with and without potential outliers is simple. Highlight the observation in a graph or in the data table. Different markers are used for regular observations and for the outlying observations. Next go to the **Rows** menu and select the toggle command **Exclude/Unexclude**. The previously selected observation row is now marked with a slash through a red circle, as shown in Figure 14.28. In this example, using the **Markers** option, the outlier also was marked with an empty circle as opposed to the filled circles of the other four observations.

Figure 14.28 Excluding a Row in JMP

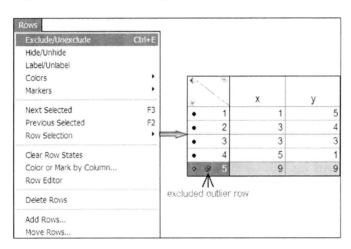

14.6.2 Hazards of Extrapolation Beyond the Range of the Data

A great advantage of fitting a model with continuous Y and X is that you can estimate mean values of Y for unobserved X. For example, in the chemical reactor yield problem you could predict Yield for a value of HP = 12.5. A horsepower of 12.5 is within the observed X range of 9.4 to 15.6. Estimating average yield for this value is called interpolation, because we use the observations from other X-values to estimate Y at X = 12.5. The analysis has shown that the linear model seems a very good fit. Interpolation is usually safe when the fitted model is adequate.

What about HP = 20? This is an extrapolation beyond the observed data range. Using the estimated straight-line equation, the average yield is estimated at 97%, a large X-value compared to those observed. Figure 14.29 shows that the model might continue as a straight line. However, Yield cannot take on values larger than 100%, so the straight-line model is inappropriate for the possible range of X-values. More likely, the model will exhibit a marginally diminishing relationship. *Extrapolation beyond the range of the data is hazardous!* Try to avoid it as much as possible.

Figure 14.29 The Hazard of Extrapolation Is the Uncertainty of the Model Form

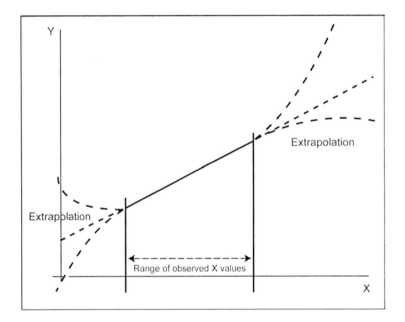

Some applications require extrapolation. For example, forecasting requires prediction of a future outcome or average. The assumption is that the model continues to hold in the future.

14.6.3 Confidence Intervals for b_0 and b_1

By right-clicking on any column of the Parameter Estimates table, 95% confidence intervals for slope b_1 and intercept b_0 estimates can be added as follows:

1. Choose the Columns option.

2. Select both the Lower 95% and Upper 95% for confidence intervals as shown on the left in Figure 14.30.

Figure 14.30 Parameter Estimates Table with Confidence Intervals

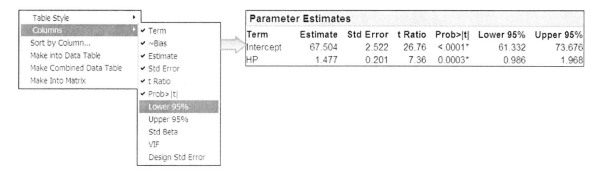

Example Yield versus HP: Figure 14.30 shows that the 95% CI for the slope b_1 is from 0.986 to 1.968. The confidence interval does not include 0, an indication that the slope is significantly different from 0 at two-sided $\alpha = 0.05$.

14.7 Summary

- Simple linear regression analysis is used with two continuous variables (X, Y). X is the independent variable, Y is the dependent variable.

- The model, estimated from the data, is a straight line with slope b_1 and intercept b_0.

- The estimated regression equation is used to give a summary of the data, to estimate and test the change in the Y-variable as a function of the X-variable, and to estimate means and predict Y values.

- The regression equation does not fit the data perfectly, except in rare cases. Therefore, the adequacy of the fit of the equation needs to be assessed.

- Inadequate model fit might be because of outlier observations, because the straight-line model does not adequately represent the relationship between Y and X, or because the variability in Y is too large for the observed range of X-values.

- The estimated slope coefficient b_1 is the estimated average change in Y per unit change in X.

- The estimated intercept b_0 is the estimated average of Y when X = 0. An interpretation of the intercept is practical only when a value of X = 0 makes sense and when X = 0 is within or near the range of the data.

- The simple regression relationship is termed statistically significant if the null hypothesis $H_0: \beta_1 = 0$ can be rejected based on the p-value associated with b_1.

- Regression can be used for estimating the average of Y at a specific value of X, or for predicting a future value of Y at X, leading to confidence and prediction intervals respectively. For the same confidence level and X, prediction intervals are much wider than confidence intervals.

- Prediction intervals are very sensitive to deviations from the Normal distribution assumption.

- Adequate fit can be assessed using RMSE, R^2, scatterplots, and residual plots.

- A good measure of the effectiveness of the regression line is to compare the RMSE (standard deviation around the regression line) with the standard deviation around the overall mean of Y. A relatively small RMSE indicates a good fit.

- RSquare (R^2) is a widely used measure of fit. $R^2 = 0$ indicates no fit, whereas $R^2 = 1$ indicates a perfect linear fit.

- In regression analysis, the presence of outliers can distort the results.

- Extrapolation beyond the range of the data is dangerous, because you cannot ascertain whether or not the model equation will be adequate outside the range of the data.

14.8 Problems

1. Describe a process in operations or production where a simple linear regression would be an applicable model.

   a. Identify the independent (X) and dependent (Y) variables.

   b. Sketch the relationship between X and Y.

c. Give an interpretation of the slope for this situation.

d. How could this regression model be used in this situation?

2. Maplecreek Estates is offering custom-built homes for sale in a suburban location. To date, homes have been built on 10 of the 25 available lots. The file *MaplecreekEstates.jmp* contains the lot number, house size (measured in square feet), and house price. Use this data to construct a simple linear regression to predict house price based on house size.

   a. Use JMP to perform the analysis and show the Summary of Fit and Parameter Estimates tables. Plot the data with the least squares regression line and show the 95% confidence and prediction intervals on the plot. Use the JMP annotate tool to identify the confidence and prediction intervals.

   b. State the regression equation. Interpret the meaning of the slope and intercept. What is the importance of the p-value of the associated t-Ratio?

   c. Justify whether the house price versus house size is a good or poor fit in as many ways as you can.

   d. Give an interpretation of R^2.

   e. Estimate the average house price for a 3000 sq. ft. home and a 3800 sq. ft. home. Which is the more precise estimate? Why? Discuss reservations you might have in using the regression model to make these predictions.

   f. Construct residual plots to verify the regression assumptions. How well are the assumptions satisfied?

3. A retail grocer is considering running a one-day special on five-pound bags of white sugar. Prior to setting the sale prices, the grocer analyzes past data on sugar sales. The data are contained in the file *sugar_sales.jmp*. Price is given in cents and sales in number of units sold per day.

   a. Estimate the regression equation.

   b. Is this linear regression an adequate model for the data? What objective evidence can you cite to justify your answer?

   c. The grocer is considering sale prices of $1.35 or $1.49. What sales could he expect at each of these prices? Discuss any reservations to these predictions.

   d. Find and interpret a 95% confidence interval for mean sales for a price of $1.35.

   e. Find and interpret a 95% prediction interval on sales for a price of $1.35.

   f. Which interval would be of most use to the grocer in this situation? Explain your choice.

4. The data file *electronic_component.jmp* contains cost data for various batch sizes of a manufactured electronic component. Use the data to develop a regression model that relates cost and batch size.

a. State the regression equation.

b. Assess the quality of the model.

c. Give a statistical interpretation of the intercept estimate.

d. Give a business interpretation of the intercept estimate.

e. Give a statistical interpretation of the slope estimate.

f. Give a business interpretation of the slope estimate.

5. The data file *fasteners.jmp* contains cost data for the production of batches of fasteners used in the assembly of cellular telephone relay towers.

a. Using JMP, perform a simple regression analysis where cost is the dependent variable (Y) and batch size is the independent variable (X). Give the regression equation. Give an interpretation of the slope estimate. Is the slope estimate significant?

b. Assess the goodness-of-fit of this model. Discuss any patterns apparent in the residual plot.

c. Discuss possible business reasons for the inadequacy of the model.

6. An official at the Winston Downs Racetrack would like to develop a model to forecast the amount of money bet (in millions of dollars) based on attendance (in thousands). A random sample of 15 days is selected, with the results given in the file *betting_attendance.jmp*.

a. Develop a forecasting model using simple regression that could be applied in this situation. Give the regression equation.

b. Assess the goodness-of-fit of this model.

c. Note on the residual plot and the X-Y data plot the point that appears to be an outlier. Examine the influence of this point on the regression estimates by excluding that point in the JMP table and refitting the model.

d. Compare the differences between the two regression models. Compare the predicted amount of money bet with an attendance of 35,000 from the two models.

e. Discuss what actions should be taken to determine whether this data point should be retained for the purposes of model fitting and prediction.

7. Data collected on cross-selling from a bank were used to estimate a simple linear regression model relating the number of households at each branch who use internet banking to the average number of products sold at that branch. The R^2 was found to be 0.16 and the slope estimate was significant (p-value < 0.0001). There were no apparent patterns in the residual plot. The data are contained in the file *cross_selling.jmp*.

a. Verify that these measures of model adequacy are correct.

b. Give possible reasons for the seemingly low R^2. Suggest possible actions to achieve a better model.

14.9 Case Study: Lost Time Occupational Injuries

The U.S. Bureau of Labor Statistics (BLS) records, among many other things, information on occupational illnesses and injuries by a number of factors including type of injury, age of worker, experience level, type of industry, etc.

Business Problem

A fast food outlet that would be classified by the BLS as a "service provider" wants to assess its recent experience with musculoskeletal injuries among workers aged 30 years or less. The most recent data available from the BLS shows that nationwide, in private industry, workers in this age group have an average of five lost workdays due to musculoskeletal injuries. The file *OccupationalInjuries.jmp* contains the fast food outlet's injury data for the calendar year.

Strategy

Construct a model using simple regression analysis that can be used to quantify the relationship between worker age and lost workdays. Additionally, use the model to determine whether the fast food outlet's number of lost workdays is significantly different from the industry average.

Tasks

Perform the following tasks:

1. Use JMP to build a simple linear regression model relating lost workdays to worker age. Show the Summary of Fit and Parameter Estimates tables. Plot the data with the least squares regression line and show the 95% confidence and prediction intervals on the plot. Label the confidence and prediction intervals.

2. Justify whether the lost workdays versus age is a good or poor fit in as many ways as you can.

3. Give the regression equation. What is the meaning of the slope coefficient? What is the importance of the associated t-Ratio?

4. What is the estimate of the mean number of lost workdays for 22-year-old workers? Give a 95% confidence interval.

5. The fast food restaurant has undertaken a recruiting effort targeted at retirees who would like part-time employment. Would this model be useful for predicting injuries for this group? Why or why not?

6. Assess the regression assumptions using residual plots.

7. Write a brief memo that summarizes your findings from this analysis.

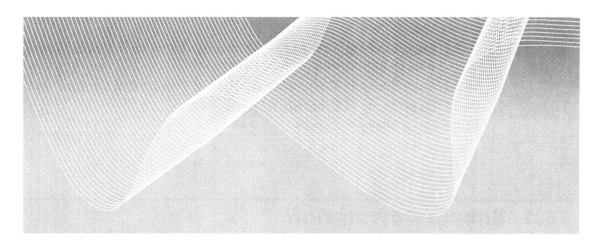

Chapter 15

Simple Regression Extensions

15.1 Simple Correlation 408
    15.1.1 Introduction 408
    15.1.2 Example: Correlation of Financial Indices 409
    15.1.3 Data Patterns and Correlation Coefficients 415

15.2 Regression and Stock Market Returns 417
    15.2.1 Introduction 417
    15.2.2 Capital Asset Pricing Model (CAPM) 421

15.3 Curvilinear Regression 426
    15.3.1 Introduction 426
    15.3.2 Quadratic Regression with Fit Polynomial 426
    15.3.3 Fitting Curves with Fit Special 430
    15.3.4 Detail Example: Price Elasticity of Demand 433

15.4 Summary 438

15.5 Problems 439
    15.5.1 Correlation 439
    15.5.2 Stock Market Returns 441
    15.5.3 Curvilinear Regression 442

15.6 Case Studies 444

    **15.6.1 Case Study 1: Comparing Correlation between Individual Stocks
and an Index 444**

    15.6.2 Case Study 2: Comparing the Risk of Two Market Sectors 444

    15.6.3 Case Study 3: Experience Curve 445

    15.6.4 Case Study 4: Engel's Law 446

15.1 Simple Correlation

15.1.1 Introduction

Simple linear regression and simple correlation analysis are related, at least numerically. Simple regression analysis examines a linear relationship between two continuous variables Y and X. X is often thought of as influencing the outcome of Y in some empirical or functional way. For example, the yield in a chemical reactor problem in the previous chapter assumes that agitation (X) increases the yield (Y) of the chemical reaction. The reasons behind the choice of the role of some variables as Y and others as X are non-statistical, typically requiring knowledge in the field of application.

What if that knowledge is not yet advanced to a stage where the cause and effect variables can be established? What if the investigation is in the early stages where the objective is merely to gain some insight into the relationships between variables? What if the study goal is to explore which variables show empirical relationships? In these situations correlation analysis is a useful tool.

Simple correlation analysis finds empirical relationship between pairs of continuous random variables Y_1 and Y_2 with both variables having an equal role. There is no need to assume that Y_1 influences Y_2 or vice versa. The goal is to determine whether or not Y_1 and Y_2 tend to increase or decrease together, move in opposite directions, or are completely unrelated.

Sample Correlation Coefficient and Sample Covariance

In many applications, you might be interested in finding associations between two variables (Y_1, Y_2). When both Y_1 and Y_2 are continuous, the Pearson correlation coefficient is a good way to measure the strength of that linear association.

For estimating the correlation coefficient, the sample covariance and the sample standard deviation of each variable are needed. The formula for the sample covariance is

$$\text{Cov}(Y_1, Y_2) = \frac{1}{n-1} \sum_{\text{all i}} (Y_{1,i} - \overline{Y}_1)(Y_{2,i} - \overline{Y}_2)$$

where the $Y_{1,i}$ and $Y_{2,i}$ represent the i-th pair of observations, and $\overline{Y}_1$ and $\overline{Y}_2$ are the sample means of each variable. Because the sample covariance is based on data, different estimates might result from different samples taken from the same population.

The sample correlation coefficient r_{12} can be viewed as a standardized covariance. It is simply the sample covariance $\text{Cov}(Y_1, Y_2)$, divided by the two sample standard deviations $\text{SD}(Y_1)$ and $\text{SD}(Y_2)$:

$$r_{12} = \frac{\text{Cov}(Y_1, Y_2)}{\text{SD}(Y_1) \cdot \text{SD}(Y_2)}$$

where $-1 \le r_{12} \le 1$.

The next section shows how to obtain sample correlations and covariances in JMP. The interpretation of sample correlation based on sample data is similar to that of the correlation coefficient of a joint probability distribution, except that one is dealing with the uncertainty of the estimate. Figure 15.1 shows that a positive correlation coefficient indicates that Y_1 and Y_2 tend to increase and decrease together, whereas a negative correlation coefficient means that as Y_1 increases, Y_2 tends to decrease and vice versa.

Figure 15.1 Patterns of Positive and Negative Correlations

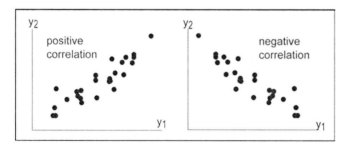

15.1.2 Example: Correlation of Financial Indices

A research study needs to explore the relationship between three widely used financial indices. The three indices are the Dow Jones Index of 30 Industrial Stocks (Dow Jones), the Standard and Poor's Index of 500 stocks (SP500), and the 10-year Treasury Note (10 year T Note).

Figure 15.2 Monthly Data of Financial Indices in *DowJSP500TNote.jmp*

	Date	Dow Jones	SP500	10 year T Note
1	01Jun2006	11260.28	1285.71	5.11
2	01May2006	11168.31	1270.09	5.11
3	03Apr2006	11367.14	1310.61	5.07
4	01Mar2006	11109.32	1294.87	4.85
5	01Feb2006	10993.41	1280.66	4.55
6	03Jan2006	10864.86	1280.08	4.53
7	01Dec2005	10717.5	1248.29	4.39
8	01Nov2005	10805.87	1249.48	4.5
9	03Oct2005	10440.07	1207.01	4.56
10	01Sep2005	10568.7	1228.81	4.33
11	01Aug2005	10481.6	1220.33	4.02
12	01Jul2005	10640.91	1234.18	4.29
13	01Jun2005	10274.97	1191.33	3.94
14	02May2005	10467.48	1191.5	4.01
15	01Apr2005	10192.51	1156.85	4.2
16	01Mar2005	10503.76	1180.59	4.5
17	01Feb2005	10766.23	1203.6	4.36
18	03Jan2005	10489.94	1181.27	4.13
19	01Dec2004	10783.01	1211.92	4.22
20	01Nov2004	10428.02	1173.82	4.36
21	01Oct2004	10027.47	1130.2	4.03
22	01Sep2004	10080.27	1114.58	4.12
23	02Aug2004	10173.92	1104.24	4.13
24	01Jul2004	10139.71	1101.72	4.47

Figure 15.2 contains 24 monthly adjusted closing prices from June 2004 to June 2006 obtained from finance.yahoo.com. The Dow Jones is a weighted subset of the SP500. The data in *DowJSP500TNote.jmp* start with the most recent observation in Row 1.

Correlation Coefficients in JMP

The easiest way to calculate correlations in JMP is to choose the **Multivariate** platform from the **Multivariate Methods** menu, as shown on the left in Figure 15.3.

Figure 15.3 Multivariate Platform Variable Selection for Stock Index
Correlations

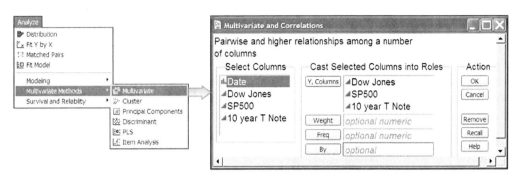

All variables are entered in the **Y, Columns** field. The right of Figure 15.3 shows the
variable specification window for the financial indices example. The nominal variable
Date is not included because the Pearson correlation coefficient is appropriate only for
continuous variables.

The Multivariate platform provides the correlation matrix and the scatterplot matrix as
default output. Figure 15.4 shows the correlations between the pairs of variables Dow
Jones and SP500, Dow Jones and 10 year T Note, as well as SP500 and 10 year T Note.
The table is a symmetric matrix. Along the diagonal of the matrix, the correlations are 1,
indicating that the correlation between a variable and itself is perfect. The off-diagonal
elements are symmetric. For example, the correlation between Dow Jones and SP500 is
the same as the correlation between SP500 and Dow Jones. In both cases, the correlation
coefficient of 0.9287 is an expectedly high correlation coefficient. The correlation
between SP500 and 10 year T Note is the lowest at 0.6782. Again this value appears
twice in the correlation matrix.

Figure 15.4 Correlation Matrix of Financial Indices Example

Correlations of 24 Monthly Stock Indices			
	Dow Jones	SP500	10 year T Note
Dow Jones	1.0000	0.9287	0.8181
SP500	0.9287	1.0000	0.6782
10 year T Note	0.8181	0.6782	1.0000

The second default output is the scatterplot matrix (Figure 15.5). It shows the strength of the correlation graphically. Each variable is assigned a row and a column. For this example, this yields a 3 by 3 matrix scatterplot, since there are three variables being correlated. The diagonal cells are used for labeling, as they would always show perfect relationships between a variable and itself. The red lines are called confidence ellipses.

The scatterplot matrix of Figure 15.5 contains six plots. However, pairs of plots marked A1 and A2, B1 and B2, as well as C1 and C2, plot the same variables but with horizontal and vertical axes reversed. Each plot shows the data points and a 95% confidence ellipse. The ellipses suggest the strength of the relationship. Narrow ellipses indicate stronger relationships than wider ellipses. Ellipses that look like circles indicate that the sample correlation coefficient is near zero. The narrowest ellipse appears in the two plots associated with Dow Jones and SP500. These are the scatterplots with reversed axes in the first row and second column (labeled A1) as well as second row and first column (A2). They are associated with a single correlation coefficient of 0.9287. All column 1 plots (A1, B1) have Dow Jones as the horizontal axis. Column 2 plots (A2, C1) have SP500 on the horizontal axis, and column 3 plots (B2, C2) have 10 year T Note as the horizontal axis. Likewise, row 1 plots (A2, B2) have Dow Jones as the vertical axis; row 2 plots (A1, C2) have SP500, and row 3 plots (B1, C1) have 10 year T Note as the vertical axis.

By clicking the red options triangle, it is possible to add features to the scatterplot. In Figure 15.5, correlation coefficients were added. It might also be useful to add histograms of the variables on the diagonal of the matrix with the **Show Histogram** option, or to change the size of the confidence ellipses.

Figure 15.5 Scatterplot Matrix from Multivariate Platform

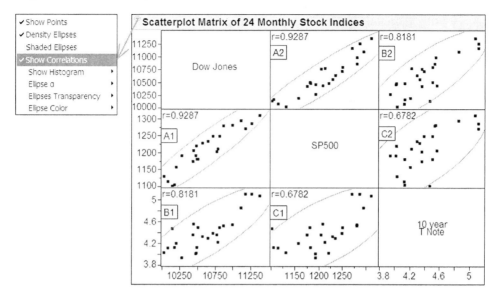

Covariances in JMP

A covariance between two variables also measures how much they vary together. Covariances are used in estimating the variance of a portfolio of stocks. Sample covariances will be used in estimating the beta of a stock in the Capital Asset Pricing Model (CAPM) (Section 15.2). They are obtained from the context menu of the Multivariate platform shown on the left in Figure 15.6. The covariance matrix, like the correlation matrix, is symmetric. It has the sample variance of each variable along the diagonal, because the covariance between a variable and itself is its variance. The first numerical entry in the matrix represents the covariance between Dow Jones and Dow Jones and is the estimated variance Var(Dow Jones) = 144256.64. Likewise, the other diagonal entries represent the estimated variances of SP500 and 10 year T Note respectively. The diagonal correlation coefficients are all 1.0000 (Figure 15.4).

The sample covariances are the off-diagonal entries. Unlike correlation coefficients, covariances take on vastly different magnitudes and tend to be less useful for interpretation of results, especially when the two variables are measured on different scales. How could you compare the Cov[Dow Jones, SP500] = 21516.69 with the Cov[SP500, 10 year T Note] = 14.20? The scale differences in the covariances make it difficult to recognize the strength of the linear relationships between these variables. Correlation coefficients give a clearer picture of linear relationships because the measurement scales have been standardized to the range from −1 to +1.

Figure 15.6 Covariances of Financial Indices Example

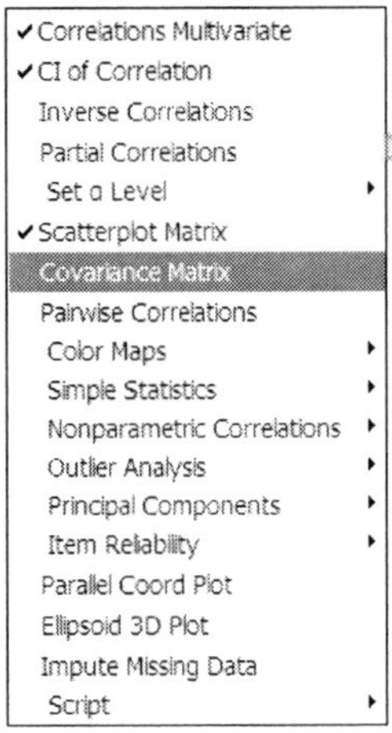

Covariance Matrix

	Dow Jones	SP500	10 year T Note
Dow Jones	144256.64224	21516.69106	106.65624
SP500	21516.69106	3721.20983	14.20219
10 year T Note	106.65624	14.20219	0.11783

Testing Significance of a Correlation Coefficient

Assume that the true correlation coefficient between Y_1 and Y_2 is the unknown ρ_{12}. However, data are available to estimate ρ_{12} with the sample correlation coefficient r_{12}. It is of interest to determine whether the data support a correlation coefficient different from zero, indicating some relationship. We use H_0: $\rho_{12} = 0$ versus H_A: $\rho_{12} \neq 0$.

The null hypothesis claims that the true ρ_{12} is zero, whereas the two-sided alternative hypothesis claims that ρ_{12} is different from zero. Select **Pairwise Correlations** from the menu on the left in Figure 15.6 to obtain correlations for all pairs shown in Figure 15.7. The Correlation column gives estimated correlation coefficients. The Count column indicates the number of pairs of observations used in estimating the correlation coefficient. Lower 95% and Upper 95% are the confidence limits for the correlation coefficients. None include zero. The column labeled Signif Prob contains individual *p*-values with the two-sided alternative hypothesis. With the three *p*-values in Figure 15.7 near zero, we would reject the null hypothesis that the correlation coefficient is $\rho = 0$ (with $\alpha = 0.05$).

Figure 15.7 Pairwise Correlations Option to Test Correlation Coefficients

Pairwise Correlations

Variable	by Variable	Correlation	Count	Lower 95%	Upper 95%	Signif Prob
SP500	Dow Jones	0.9287	24	0.8399	0.9690	<.0001*
10 year T Note	Dow Jones	0.8181	24	0.6189	0.9184	<.0001*
10 year T Note	SP500	0.6782	24	0.3784	0.8493	0.0003*

15.1.3 **Data Patterns and Correlation Coefficients**

The Pearson correlation coefficient measures the strength of a linear relationship between two variables. Figure 15.8 shows increasingly correlated patterns of 26 random observations. Each pair is represented by a point in these two-dimensional scatterplots. Low correlation coefficients show a broad scatter of points, whereas with high correlation coefficients the data points are more concentrated around an imagined line.

Figure 15.8 Random Patterns of 26 Observations with Various Correlation Coefficients

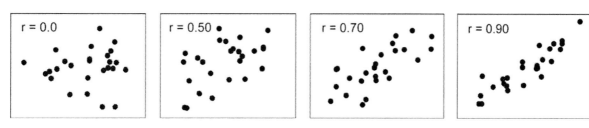

Correlation coefficients can be affected by data outliers. Figure 15.9 shows three graphs. The left graph shows baseline data common to all three graphs. The middle and right graph add a single outlier. Depending on the location of the outlier relative to the original data, the outlier increases the correlation coefficient as in the middle graph and lowers it as in the right graph. The baseline data have an estimated correlation coefficient r = 0.5. In the middle graph, a single outlier observation in the lower left raises the correlation coefficient to 0.70. This observation strengthens the relationship that exists in the left plot. In the right graph, the outlier added below the base line pattern weakens the linear relationship by increasing vertical variability resulting in r = 0.30.

Figure 15.9 Outlier Influence on Correlation Coefficients

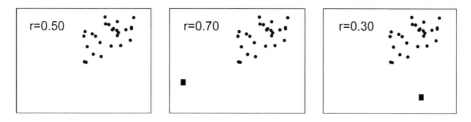

Perfect relationships do not always result in perfect correlation coefficients. The correlation coefficient measures the strength of a *linear* relationship. A perfect linear relationship exists when all data fall on a straight line. In this case, the correlation coefficient is either +1 or −1. The left graph of Figure 15.10 shows a perfect linear relationship for 10 data points. The right graph of Figure 15.10 also represents a perfect curvilinear relationship. However, the correlation coefficient is not a perfect +1, but r = 0.9573, because this is not a straight-line relationship.

Figure 15.10 Imperfect Correlation for Perfect Quadratic Relation

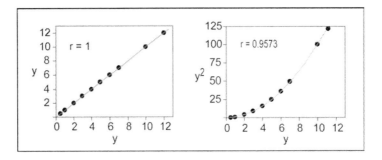

Relation of Correlation Coefficient r with Simple Regression R^2 and Slope b_1

The correlation coefficient r_{xy}, the coefficient of determination R^2, and the slope coefficient b_1 are closely related. The formulas in Table 15.1 define these relationships. Note that these formulas apply only to simple regression.

Table 15.1 Correlation Coefficient, Simple Regression Slope and R^2

$$b_1 = r_{xy} \cdot \frac{s_y}{s_x} = \frac{Cov(X, Y)}{s_x^2}$$	The slope equals the correlation coefficient multiplied by the ratio of the standard deviation of Y divided by the standard deviation of X. The slope is also equal to the Cov(X, Y) divided by the variance of X.
$$R^2 = r_{xy}^2 \quad \text{or} \quad r_{xy} = \text{sign}(b_1)\sqrt{R^2}$$	R^2 is the square of the correlation coefficient. Alternately, the correlation coefficient is the square root of R^2 multiplied by the sign (either + or −) of the slope coefficient.

15.2 Regression and Stock Market Returns

15.2.1 Introduction

A common task in financial analysis is to compare returns of stocks and their volatility. What is the return of a stock? A total return on a stock during an interval of time (t, t + 1) consists of dividend income plus capital gain or loss. In comparing several stocks, rather than comparing returns in dollar amounts, it is more convenient to deal with proportional (or percentage) returns. For this we need to define several terms:

- t is a point in time, usually a date such as the market closing of 7/26/2006.

- (t, t + 1) is the time period for which the return is sought. In the CAPM discussed later, daily returns are frequently used. An example of a daily interval is (7/26/2006, 7/27/2006)—in other words, the closing of two consecutive business days.

- P_t = price of the stock at time t.

- Div_{t+1} = dividends paid on a stock during time period (t, t + 1).

- Dividend yield = $\dfrac{Div_{t+1}}{P_t}$ = dividends paid during period (t, t +1) divided by price at the beginning of the period.

- Capital gain = $\dfrac{P_{t+1} - P_t}{P_t}$ = price change during period (t, t + 1) divided by price at t.

- Total proportional return on stock = Dividend Yield + Capital Gain =
$$R_{t+1} = \frac{Div_{t+1}}{P_t} + \frac{P_{t+1} - P_t}{P_t}$$

- Market Return = the SP500 return, often taken as a substitute for the market return.

- Risk = the uncertainty of a stock's return, often the standard deviation of the returns.

Returns on stocks or stock portfolios are compared to the returns on risk-free assets such as 13-week Treasury Bills. Their return is quoted as an annual percentage. To convert it into a daily proportional return, divide by 100, because the yield is given as a percentage, and by 360 to convert the annual return to yield the

$$\text{daily return on 13 - week Treasury Bill} = \frac{\text{Annual Return in \%}}{100 \cdot 360}.$$

Example: Comparing the Risk of General Electric, SP500 and 13-week Treasury Bill

In this example we show the calculations of the daily returns for General Electric, the SP500, and the 13-week Treasury Bill (T-Bill) for a three-month period from 5/1/2006 to 7/31/2006.

Figure 15.11 Data Excerpt for Three Investments: GE, SP500 index, 13-week T-Bill

	Date	GE Adj. Close*	SP500	^IRX	GE standardized	SP500 standardized	^IRX standardized
1	07/31/200	32.69	1276.6	4.93	-0.0099939	-0.0014782	0.0001369
2	07/28/200	33.02	1278.5	4.92	0.01133231	0.01215168	0.0001366
3	07/27/200	32.65	1263.2	4.95	-0.000918	-0.0040997	0.0001375
4	07/26/200	32.68	1268.4	4.97	-0.0006116	-0.0003783	0.0001380
5	07/25/200	32.7	1268.8	5	0.00245248	0.00632083	0.0001388
6	07/24/200	32.62	1260.9	4.95	0.01147287	0.01662514	0.0001375
7	07/21/200	32.25	1240.2	4.95	-0.0070813	-0.0070769	0.0001375
8	07/20/200	32.48	1249.1	4.95	-0.0121655	-0.0084775	0.0001375
9	07/19/200	32.88	1259.8	4.97	0.012939	0.01855505	0.0001380
10	07/18/200	32.46	1236.8	5	0.00309023	0.00191982	0.0001388
11	07/17/200	32.36	1234.4	4.94	0.00778574	-0.0013833	0.0001372
12	07/14/200	32.11	1236.2	4.91	-0.0171411	-0.0048942	0.0001363

GESP500TBill3months

GESP500TBill3months

Columns (7/0)
Date
GE Adj. Close*
SP500
^IRX
GE standardized
SP500 standardized
^IRX standardized

Rows
All rows 64
Selected 0

Data Requirements: Figure 15.11 shows 12 of 64 daily returns for General Electric (GE), the SP500, and the 13-week Treasury Bill (^IRX) for May 1 to July 31, 2006. The complete data for a three-month period are given in the file *GESP500TBill3month.jmp*. Column 1 contains the date in descending order, because this is the way the data were imported from finance.yahoo.com. Column 2 lists the adjusted closing price of GE. Column 3 gives the SP500 index of each day. Column 4 is labeled by the symbol ^IRX and contains the 13-week T-Bill annual rate of return.

Standardized daily returns are needed for a comparison. Column 5 of Figure 15.11 contains GE standardized returns, representing the proportional changes in price from the previous day's price. Column 6 contains SP500 standardized, representing the proportional changes in the SP500 index from that of the previous day. Column 7 is labeled ^IRX standardized, representing the daily returns of the 13-week T-Bill. JMP formulas for calculating these daily returns are in Table 15.2. The formulas for GE standardized and SP500 standardized involve differences. As a result, only 63 instead of 64 values appear in those two columns.

Table 15.2 JMP Formula to Generate Standardized Daily Returns

GE standardized: Proportional change in price from previous day's price	SP500 standardized: Proportional change in SP500 index from the index of previous day	^IRX standardized: Daily return of the 13-week T-Bill
$$\dfrac{GE\ Adj.\ Close^*_{\text{Row}()} - GE\ Adj.\ Close^*_{\text{Row}()+1}}{GE\ Adj.\ Close^*_{\text{Row}()+1}}$$	$$\dfrac{SP500_{\text{Row}()} - SP500_{\text{Row}()+1}}{SP500_{\text{Row}()+1}}$$	$$\dfrac{\text{^IRX}}{100*360}$$

Asset Variability as a Measure of Total Risk

The historical volatility of a single stock can be assessed by comparing daily returns for the last three months with returns to the broad market and a risk-free investment. First examine the histograms of each investment to assess the distributions of the returns. A relatively larger spread of returns indicates an increased level of total risk.

Table 15.3 Comparison of Variability of Returns

Asset	Histograms of daily returns from 5/1/06 to 7/31/06	Data Summary	Comment
^IRX standardized	^IRX standardized histogram (−0.02 0 0.01 0.03)	Mean 0.0001333 Std Dev 2.8659e-6 Std Err Mean 3.5823e-7 upper 95% Mean 0.000134 lower 95% Mean 0.0001326 N 64	T-Bills show little variability compared to GE or the SP500 returns.
SP500 standardized	SP500 standardized histogram (−0.02 0 0.01 0.03)	Mean −0.000312 Std Dev 0.0088699 Std Err Mean 0.0011175 upper 95% Mean 0.0019217 lower 95% Mean −0.002546 N 63	The SP500 index shows considerable variability. The average return is negative for this period.
GE standardized	GE standardized histogram (−0.02 0 0.01 0.03)	Mean −0.000656 Std Dev 0.0074944 Std Err Mean 0.0009442 upper 95% Mean 0.0012311 lower 95% Mean −0.002544 N 63	GE shows a negative average return. The standard deviation is close to that of the SP500.

The horizontal scales of each histogram should be made all the same, so the histograms can be directly compared. For such a comparison use the **Uniform Scaling** option in the Distribution platform of JMP as shown in Figure 15.12.

Figure 15.12 Uniform Scaling Option in Distribution Platform

Comparing the histograms in Table 15.3, the 13-week T-Bill (^IRX) has the smallest variability, while the variability of the SP500 and GE are similar. The spread is also measured by the standard deviations of monthly returns. ^IRX has the lowest standard deviation, while GE and SP500 have very similar standard deviations, at least in this three-month period. For this period of three months, T-Bills have on average the highest and GE the lowest daily return.

15.2.2 Capital Asset Pricing Model (CAPM)

The CAPM is widely used in finance. The CAPM produces the beta coefficient to evaluate the specific risks of individual stocks. The variance or standard deviation of the return measures the total risk for holding a single asset. The beta coefficient in the regression model measures the risk of a single stock relative to a well-diversified portfolio. Total risk of holding a single asset is the sum of the market risk and the firm-specific risk. Total risk associated with a single stock can be assessed using the standard deviation (or variance) of the returns from historical data. The total risk of a stock is compared to returns from an index, such as the SP500, as a proxy for the broad market. The total risk of a single stock can also be compared to the returns from a risk-free investment, such as the 13-week T-Bill.

Expected Return on Market

It is usually assumed that the expected return on an asset is positively related to its risk. We have seen that even a broad-based portfolio such as the SP500 is not free of risk. You might be willing to invest in the SP500 only if the average return is higher than the risk-free T-Bill. This can be expressed as $R_{Market} = R_{Risk-free} + $ Risk Premium, where $R_{Risk-free}$ is the return on T-Bills and the Risk Premium is the excess return of a risky asset over a risk-free asset. The market risk premium is the difference between the market return and the risk-free return: Risk Premium $= R_{Market} - R_{Risk-free}$.

Figure 15.13 Daily Risk Premiums of SP500 to 13-week T-Bill

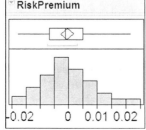

RiskPremium

-0.02   0   0.01 0.02

Using the Distribution platform on the data in Figure 15.11, the daily Risk Premium, as measured by the difference in the mean daily returns for the T-Bills and the SP500 for the three-month period, is $-0.000312 - 0.000133 = -.000445$. Figure 15.13 shows a histogram of the daily risk premiums of the SP500 relative to the 13-week T-Bill for three months. The risk premiums show considerable variability, including both negative and positive values.

Expected Return on Individual Stock

The CAPM for an individual stock is expressed as a linear equation:

$$R_{Individual} = R_{Risk\text{-}free} + \beta \cdot (R_{Market} - R_{Risk\text{-}free}) = R_{Risk\text{-}free} + \beta \cdot (Risk\ Premium)$$

$R_{Individual}$ is also called the firm-specific risk. The coefficient β, called beta, is a widely used measure of risk of a security with the following interpretations:

- A stock with beta $= 0$ has the same risk as a T-Bill.
- A stock with beta $= 1$ has the same risk as the market (here the SP500).
- A stock with beta < 1 has less risk than the market.
- A stock with beta > 1 has more risk than the market.

Historical returns from an individual stock and the market can be used to estimate the stock's beta from a given time period.

Example: GE and Its Three-Month Daily Beta for 2006

We use the data in Figure 15.11 to estimate the beta of GE for the three-month period in 2006. A convenient method to estimate the beta of a stock is simple linear regression between the standardized daily returns of the stock as Y and the standardized daily return of the market as the X-variable. The resulting slope is the beta coefficient of the CAPM.

Using either the Fit Y by X or the Fit Model platform, for GE daily returns this equation is

GE standardized $= -0.000456 + 0.6413726$ SP500 standardized

The estimate of GE's beta is the slope estimate $b_1 = 0.64137$ (see Figure 15.14). This means that during May, June, and July of 2006, GE had a less variable return than the SP500 index and is therefore a stock with less risk than the market.

Figure 15.14 Scatterplot of Daily Returns of GE versus SP500

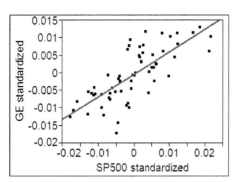

Interpreting the Slope Coefficients

What does the slope coefficient measure? The scatterplot of Figure 15.14 shows the linear relationship between daily proportional changes in value between the SP500 and GE. The fitted line is an estimate of the average change in GE for a given change in the SP500. The slope coefficient b_1 of the fitted line measures the average proportional change in GE stock per unit proportional change in the SP500. A slope $\beta_1 = 1$ indicates that the average proportional change in value of GE is the same as the average proportional change of the SP500. The left panel of Figure 15.15 shows this relationship. A change in value of the SP500 (from SP_1 to SP_2) at two different times $t = 1$ and $t = 2$ is associated with the same average proportional change in the stock (from $Stock_1$ to $Stock_2$). A stock with beta = 1 on average has the same volatility as the market.

Figure 15.15 Relationship of Beta Coefficient Stock Returns

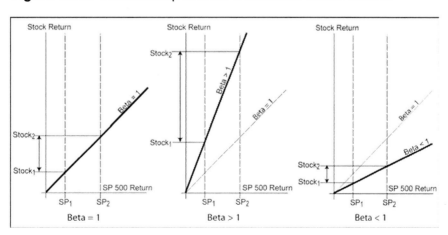

The middle panel of Figure 15.15 shows the relationship when beta is greater than 1. A slope coefficient greater than 1 indicates that a fixed average proportional change of the SP500 is associated with a greater average proportional change in the value of the stock. For every unit change in Market Return (SP_1 to SP_2) the Stock Return changes more. This is shown by the length of the double arrow in the middle panel from $Stock_1$ to $Stock_2$.

The right panel of Figure 15.15 shows the relationship when beta is less than 1. A slope coefficient less than 1 indicates that the average proportional change in a stock is less than the SP500. For every unit change in Market Return (SP_1 to SP_2) the Stock Return changes less. This is shown by the shorter length of the double arrow in the right panel from $Stock_1$ to $Stock_2$. GE's beta is indeed less than 1, at least for the three-month period for which data were analyzed. GE's slope for the given three-month period was 0.64 (rounded to two decimals). This indicates considerably less volatility in the value of GE than that of the SP500.

A Hypothesis Test Associated with Beta

The slope estimate $b_1 = 0.6413726$ is the estimate of GE's beta based on daily data for the three-month period starting with May 1, 2006 and ending with July 31, 2006. This implies that GE is a less risky stock than the SP500. However, the difference from 1 might be due to sampling error. A statistical test of hypothesis can be performed to determine whether GE's beta is significantly lower than the market. The appropriate hypotheses are H_0: beta = 1 versus H_A: beta < 1.

This is a one-sided hypothesis, because we want to test whether GE is less risky than the market. (The *p*-value from the JMP Parameter Estimate output in Figure 15.16 is not useful because it tests that H_0: beta = 0.)

Figure 15.16 Parameter Estimates for Linear Fit of GE versus SP500 Daily Returns

Parameter Estimates				
Term	Estimate	Std Error	t Ratio	Prob>|t|
Intercept	-0.000456	0.00062	-0.74	0.4647
SP500 standardized	0.6413726	0.070424	9.11	<.0001*

The t-Ratio and *p*-value can be calculated for the appropriate null hypothesis using the Std Error of the slope $s(b_1) = 0.070424$. The t-Ratio becomes

$$t = \frac{b_1 - 1}{s(b_1)} = \frac{0.6413726 - 1}{0.070424} = -5.092$$

Because df = 61 (63 – 2) is very large, we can use the Normal distribution to calculate the
p-value. Looking up z = –5.09 gives p < 0.0000 < α = 0.05. So, there is sufficient
evidence to reject the null hypothesis that GE has the same beta risk as the SP500 on a
daily basis. GE is less volatile than the SP500 index during this period assumimg that the
daily proportional changes are random.

A specific company's beta depends on several factors and can change over time. For
example, the length of the data series on which the beta is based might be different from
the three months considered in the GE example. The risk of a company might vary over
time. In fact, during several three-month periods during the last decade, GE has shown a
beta larger than 1. The current beta is influenced by changes in financial leverage and
business risk.

A three-month period of daily returns is common to estimate beta. In actual financial
analysis, these problems often require more detailed methodology and sophisticated tools
than the ones presented here. But the basic idea should help your understanding of this
important concept in finance.

Using Covariance and Variance of Returns to Estimate Beta

A second method to estimate beta is also found in finance texts. This method uses the fact
that the beta of a stock is simply the covariance between the returns of the stock and the
returns of the market, divided by the variance of the market returns. The values of the
covariance matrix can be obtained from a JMP Multivariate analysis and are shown in
Figure 15.17 for the GE data.

Figure 15.17 Covariance Matrix of Three Financial Investments

Covariance Matrix			
	^IRX standardized	SP500 standardized	GE standardized
^IRX standardized	0.00000000	0.00000000	0.0000000
SP500 standardized	0.00000000	0.00007868	0.00005046
GE standardized	0.0000000	0.00005046	0.00005617

Using the values from the covariance matrix, the estimated beta is

$$beta = \frac{Cov(R_{Stock}, R_{Market})}{Variance(R_{Market})} = \frac{0.00005046}{0.00007868} = 0.6413$$

The difference in the last decimals may be due to rounding.

15.3 Curvilinear Regression

15.3.1 Introduction

Straight-line equations are a first-order approximation to a relationship between Y and X. Not all relationships between continuous variables are linear. Some relationships are better modeled by quadratic or cubic polynomials. In some models, a transformation of either Y or X can linearize the relationship. The Fit Y by X platform offers several choices to fit curves to data. This section shows two approaches that fit linear equations to curvilinear data.

15.3.2 Quadratic Regression with Fit Polynomial

A quadratic regression of Y on X fits an equation of the form $Y = b_0 + b_1X + b_{11}X^2$. The simple regression model, containing an intercept and a linear term, is augmented by a quadratic term. The third parameter b_{11} is the coefficient of the quadratic term. With this quadratic equation, a great variety of curvilinear relationships can be adequately modeled. Figure 15.18 shows two representations of quadratic curves. In the graph, the coefficients b_0, b_1, and b_{11} are assumed to be positive. The + or − signs indicate their actual value. Thus in the left panel, the sign of b_1 is negative and of b_{11} is positive, while in the right panel the signs are just the opposite. Figure 15.18 shows how coefficients influence the appearance of a quadratic model.

Figure 15.18 Examples of Quadratic Curves

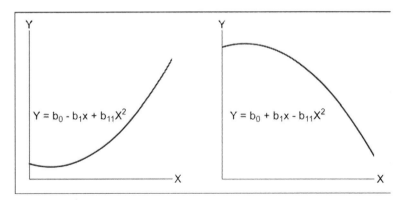

How do we know when to use a quadratic regression? Examining a scatterplot of the data can reveal the nature of the relationship. Fitting both a straight line and a quadratic might help in deciding which model is better. As always, model terms need to be affirmed with *p*-values. With a quadratic model, there will be three *p*-values in the Parameter Estimates table, one for each parameter. The *p*-value associated with the quadratic term tests the significance of adding a quadratic term to a linear model. The quadratic term is deemed significant if the *p*-value is smaller than a specified α value.

Example: Detecting Possible Voting Irregularities

On April 11, 1994, The New York Times covered a case of alleged voting fraud in a special 1993 election to fill a vacant seat in the Second State Senatorial District in Philadelphia. The Republican candidate Bruce Marks received 19,691 votes on voting machines. His Democratic opponent William Stinton received only 19,127. However, on absentee ballots Marks received only 366, whereas Stinson received 1,391. This gave Stinson a 461 vote margin of victory. Marks contested the election at the Federal District Court in Philadelphia, claiming that absentee ballots had been obtained and processed illegally by the Democratically controlled Board of Elections.

Professor Orley Ashenfelter of Princeton University was called as an expert witness. Ashenfelter looked at the historic pattern of machine votes versus absentee ballots for the even election years from 1982 to 1992. In each year, either 3 or 4 Senate Districts voted. He compared the 1993 election to the empirical model based on 21 district elections.

The data file *Vote.jmp* contains 22 rows of historical data and the outcome of the 1993 election. The four original variables are as follows:

- absentee votes for Democratic candidate
- absentee votes for Republican candidate
- machine votes for Democratic candidate
- machine votes for Republican candidate

These four variables were transformed into two derived variables:

- the difference between absentee votes for the Democratic and Republican candidates
- the difference between machine votes for the Democratic and Republican candidates

The scatterplot of Figure 15.19 shows the data with the election outcome of 1993 highlighted. The statistical model consists of the two derived variables and a straight-line relationship between those two variables. The straight-line relationship is important because it claims that as differences in machine votes between Democrats and Republican increase, the differences in absentee votes tend to increase as well.

Figure 15.19 Scatterplot with Straight-Line Fit

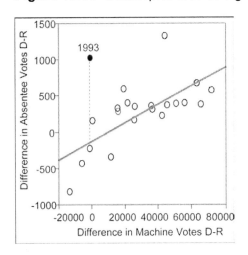

The model shows that the 1993 election does not fit that pattern. For 1993, the difference in absentee votes is much too high in favor of the Democrat, given the difference in the machine vote. This is why the solid circle is far above the straight line. Another election, the recent 1992 election in District 1, is also much above the line, also in favor of the Democratic candidate. You could argue that this excessive difference in absentee ballots was not decisive for the outcome of the election. The graph is intended to show that the 1993 election result was an aberration, if you believe that historical patterns continue to hold in the present. Professor Ashenfelter calculated that if voting history is an indicator of how people vote currently, there is a 94 percent chance that irregularities in the absentee ballots gave the election to the Democratic candidate.

What was the outcome of the court case? Judge Clarence Newcomer of the Federal District Court ruled that absentee ballots had been improperly obtained as a scheme to influence the outcome of the election. He declared Mr. Marks the winner of the election. A Federal Appellate Court ruled that the election was indeed fraudulent, but ordered Judge Newcomber to reconsider his declaring Mr. Marks the winner of the election.

There was much controversy about the statistical model and its conclusions. Other statisticians who were engaged as expert witnesses disputed Professor Ashenfelter's conclusions. They claimed that a statistical model is merely a representation of historical facts. It does not and cannot guarantee that a relationship that was true in the past also holds in the future. Another argument was that the Democratic candidate could have made a special effort to attract potential absentee voters and thus produced an aberrant result in the 1993 election. Yet another interpretation was that even if the model was correct, the outcome could have happened by chance.

Would a quadratic model provide a better fit? For this we reanalyze the data in *Vote.jmp*. In order to streamline the following analysis, the outlier observations 18 and 22 are excluded. The panel on the left in Figure 15.20 shows a plot of AbsenteeDiff versus MachineDiff with the straight regression line that was obtained using the Fit Line command on the Fit Y by X platform. The straight-line model of the type $\hat{Y} = b_0 + b_1 \cdot X$ fits the data well with $R^2 = 0.57$ and RMSE = 253.460. However, the scatterplot reveals a slight curvature. It is therefore prudent to fit a quadratic model and examine the *p*-value of the quadratic term.

Figure 15.20 Linear and Quadratic Fit to *Vote.jmp*

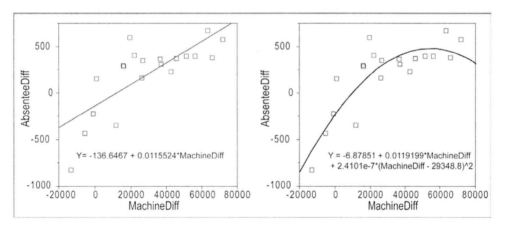

On the Fit Y by X platform, a quadratic equation can be fitted by selecting the **2,quadratic** option from the **Fit Polynomial** menu, shown in left panel of Figure 15.21. This produces the graph in the right panel of Figure 15.21. Selecting **Fit Polynomial →** **2,quadratic** fits a quadratic equation of the form $\hat{Y} = b_0 + b_1 \cdot + b_{11} \cdot \left(X - \overline{X}\right)^2$. The quadratic term is here expressed as the squared deviation of X from its average, $\overline{X}$. In the voting example, $\overline{X} = 29348.8$. These are centered polynomials and are preferred in JMP because of certain computational advantages. The polynomial also can be rewritten in the original equation form.

It appears that the quadratic provides a better fit than the linear fit. This is supported by a higher $R^2 = 0.72$ and a lower RMSE = 211.076. The *p*-value of the quadratic term in the

Parameter Estimates table of Figure 15.21 shows that the quadratic term is significantly different from 0, because its *p*-value is p = 0.0082.

Figure 15.21 Parameter Estimates of Quadratic Fit for *Vote.com*

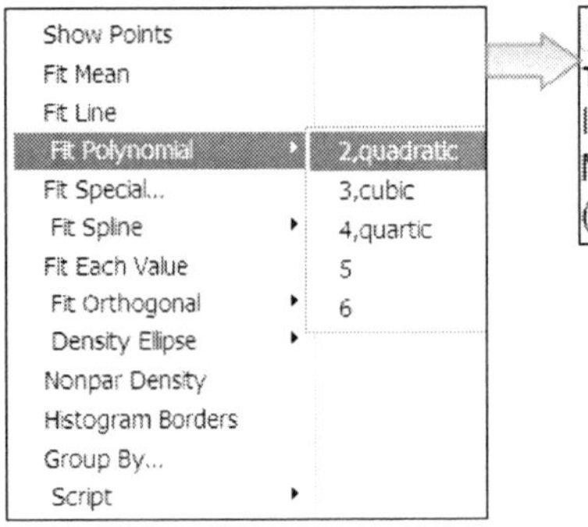

Parameter Estimates Polynomial Fit Degree=2			
Term	Estimate	Std Error	Prob>|t|
Intercept	-6.878951	86.0056	0.9372
MachineDiff	0.0119199	0.001958	<.0001*
(MachineDiff-29348.8)^2	-2.41e-7	8.051e-8	0.0082*

Menu items: Show Points, Fit Mean, Fit Line, Fit Polynomial ▸ 2,quadratic, Fit Special..., 3,cubic, Fit Spline ▸ 4,quartic, Fit Each Value, 5, Fit Orthogonal ▸ 6, Density Ellipse ▸, Nonpar Density, Histogram Borders, Group By..., Script ▸

The quadratic model has significant statistical advantages over the straight-line model. However, do they matter in this specific application where one is concerned with predicting absentee differences from machine differences? In this case, both models make a similar point, although the quadratic model appears to be more precise. Note, for example, that the parameters of the linear term of both equations are close in value (0.01155 for the straight line and 0.01192 for the quadratic equation). The quadratic term could be considered a correction to the linear model. In this example it is a small, but noticeable correction.

15.3.3 Fitting Curves with Fit Special

The Fit Special option on the Fit Y by X platform enables the user to transform either Y, X, or both. Transformations are re-expressions of either Y or X. The main reasons for transformations in simple regression are (1) to linearize a curvilinear relationship between Y and X, and (2) to eliminate differences in the variability of Y at different values of X. There are other reasons for using transformations, but they are beyond the scope of this book.

Five transformations are available in Fit Special:

- Natural Logarithm (log(y))
- Square Root ($\sqrt{y}$)
- Square (y^2)
- Reciprocal ($\frac{1}{y}$)
- Exponential (e^y)

These transformations can be applied to both Y and X. Fit Special transforms the variables internally without creating new, transformed variables in the data file. For transformations other than the ones available in Fit Special (shown in Figure 15.26), you must create separate columns for the transformed variables with the aid of functions available in the formula editor.

Transformations change the units of the transformed variable. For example, taking the log of a variable measured in inches changes its unit to log(inches). You must be careful in comparing equations with different Y units as the result of different transformations. The units of the transformed Y variables affect the units of their RMSE and their residuals. RMSEs and residuals of models with different Y units are not directly comparable.

How do we decide which transformation to use? The short answer is that we choose the transformation suggested by theory or, if there is no theory known, choose whichever transformation works best, i.e., has the best fit and is easy and practical in its use. In deciding on the transformation, it is useful to keep in mind the functional form of the transformation.

First consider the equation $Y = b_0 + b_1 \cdot \log(X)$. In this equation, the X variable is transformed by taking its natural logarithm. This equation can be rewritten as $(Y - b_0)/b_1 = \log(X)$; it will follow a curve similar to the one in the left panel of Figure 15.22. Log(X) is a good transformation when the Y values increase at a disproportionately lower rate than the X values. Log(X) keeps on increasing, albeit at an increasingly slower rate than X.

Figure 15.22 Logarithmic and Exponential Functions

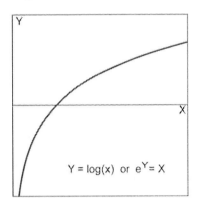
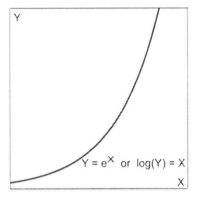

Next consider the equation $\log(Y) = b_0 + b_1 \cdot X$, where the logarithm is taken of Y, and X remains untransformed. This equation can be rewritten as $Y = e^{b_0 + b_1 \cdot X}$. This is the exponential function of the right panel of Figure 15.22, which is useful when the increase in Y tends to be disproportionately larger than that of X. This function is used in growth curves. Taking the log of Y or of X leads to different curves with different applications.

The logarithmic transformation of X might also lead to graphs as shown on the left in Figure 15.23. The regression model is the same as with the left panel of Figure 15.22 (except for the constant term). The difference is the sign of the slope coefficient, which was positive in Figure 15.22 and is negative in Figure 15.23.

Figure 15.23 Equations with Negative Slope on Log(X)

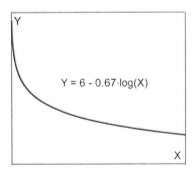
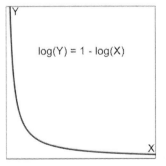

Lastly, it is possible to transform both the X and the Y variable. This transformation leads to the model $\log(Y) = b_0 + b_1 \cdot \log(X)$. This equation is used to estimate the price elasticity coefficient of demand. It is also used in the experience curve, a tool used in strategic planning and technology applications. On the right of Figure 15.23, the vertical and horizontal axes are in Y and X units respectively. If they were both drawn on a logarithmic scale, the curve would turn into a straight line with negative slope. Some transformations, such as 1/y, are very similar to the logarithmic transformation within a restricted range. The squared transformation is similar to the quadratic polynomial of the previous section.

15.3.4 Detail Example: Price Elasticity of Demand

A firm needs to know how responsive demand for their goods is to changes in price. Sometimes lowering a price (P) increases demand (Q) sufficiently to make up for lost revenue on an individual unit and increases total revenue. For other goods, a reduction in price P does not increase demanded quantity Q sufficiently and so total revenue goes down. The responsiveness of demand to changes in price is measured by the *elasticity coefficient*. Economists distinguish between two types of elasticity. Point Price Elasticity is the responsiveness at a particular point on the demand curve. Arc Price Elasticity is the responsiveness between two distinct points on the demand curve.

Point Price Elasticity E_P

For a particular good, Point Price Elasticity (E_P) is the proportional change in the quantity demanded divided by the proportional change in its price, holding constant all other variables. In mathematical terms this is,

$$E_{\mathrm{P}} = \frac{\text{Proportional change in Quantity } Q}{\text{Proportional change in Price } P} = \frac{\Delta Q/Q}{\Delta P/P} = \frac{\Delta Q}{\Delta P} \cdot \frac{P}{Q}$$

where ΔQ and ΔP are the change in quantity and the change in price respectively.

As an example, assume that a 10% increase in the price of chicken results in a 2% decrease in the quantity of chicken sold. Since total revenue $\text{TR} = \text{Price} \cdot \text{Quantity} = P \cdot Q$, total revenue after the price change is $\text{TR}_{\text{new}} = (1.1 \cdot P) \cdot (0.98 \cdot Q) = 1.078 \cdot P \cdot Q$, where P and Q are price and quantity that determined the old revenues. The 10% increase in price increased TR by 7.8%, even though the quantity Q sold decreased by 2%. The elasticity coefficient is

$$E_p = \left| \frac{-0.02 \ \text{change in quantity}}{0.10 \ \text{change in price}} \right| = \left| -0.20 \right|$$

Meaning of E_P

Elasticity coefficients are naturally negative, although there are rare situations when a price increase leads to an increase in quantity demanded. Typically, the absolute value of E_P is used. The interpretation of E_P is summarized in Table 15.4.

Table 15.4 Interpretation of Elasticity Coefficients

$|E_P|$	Elasticity	Effect
$|E_P| = 1$	Unitary demand	A small price change leads to no change in revenue.
$|E_P| < 1$	Inelastic demand	A small price increase (decrease) leads to an increase (decrease) in revenue.
$|E_P| > 1$	Elastic demand	A small price increase (decrease) leads to a decrease (increase) in revenue.

Unitary demand exists whenever a price change leads to no change in total revenue. Increasing the price by a small amount reduces the purchased quantity proportionately, so that total revenues remain the same. Examples of goods and services that have been found to have approximately unitary demand are owner-occupied housing, movies, and certain types of foods such as shellfish.

Inelastic demand exists whenever a price increase has the effect of decreasing the demanded quantity by a proportionally lesser amount. The result is that total revenues are increased. An example of a good with an inelastic price is gasoline. Small changes (a few cents) in price hardly change the demand for gasoline, with the result that total revenues for oil companies are also increased.

Elastic demand exists whenever a price increase has the effect of reducing the demanded quantity by a proportionally larger amount. The result is that total revenues are decreased. For example, demand for mortgages is elastic if a 5% price increase in the cost of mortgages results in a 15% decrease in the quantity demanded. The elasticity coefficient is $|E_P| = 15/5 = 3$. This is an elastic demand. Luxury goods, non-essential goods for the conduct of your life, or goods where adequate substitutes are available are associated with elastic demand. Restaurant meals, foreign travel, airline travel, and certain fresh fruits or vegetables fall into this category.

Economic factors affecting $|E_P|$ are varied and include the following:

- the ease of substitution of another good for the one in question
- the proportion of income spent on a good
- necessary goods versus luxury goods
- time since price change

Elasticity and Regression

Figure 15.24 shows a non-linear price-quantity relationship. The curve shows that an increase in price from P_0 to P_1 results in a disproportional decrease in the quantity from Q_0 to Q_1. Despite this curvilinear relationship, you can use linear regression with transformed variables to evaluate elasticity.

Figure 15.24 Demand Curve Relating Price and Quantity Demanded

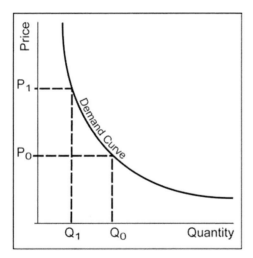

Elasticity under Unitary Demand

The movement of price P versus quantity Q can be represented as a curve. For price elasticity of demand, one looks for changes in demand as a result of changes in supply and their combined effect on total revenue. You would expect price to be plotted along the horizontal axis and quantity on the vertical axis. However, in economics just the reverse is common practice: price is the vertical axis and quantity is the horizontal axis, despite the fact that Q is the variable that needs to be explained and is thus the Y variable from a regression viewpoint. Get used to this tradition and don't be confused by it. In statistical analysis of price elasticity of demand data, quantity must represent the Y-variable and price the X-variable.

The three important quantities in determining the price elasticity of demand are TR = Total Revenue, Q = Quantity Demanded, and P = Price. As mentioned before, with unitary demand the elasticity coefficient E_P = 1, and the total revenue does not change as the price changes, i.e., TR = $P \cdot Q$ = constant. To express the demand as a function of price, you need to take logarithms, because the multiplication turns into simple addition when taking logs. Thus, log(TR) = log(Q) + log(P). Solving this for log(Q) (as the Y-variable), you get the equation under unitary demand: log(Q) = log(TR) – log(P), or more generally, log(Q) = b_0 + $b_1 \cdot$log(P), where b_0 = log(TR) and b_1 = elasticity coefficient = –1 (when the demand is unitary as assumed).

The relationship between quantity Q=Y and price P=X, both on the log scale, is linear. The elasticity coefficient b_1 can be estimated with simple linear regression. In order to test whether a demand is unitary or not, the appropriate null hypothesis is H_0: β_1 = 1 versus H_A: $\beta_1 \neq$ 1. To test this null hypothesis, calculate the *p*-value manually or rely on two-sided confidence intervals for the slope coefficient. The null hypothesis cannot be rejected if the confidence interval for b_1 includes –1 as a possible value.

Example: Demand for Chicken

Examine the demand for chicken in pounds (lbs) per household. The data are contained in *ChickenPriceElasticity.jmp* and are given in Figure 15.25. Column 1 contains the price per pound of chicken in dollars (price/lb). Column 2 contains the demand of chicken in pounds at the given price (demand in lbs). Column 3 is Total Revenue, the product of price/lb and demand in lbs.

Figure 15.25 Data File *ChickenPriceElasticity.jmp*

		price/lb	demand in lbs	Total Revenue
•	2	0.79	37.1	29.309
•	3	0.81	37.4	30.294
•	4	0.84	34.9	29.316
•	5	0.89	32.8	29.192
•	6	0.86	35.5	30.53
•	7	0.94	30.6	28.764
•	8	0.91	34.2	31.122
•	9	0.96	31.2	29.952
•	10	0.99	29.8	29.502

ChickenPriceElasticity

ChickenPriceElasticit

Columns (3/0)
⊿ price/lb
⊿ demand in lbs
⊿ Total Revenue +

Rows
All rows   10
Selected   0
Excluded   0

The Fit Special option is used with the untransformed variables demand in lbs and price/lb. JMP creates the appropriate variable internally. On the Fit Y by X platform, select the price/lb as the **X, Factor** and demand in lbs as the **Y, Response**.

Figure 15.26 Choices of Y and X Transformations for Fit Special

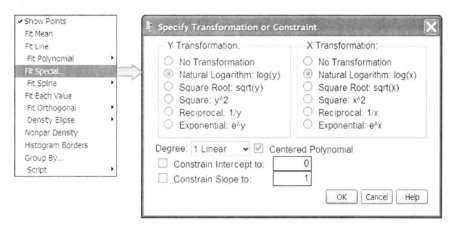

The result is a scatterplot with the original Y and X axes. Select **Fit Special** from the red triangle on the bar above the scatterplot (Figure 15.26). A window appears in which you select the axis transformations for both Y and X. As shown, select **Natural Logarithm** for both the X and the Y. Click **OK**. Regular regression output appears, except that the scatterplot is on the original P and Q scale. Because the model in Figure 15.27 uses the original instead of the log-transformed variables, the fitted model has a slightly curved appearance. Notice also that the plot is Q=Y versus P=X and not with the axes reversed as they are in economics. Figure 15.27 gives the Parameter Estimates table with the 95% confidence limits for each regression coefficient (rounded to four decimal places).

Figure 15.27 Log-Log Fit to Original Data with Parameter Estimates for Y = Log(demand) and X = Log(price)

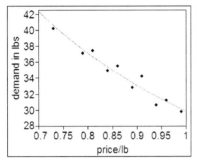

Parameter Estimates					
Term	Estimate	Std Error	Prob>|t|	Lower 95%	Upper 95%
Intercept	3.3943	0.0150	<.0001*	3.3597	3.4289
Log(price/lb)	-0.9838	0.0897	<.0001*	-1.1906	-0.7770

The resulting equation is Log(demand) = 3.3943 – 0.9838*Log(price). The absolute value of coefficient $b_1 = -0.9838$ is near 1 and suggests unitary demand. Testing for unitary demand requires the hypotheses $H_0: \beta_1 = 1$ versus $H_A: \beta_1 \neq 1$. (The default test in the Parameter Estimates table is $H_0: \beta_1 = 0$ versus $H_A: \beta_1 \neq 0$.) For a test of unitary demand, note that the 95% confidence interval for β_1 is from –1.1906 to –0.7770 and includes the value –1 associated with unitary demand.

15.4 Summary

- Correlation analysis is a widely used method for empirical relationships between variables.

- Simple correlation analysis examines pairs of variables.

- The simple correlation coefficient measures linear association between two continuous variables.

- The sample correlation coefficient is also called the Pearson correlation coefficient.

- The correlation coefficient can be viewed as a standardized covariance between two variables. Correlation coefficients are restricted to values between –1 and +1.

- A correlation coefficient of 0 indicates that no linear relationship exists. Values of either –1 or +1 suggest perfect linear relationships. The sign determines the orientation of the relationship.

- Outlier observations can influence the value of the estimated correlation coefficient.

- The simple correlation coefficient is related to the slope and the R^2 of simple regression.

- The Capital Asset Pricing Model (CAPM) is used to evaluate the risk of a financial investment.

- The total risk of holding a single asset is Market Risk + Firm-Specific Risk.

- Risk is related to the volatility of an investment as measured, for example, by the standard deviation of its historical returns.

- Return consists of dividends and capital gain or loss.

- In the CAPM, the beta coefficient measures the risk of a stock relative to the market risk. Market risk can be measured by a broad index such as the S&P 500.

- The typical data for a beta coefficient require daily returns for the most recent three-month period. Other periods are possible.

- The beta coefficient can be calculated with a simple regression of standardized returns of the stock as the Y variable and standardized market return as the X variable.

- A beta coefficient of 1 indicates that a stock has the same risk as the market. A beta greater than 1 indicates a risk greater than the market; a beta less than 1 indicates a risk less than the market.

- Elasticity concerns the responsiveness of demand to price changes. Point Price Elasticity is the proportional change in quantity demanded divided by the proportional change in its price.

- Depending on its effect on total revenues, elasticity is unitary, inelastic or elastic.

- The basic elasticity model explains the logarithm of the quantity purchased to the logarithm of the price, i.e., $\log(Q) = b_0 + b_1 \cdot \log(P)$.

- The slope coefficient b_1 is the elasticity coefficient.

15.5 Problems

15.5.1 Correlation

1. Each week a newspaper prints a table showing interest rates for various products at local banking institutions. The data for three-month certificates of deposit, home equity loans, three-year auto loans, and money market accounts are contained in the file *BankProducts.jmp*.

   a. Describe the data.

    b. Perform a correlation analysis and determine which bank products are significantly correlated.

    c. Use the scatterplot matrix to identify outliers. Discuss any relationships you observe between these outliers.

2. A regional economic development organization in a small state is considering an initiative to attract high technology companies. The director of research obtains data on demographic and technology indicators for each of the 50 states and the District of Columbia from the U.S. Census website. Specifically, the data to be examined are as follows:

Gross state product	$ million
Total research and development investment	$ million
Area	Square miles
Population density	Thousands per square mile
Population	Thousands
Number of patents issued to state residents	-

The data can be found in *HiTechInitiative.jmp*.

    a. Which variables are significantly correlated? Comment on the strength of the correlations.

    b. For the purposes of this analysis, a small state is one with a population of 2,000,000 or less. Separate the data into two JMP tables—one for small states (population less than 2,000,000) and one for large states (population over 2,000,000). Perform a correlation analysis on each of the tables. Summarize the significant correlations for each. Discuss possible reasons for the differences between small and large states.

3. A corporation is considering opening a chain of sports-themed restaurants in major American cities and has gathered information on the demographics of the cities and their professional sports teams' attendance and performance. The data are in *sports_attendance.jmp*. An analysis of this data will be included in the business case analysis to support initiating this business venture.

    a. Describe the data.

    b. Perform a correlation analysis.

    c. Which correlations are significant? Give 95% confidence intervals for the significant correlations and interpret them.

    d. Prepare a short statistical summary of your findings and discuss how this information will be useful to support the business case for opening the restaurants.

4. Download the adjusted weekly closing values for the last year for the following market indices—Dow Jones Industrial Average, NASDAQ, S&P 500, FTSE 100, Nikkei, and Hang Seng. Also download the adjusted weekly closing for the U.S. 10-year Treasury Notes. Be sure to align the dates correctly to accommodate for the differences in the days the various markets are closed. Perform a correlation analysis. Determine which correlations are significant and give the *p*-values. Note any outliers and their effect on the correlations. Discuss your results in terms of current world events and differences between the various stock exchanges.

15.5.2 Stock Market Returns

1. The file *pepsi.xls* contains the weekly adjusted closing prices for the Pepsi Bottling Group common stock, Vanguard Windsor mutual fund, the Dow Jones Industrial Average, and the 13-week U.S. Treasury Bill from January 2004 to January 2005.

    a. Use the JMP formula editor to create the standardized returns for each investment.

    b. Compare and comment on the standard deviation and histogram associated with each of these investments and the Dow Jones index.

    c. Summarize your findings in a table and write a paragraph comparing the risks of these investments from a business viewpoint.

2. The daily adjusted closing price for Boeing stock is given in the file *boeing.xls* along with the corresponding adjusted closing value of the S&P 500 index for the period from June 1, 1986, to September 8, 1986.

    a. Create the standardized returns and compare the risks of Boeing and the S&P 500.

    b. Fit the CAPM to the Boeing data.

    c. What is the estimate of the beta risk for Boeing compared to the broad market as measured by the S&P 500? Is Boeing significantly riskier than the S&P 500?

3. Download the daily adjusted closing prices for GE stock and the S&P 500 index for the same three-month period of 1993 and 2003. Use the CAPM to estimate and compare GE's beta coefficients for each period.

4. The file *Stock_Comparison.jmp* contains adjusted closing prices for the S&P 500 index and Apple Computer, Walmart, and Caterpillar stocks. Use the CAPM to evaluate the risk of each of the three stocks relative to the S&P 500. Based on the estimate of the beta coefficient for each model, perform

statistical tests of hypothesis to classify the risk of each stock as one of the following:

    a. same risk as the market (S&P 500)

    b. less risk than the market

    c. greater risk than the market

5. Another measure that is used in portfolio management is Jensen's alpha, which gives the excess return of a stock over what is expected for the market. The simple regression model,

$$R_{individual} - R_{Risk-free} = \alpha + \beta(R_{Market} - R_{Risk-free}) + \varepsilon$$

can be fitted from the historical returns for a single stock, the market, and a risk-free investment. Select a stock of interest and obtain estimates of α and β.

    a. Download the adjusted daily closing price for the stock along with the corresponding values of the S&P 500 and 13-week Treasury Bill for the past three months.

    b. Create the standardized daily returns for the three investments.

    c. Create two new columns using the JMP formula editor. In one column, subtract the value of the daily 13-week Treasury Bill return from the standardized stock return. Label this column StockReturn—T-Bill. In the other column, subtract the value of the daily 13-week Treasury Bill return from the standardized S&P 500 return. Label this column MarketReturn—T-Bill.

    d. Fit a simple regression using StockReturn—Tbill as the Y and MarketReturn—Tbill as the X. The resulting intercept is the estimate of Jensen's alpha. The slope estimates the beta risk.

    e. (Optional) Write a paragraph summarizing the performance of the stock in terms of the estimated α and β.

15.5.3 Curvilinear Regression

1. A supermarket is considering lowering the price of a dozen store-baked gourmet cookies for the months of September and October. The file *cookie_demand.txt* contains a randomly selected sample of price and quantity demanded over the last two years.

    a. Estimate the price elasticity of the cookies using a simple regression of log(Price) and log(Quantity).

    b. Find the 95% confidence interval on elasticity (the slope coefficient) and use it to determine whether the demand for the gourmet cookies is elastic, unitary, or inelastic.

2. A fast food restaurant is introducing a new line of flavored coffee targeted at the breakfast market. The coffee was test marketed in 10 different regions. The results are given in *coffee_test_markets.jmp*. Calculate the elasticity of the new coffee line by relating Demand in Cups to Price. Write a brief memo that summarizes your results and the implications for extending the new coffee line nationally.

3. An amusement park has recently built an indoor water park. At present, there are separate admission tickets, one for each park. Park management is considering offering a two-day ticket at a discounted rate that allows a one-day admission to each park. The data found in *AmusementPark.jmp* show daily admissions to the indoor water park for various prices at the amusement park. Fitting an elasticity model where Indoor Water Park Admission is the X variable and Amusement Park Price is the Y variable estimates what is referred to as cross-price elasticity. This is the responsiveness of demand for one good to a change in the price of another. The cross-price elasticity will be positive if the goods are substitutes and negative if the goods are complements.

   a. Find the cross-price elasticity. Is the water park a substitute or complement to the amusement park?

   b. Based on your analysis, would you recommend offering the two-day ticket? Explain how you arrived at your answer.

4. A national restaurant chain is running a television advertising campaign during primetime. The file *PrimetimeAdvertising.jmp* contains the number of times the 30-second commercial is broadcast per day and sales (in millions of dollars) for the following day.

   a. Fit a simple linear regression to the data and evaluate the goodness-of-fit.

   b. Fit a quadratic regression to the data and evaluate the goodness-of-fit.

   c. Use the quadratic regression to predict sales when the commercial is shown nine times.

   d. Compare the linear and quadratic models. Which model would you recommend? Why?

5. The file *pickup_truck_mpg.jmp* contains fuel economy data on 17 pickup trucks. Fuel economy is measured in miles per gallon (mpg) and is related to the weight of the vehicle. Curb weight, recorded in pounds, is defined as the total weight of a vehicle with all standard equipment, all necessary operating fluids (oil, coolant, etc.) and a full tank of fuel. Curb weight does not include the weight of passengers or cargo.

   a. Fit two curvilinear models, one for highway driving and one for city driving, to predict mpg from curb weight using a $1/y$ transformation and no transformation for X.

   b. Evaluate the goodness-of-fit for the two models.

   c. Predict the city and highway fuel economy for a pickup truck with a curb weight of 4500 pounds.

15.6 Case Studies

15.6.1 Case Study 1: Comparing Correlation between Individual Stocks and an Index

The objective of this case is to examine the correlation between the Dow Jones Utility Average (DJUA) and the individual stocks that comprise it.

1. Find the 15 individual stocks that constitute the DJUA. Download the daily adjusted closing prices for the last three months and calculate the standardized returns.
2. Give a statistical summary of the returns and risk associated with each stock and the index.
3. Find the correlations between the individual stocks and the DJUA. Determine which correlations are significant.
4. Write a summary, not to exceed one page, of your analysis. Include in your discussion a comparison of the associations between the stocks and the DJUA. Also discuss the correlation between and within stocks of similar utilities (e.g., electric, diversified). Interpret the results in the context of recent political, natural, and financial events.

15.6.2 Case Study 2: Comparing the Risk of Two Market Sectors

The objective of this case is to use the CAPM to compare the risk within and between two market sectors.

1. Select two stocks from the healthcare/pharmaceutical industry and two stocks from the energy sector. Also, select an index that represents the broad market.
2. Download the daily adjusted closing prices for the last three months from the Internet. Calculate the standardized returns.
3. Give a statistical summary of the returns and risk associated with each stock and the index.
4. Estimate the beta coefficient for each stock and determine those that are significantly different from the broad market.
5. Write a one-page summary of your analysis. Include in your discussion a comparison of the risk within and between sectors. Interpret the results in the context of the political, natural, and financial events that have occurred in the last year.

15.6.3 Case Study 3: Experience Curve

The experience curve is a useful management tool for strategic and production planning. In the 1960s, the Boston Consulting Group observed a relationship between the cumulative number of units produced and the cost per unit. They found that for each doubling in cumulative production volume, unit costs fell by a constant percentage. This relationship, referred to as Henderson's Law, can be expressed mathematically as

$$C_n = C_1 X^{-a}$$

where

C_n = cost of the nth production unit
C_1 = cost of the 1st production unit
X = cumulative volume of production
a = elasticity of cost with respect to output

This model can be formulated as a simple linear regression by taking the log of Henderson's Law:

$$\log C_n = \log C_1 - a \log X$$

The slope a can be interpreted as the rate at which experience is translated into cost savings. Experience curves are often characterized by the experience rate, $r = e^{-a(\log 2)}$.

$(1 - r)$ is interpreted as the percentage cost reduction for each doubling of output. For example, an experience curve with $a = -0.152$ will have a 10% reduction in unit cost for each doubling of output. This is also referred to as a 90% experience curve.

The experience curve considers all value-added costs such as marketing, distribution, and production. The closely related learning curve considers only the effect of direct labor on unit cost.

A company that creates video games has recorded the cost of producing their games. The number of games developed and the cost for the most recently developed game are given in the file *video_game_production.jmp*.

 a. Fit an experience curve to the data using Cost as the Y and Games Produced as the X.

 b. Find the experience rate for the video game production.

 c. What cost would be expected for their 50th game?

 d. What cost would be expected for their 800th game?

 e. Discuss how the experience curve could benefit the company's strategic planning.

15.6.4 Case Study 4: Engel's Law

Income Elasticity of Demand (IED) is the responsiveness of demand for a commodity to changes in income. The values of income elasticity can be classified as shown in the following table:

	IED Value	Type of good
Income elastic	IED > 1	Normal good
Income inelastic	IED < 1 but > 0	Normal good
Negative income elasticity	IED < 0	Inferior good

Goods with positive income elasticity are referred to as normal goods, while those with negative income elasticity are referred to as inferior goods. Factors affecting income elasticity of demand include the following:

- **Degree of necessity of a good:** The more necessary a good, the more people will want to buy it whatever their level of income, and therefore the more income inelastic it will tend to be.

- **The rate at which the desire for a good is satisfied as consumption increases:** If desire for the good is not quickly satisfied, then people will want to go on buying the good as their income level increases, and therefore the more income elastic the good will tend to be.

- **The level of income of consumers:** As income levels change, so the pattern of spending will change. The higher the level of income, the more goods will tend to become income elastic.

Ernst Engel, director of the Bureau of Statistics for Prussia, observed in 1857 that as income increases, the proportion spent on food decreases even if the actual amounts spent on food increase. This is known as Engel's law.

A purveyor of gourmet whole and natural foods collected data on annual income and weekly fresh produce expenditure from the geographic region they serve. The data can be found in *ProduceExpenditures.jmp*.

1. When Engel's law holds, what values of income elasticity can be expected?
2. Estimate the income elasticity of demand (Y = Produce Expenditures, X = Income) and determine whether Engel's law applies in this situation? Cite the statistical evidence used to support your conclusion.
3. Write a brief report for the gourmet food merchant summarizing your results, (especially the income elasticity), and provide recommendations on how this information is useful in business strategies (e.g., marketing, pricing).

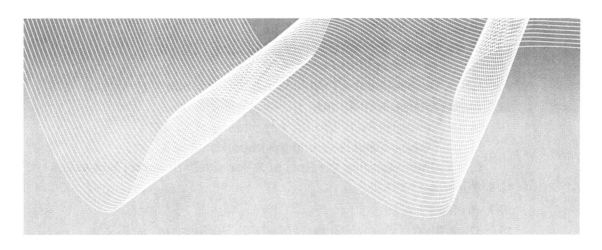

Chapter **16**

Multiple Regression Analysis

16.1 Introduction   448

16.1.1 Data Requirements for Multiple Regression   449

16.1.2 Multiple Regression Analysis Process   450

16.1.3 Multiple Regression Model   450

16.2 Detail Example: Profits of Bank Branches   451

16.3 JMP Analysis of Bank Branch Profits Example   453

16.3.1 Preliminary Fitting of Single X-Variables to Y   453

16.3.2 Fitting Several X-Variables to Y   456

16.3.3 What Do t-Ratios Measure?   459

16.3.4 Conclusion from the Three X-Variable Model   460

16.3.5 Regression Model Using Total Sales per Year = X_1
        and Total Sqft = X_2 460

16.4 Evaluating Model Assumptions and Goodness of Fit   462

16.4.1 Residuals e_i 462

16.4.2 Residuals e_i and RMSE   463

16.4.3 Residuals and R^2 464

16.4.4 R^2 and RMSE   464

16.4.5 R^2_{adj} 465

16.4.6 Plot of Residuals versus Predicted Y (Yhat)
        or versus Time Order 465

16.4.7 Durbin-Watson Test for the Independence of Residuals 466

16.5 Model Interpretation 468

16.5.1 Leverage Plots and the Importance of X_i 469

16.5.2 Standardized Beta and the Importance of X_i 472

16.5.3 Understanding the Role of X-Variables: Column Diagnostics 473

16.6 Summary 475

16.7 Problems 476

16.8 Case Study: Forbes Global 2000 High Performers 479

16.9 References 480

16.1 Introduction

Multiple regression is a widely used tool in business and economic analysis, because it allows the exploration of complex relationships between variables. For example, multiple regression is used to do the following:

- explain the value of a business Y as a function of accounting data (assets, debt, sales, inventory turnover, risk, etc.)

- explain the demand for a product (Y) in terms of its price, advertising expenditure, income of purchaser, etc.

- explain the cost of a project (Y) from component costs of past projects

Multiple regression extends the ideas in simple regression analysis to establish relationships between a single continuous Y-variable and several X-variables. Multiple regression is more complex than simple regression, because it fits a set of X-variables simultaneously. Consequently, you need to sort out which subset of X-variables forms the best model to explain Y. A further complication is that the X-variables might be correlated.

The following are typical tasks in multiple regression analysis:

- define the X-variables according to the importance of their effect on Y

- estimate the average value of Y for a particular value of $(X_1, X_2, \ldots, X_{p-1})$

- find the values (settings) of $(X_1, X_2, \ldots, X_{p-1})$ that optimize the value for Y with respect to some performance criterion
- predict a future value of Y at $(X_1, X_2, \ldots, X_{p-1})$

The X-variables can be classified into three different categories that affect how these variables are selected and recorded:

1. *Input variables* measure inputs into a process. Selection of different values for these inputs is usually within the control of the system operator. Examples include measurements on raw materials, parts, etc.

2. *Noise variables* are extraneous variables that act as inputs, but cannot be controlled or influenced by the system operator. This group includes weather-related variables, general economic conditions, etc.

3. *Process variables* affect the operation of the system. This group includes process temperature, type of machine used to process, and process layout. Process variables can be difficult to change because of fixed-process technology.

All three categories are treated as standard X-variables. However, it is important to ascertain which variables of the model can be directly influenced and which are beyond control. You need to assess the actions that can be taken in order to change the output variable Y in the desired direction. Whatever the mix of variables, all important variables need to be represented in the model, even though they might be beyond direct control.

16.1.1 Data Requirements for Multiple Regression

Multiple regression requires *one continuous Y-variable and several X-variables*. The X-variables are continuous, although JMP allows both nominal and continuous variables in the model. The next chapter discusses how nominal variables are recoded by JMP into continuous variables. A multiple regression data table consists of n rows and p columns. The number of observations is *n* and corresponds to the number of rows. Any multiple regression model contains p variables. There are (p − 1) X-variables and a single Y-variable. Of course, a JMP data table might contain extraneous variables not included in the analysis and not specified in the model statement. The X- and Y-variables have the same names as in simple regression analysis (see Chapter 14, Table 14.1).

16.1.2 Multiple Regression Analysis Process

Multiple regression requires a more subtle analysis and interpretation than simple regression analysis. Multiple regression also requires additional tasks. The first group of tasks concerns the finding of a good multiple regression model:

1. Determine which X-variables should be included in the equation. Eliminate unnecessary X-variables. Look for X-variables that are not yet available for analysis, but are potentially useful.

2. Estimate the coefficients of the multiple regression model.

3. Characterize the quality of the fit of the regression model.

4. Prioritize the X_i in terms of importance in the model.

For example, in an application that seeks to explain the differences in profits among branches of a bank corporation, three X-variables are available. You need to find which of these three X-variables help explain variation in profits. All three X-variables might not be needed to explain the variation in Y, because some might prove to be statistically non-significant. Non-significant X-variables can be eliminated from the model. After we know which of the X-variables are important, we will have to estimate the corresponding model coefficients and perform further analysis.

Another group of tasks concerns assessing the quality of the regression fit and how the model meets the assumptions of regression analysis:

1. Look for outlier observations and their effect on the model.

2. Look for multi-collinearity (dependence) among the X-variables and their effect on the interpretation of the model.

3. Look for unwanted correlations in the data.

16.1.3 Multiple Regression Model

In multiple regression analysis, several X-variables are included in the model simultaneously. Table 16.1 explains the terms of the standard multiple regression model, written as

$$Y_i = \beta_0 + \beta_1 X_1 + \beta_2 X_2 + \cdots + \beta_{p-1} X_{p-1} + \varepsilon_i$$

Table 16.1 Components of the Multiple Regression Model

Y_i	Response of observation at $(X_{1i}, X_{2i}, \ldots, X_{p-1,i})$, where the subscript i identifies the i-th observation.
β_0	Intercept of model with Y-axis at $(X_1, X_2, \ldots, X_{p-1}) = (0,0, \ldots ,0)$ representing the average value of Y when all $X_i = 0$.
$\hat{\beta}_k$	Slope coefficient of X_k, representing the estimated average change in Y per unit change in X_k, assuming that all other X-variables remain constant.
$(X_1, X_2, \ldots, X_{p-1})$	Array of all X-variables at which Y is evaluated.
$\beta_0 + \beta_1 X_2 + \cdots + \beta_{p-1} X_{p-1}$	$\mu_{Y\vert X}$ = Deterministic or average component representing the mean of Y at $(X_1, X_2, \ldots, X_{p-1})$.
ε_i	Random error term with $E(\varepsilon_i) = 0$ and $V(\varepsilon_i) = \sigma^2$, $Cov(\varepsilon_i, \varepsilon_j) = 0$ for $i \neq j$. It is usually assumed to be normally distributed (with mean = 0 and standard deviation = σ)

16.2 Detail Example: Profits of Bank Branches

Situation: Regional management of a large retail bank needs to look at the profitability of bank branches. They want to find variables that explain the considerable differences in profitability between branches. Many variables might play a role in determining the profits. Management decides to select three X-variables to explain the single Y-variable. All four variables are continuous:

- X_1 = Total Sales per year is related to business activity. It simply counts all sales, such as loans, insurance, and investments executed in a branch.

- X_2 = Total Sqft measures the physical size of the branch. It represents the total usable square footage of the branch, excluding foyers and other areas that do not contribute to business activity.

- X_3 = FTE Managers measures the number of bank managers that are assigned to the bank branch. They are persons with a level of knowledge and experience to increase profits.

- Y = Branch Profit (per year in $) is the dependent or performance variable that management is trying to understand.

Purpose of Analysis: Construct a model to predict the profits of a bank branch. Determine which variables affect branch profits and by how much.

Data for Branch Profits: For these variables, data are available on 18 branches in a geographical region. Each of the four variables needs a column. Each branch represents an observation with a row in the data table. The JMP data table *BranchProfits.jmp* with four columns (X_1, X_2, X_3, Y) and 18 rows is shown in Figure 16.1.

Anticipated Results: Management expects that all three variables contribute positively to branch profits.

Figure 16.1 Data Table for Branch Profits Example

	Total Sales per year	Total Sqft	FTE Managers	Branch Profits
1	10591	2474	2	821298
2	9069	4262	3	718488
3	7511	7608	3	630752
4	1541	3248	2	-40582
5	7569	5095	2	606998
6	6014	6455	2	474231
7	10506	3628	3	861916
8	4588	2818	1.5	246162
9	6058	2481	2	386684
10	3073	7027	3	187208
11	12021	4052	2.5	1031091
12	7541	1697	2	504632
13	3090	3023	2.5	100223
14	7594	3483	3	561795
15	10587	5646	2.5	901631
16	1529	5813	3	29199
17	6053	1821	2	346304
18	7578	2872	2.5	532738

BranchProfits

Columns (4/0)
- Total Sales per year
- Total Sqft
- FTE Managers
- Branch Profits

Rows
All rows — 18
Selected — 0
Excluded — 0
Hidden — 0
Labelled — 0

16.3 JMP Analysis of Bank Branch Profits Example

Two different facets of the analysis are discussed in this section. First is a preliminary review of the data. It examines the individual relationships of each X-variable with the Y-variable with simple correlations and regressions as well as a scatterplot matrix. This approach is called the "one X-variable at a time" approach. Fitting one X-variable at a time to the Y-variable has its limitations in interpreting complex relationships. *This step is not required for multiple regression analysis, but is taken to familiarize the analyst with the data.*

The second step is multiple regression analysis proper. Variables are included and analyzed in the model simultaneously, an approach that is often more effective in identifying all sources of variation or in deciding which variables are important for determining branch profits.

16.3.1 Preliminary Fitting of Single X-Variables to Y

This preliminary step is not part of multiple regression analysis and could be skipped. It first looks at simple correlations between Xs and Y. Then it fits a separate simple regression between each X- and the Y-variable. No final conclusions are drawn from this step.

Simple Correlations

Simple correlations provide a quick overview of obvious pairwise relationships between the variables. Look for correlations between the Xs and Y as well as between the Xs. Simple correlations were introduced in Chapter 15. Use the Multivariate platform to calculate simple correlation coefficients between all four variables. Correlation analysis treats all variables as Y-variables.

For this analysis the Pairwise Correlation option gives, in addition to correlation coefficients, confidence intervals and p-values for a test of zero correlation for each pair of variables. Figure 16.2 shows the Pairwise Correlation results. The first two columns list the two variables for which the correlation coefficients are calculated. The third column shows estimated correlation coefficients. The fourth column gives a count of observations for which values on the two variables are available. (In case of missing values, the Count column is different from the number of rows.) The fifth and sixth columns give 95% confidence limits for the actual correlation coefficient. The last column, called Signif Prob, gives the p-value for the null hypothesis that the correlation coefficient is zero. A bar chart showing the magnitude of correlation coefficients is given in the last column.

Simple correlations measure only pairwise linear relationships. As can be seen from Figure 16.2, two pairs of variables have a p-value less than $\alpha = 0.05$. The correlation coefficient (r = 0.9897) between Total Sales per year and Branch Profits is significantly different from zero, with a p-value < 0.0001 and the 95% confidence interval for ρ ranging from 0.9718 to 0.9962. FTE Managers is correlated with Total Sqft (r = 0.5051) and a p-value = 0.0325. Neither Total Sqft nor FTE Managers are significantly correlated with Branch Profits.

Figure 16.2 Pairwise Correlations with p-Values in Branch Profits Example

Pairwise Correlations							
Variable	by Variable	Correlation	Count	Lower 95%	Upper 95%	Signif Prob	-.8-.6-.4-.2 0 .2 .4 .6 .8
Total Sqft	Total Sales per year	-0.0991	18	-0.5410	0.3856	0.6956	
FTE Managers	Total Sales per year	0.0981	18	-0.3864	0.5402	0.6985	
FTE Managers	Total Sqft	0.5051	18	0.0500	0.7865	0.0325*	
Branch Profits	Total Sales per year	0.9897	18	0.9718	0.9962	<.0001*	
Branch Profits	Total Sqft	0.0402	18	-0.4349	0.4977	0.8743	
Branch Profits	FTE Managers	0.1698	18	-0.3227	0.5899	0.5006	

According to pairwise p-values, Total Sales per year is the most significant X-variable in explaining Y if X-variables are examined one at a time.

Scatterplot Matrix

The interpretation of the correlation coefficient can be enhanced by examining the scatterplot matrix, a two-dimensional array of scatterplots of all possible pairs of variables in a matrix format. Figure 16.3 shows a scatterplot matrix for the branch profits example. The variable names are shown in the cells of the main diagonal. Correlation coefficients are optionally added. For example, the first cell in the first row is Total Sales per year. This indicates that in row 1, the vertical axes of all scatterplots are for Total Sales per year. Similarly, the horizontal axes of all scatterplots of column 1 are for Total Sales per year. The same applies analogously to the other variables and axes. The scatterplot also shows 95% density ellipses. Narrow ellipses suggest high correlation. Circular ellipses suggest no correlation. Of course, all interpretations must be made with the caution that they are based on a random sample.

Like the correlation coefficients, the scatterplot shows strong pairwise relationships only between the Total Sales per year and the Branch Profit variables. The scatter between Total Sqft and FTE Managers is along the main axis of the ellipses, but the ellipse is much wider, corresponding to a estimated correlation coefficient of 0.5051. The scatterplot between Total Sqft and Branch Profits shows nearly round confidence ellipses, corresponding to a low correlation coefficient of 0.0402.

Scatterplots can be examined for the presence of outliers or unusual patterns. Except for Branch Profits versus Total Sales per year, the 95% contour ellipses form near circles, a fact that indicates no linear relationship.

Figure 16.3 Scatterplot Matrix of Variables

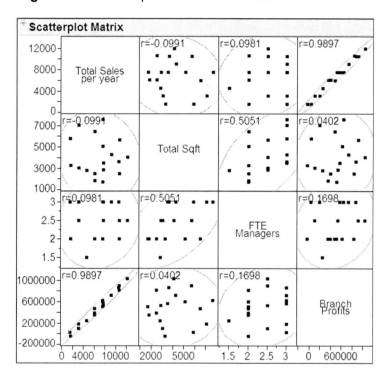

Simple Regression Analysis with Individual Independent Variables

Simple regression and simple correlation are inadequate for analyzing data with many X-variables, because these methods treat each X-variable in isolation, as if the other variables did not exist. (This is similar to treating two-way ANOVA by two separate one-way ANOVAs.) In multiple regression, the inclusion of more than one X-variable might reduce the RMSE to a point where even weaker X-variables might be significant. Joint explanations of Y, not possible in simple regression, are the very purpose of multiple regression.

Simple regression of each X-variable with the Y-variable provides significant results only between Total Sales per year and Branch Profit. The slopes taken from the Parameter Estimates table for each of the three X-variables can be used to test the significance of each Y versus X relationship. The *p*-values of the slopes are the same as those for the pairwise correlation coefficients between the respective X-variable and Y. Simple correlation and simple regression both produce similar conclusions (*p*-values). The slopes of each of the three simple regressions are given in Table 16.2.

Table 16.2 Simple Regression Results for Y-Branch Profits

X-Variable in Model	Slope Coefficient	Interpretation
X_1 = Total Sales per year	b_1 = 97.68 (rounded) dollars per Sale	Significantly different from 0 with $p < 0.0001$.
X_2 = Total Sqft	b_2 = 6.94 dollars per Square Foot	NOT significantly different from 0 with $p = 0.8743$.
X_3 = FTE Managers	b_3 = 106779.43 dollars per FTE Manager	NOT significantly different from 0 with $p = 0.5006$.

Judging from simple regression or simple correlation, only the X-variable Total Sales per year is significant in explaining Branch Profits; Total Sqft and FTE Managers are not. You might be tempted to drop the non-significant X-variables from further consideration. But in multiple regression things are a bit more complicated.

- Non-significant X-variables in simple regression might be significant in multiple regression.
- Significant X-variables in simple regression might be non-significant in multiple regression.

Of course, an X-variable might be significant in both simple regression and in multiple regression. In this example this is true for Total Sales per year.

16.3.2 Fitting Several X-Variables to Y

Multiple regression examines the relationship of Y with a combination of X-variables. In the branch profits example we start with fitting all three X-variables to Y.

Multiple Regression Model with All Three Independent Variables

Multiple regression analysis is a process that can be characterized by five steps.

1. Estimate the coefficients of $\hat{Y} = b_0 + b_1 X_1 + b_2 X_2 + b_3 X_3$

2. Test whether they are significantly different from 0. If they are, then the variable contributes to explaining the variation in the Y-variable, given the current equation. If they are not, consider eliminating one non-significant variable at a time.

3. Examine whether this is a good equation.

4. Revise the model if necessary.

5. Interpret the final model.

Figure 16.4 Variable Selection Window for Fit Model

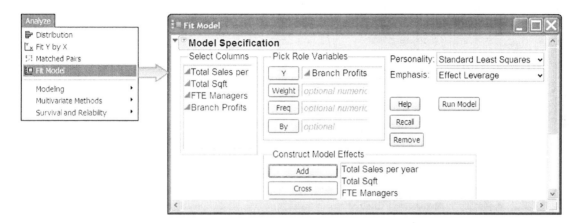

In JMP, Fit Model is the proper platform for multiple regression analysis. From the **Analyze** menu, choose **Fit Model**. As model effects we use the X-variables. Not all available X-variables need to be included in the model. A regression model that contains all reasonable X-variables is called a *full model*. Figure 16.4 shows the selection of variables according to their role in the branch profits example.

The three X-variables (Total Sales per year, Total Sqft, and FTE Managers) can be entered simultaneously by highlighting them in the **Select Columns** list and then clicking the **Add** button. Branch Profits is added by clicking the **Y** button.

Summary of Fit from Fit Model

The Fit Model platform provides extensive output. Summary of Fit and Parameter Estimates are the most basic outputs. The Summary of Fit shown in Figure 16.5 provides basic information about the model. For the branch profits example, it includes the Mean of Response = 494487.1 dollars profit, representing the average profit of all 18 branches. It also includes the Root Mean Square Error or RMSE = 12194.79 dollars and the R^2 = 0.9987 as the proportion of the total sum of squares explained by the regression model.

The Summary of Fit results can be compared with the baseline model, which only fits the average of the Y. The RMSE = 12194.79 dollars is a considerable reduction when compared to the Standard Deviation = 309775.14 of the Y-variable without the X-variables in the baseline model. The conclusion that the multiple regression model explains most of the Y variation is supported by the very high R^2 = 0.9987.

Figure 16.5 Summary of Fit of Three X-Variable Model

Summary of Fit	
RSquare	0.998724
RSquare Adj	0.99845
Root Mean Square Error	12194.79
Mean of Response	494487.1
Observations (or Sum Wgts)	18

Parameter Estimates Table

The Parameter Estimates table in Figure 16.6 summarizes many important characteristics of an equation. The format is the same as in simple regression. Standard output includes the Estimate column with the intercept estimate b_0 and all the slope coefficient estimates b_i. This is the column of coefficients for the regression equation. The second numerical column gives the standard errors of the regression coefficients. The next column lists the t-Ratios representing the Estimate column divided by the Std Error column. The fourth column contains Prob > |t|, the *p*-values of the t-Ratio.

Figure 16.6 Parameter Estimates Table of Three X-Variable Model

Parameter Estimates				
Term	Estimate	Std Error	t Ratio	Prob>|t|
Intercept	-279354.9	15620.28	-17.88	<.0001*
Total Sales per year	99.031027	0.961476	103.00	<.0001*
Total Sqft	24.027674	1.940939	12.38	<.0001*
FTE Managers	700.25959	7062.842	0.10	0.9224

The estimated regression coefficients in the Estimate column allow us to write the estimated regression equation. In the branch profits example, the (rounded) equation is as follows:

$$\hat{Y} = -279354.9 + 99.03 \cdot \text{Total Sales per year} + 24.03 \cdot \text{Total Sqft} + 700.26 \cdot \text{FTEManagers}$$

16.3.3 **What Do t-Ratios Measure?**

The estimated slope coefficient b_i of the multiple regression model represents the estimated average change in Y per unit change in X_i, assuming that all other X-variables in the equation remain constant. The significance of the slope of an X-variable in multiple regression depends on both the data and the model (equation). The t-Ratio of each X_i-variable tests H_0: $\beta_i = 0$ versus H_A: $\beta_i \neq 0$. Large t-Ratios lead to small *p*-values and a rejection of the null hypothesis. However, the test assumes that *all* other variables in the equation remain in the equation. The t-Ratio and its *p*-value in the Prob > |t| column are valid only for the current equation.

The interpretation of a t-Ratio with a significant *p*-value is that the corresponding X-variable has a slope significantly different from 0, given that all other X-variables remain in the equation. The t-Ratio of b_k measures whether adding X_k to an equation already containing all other variables (X_1, ... , X_{k-1}, X_{k+1}, ... , X_{p-1}) leads to a significant reduction of error in Y. *Thus the t-Ratio measures the significance of X_k above and beyond all other X-variables currently in the equation, as if it had been added last.* In multiple regression, the t-Ratio $t_k = b_k/s(b_k)$ is not an absolute test of the variable X_k, but conditional on the other X-variables in the equation.

In the branch profits example, the t-Ratio = 12.38 of Total Sqft can be explained in several equivalent ways:

- The t-Ratio measures the significance of the additional contribution Total Sqft makes, given that Total Sales per year and FTE Managers are already in the equation.

- The t-Ratio $t_{X_2} = t_{X_2|X_1,X_3} = 12.38$ tests a conditional hypothesis H_0: $\beta_2 = 0$, given X_1 and X_3 are also in the equation.

- The t-Ratio measures whether the slope b_2 is significantly different from 0, given that X_1 and X_3 are in the equation.

The t-Ratio = 0.10 for FTE Managers with its *p*-value = 0.9224 indicates that the slope of X_3 is not significantly different from 0 at $\alpha = 0.05$, given that Total Sales per year and Total Sqft are in the model.

Because the value of the t-Ratio (and the *p*-value) depends on the other X-variables in the model, non-significant X-variables should be eliminated only *one at a time*. At each step, the significance of a variable might change. This approach is used in stepwise regression discussed in Chapter 18.

16.3.4 Conclusions from the Three X-Variable Model

Both b_1 and b_2 are significantly different from 0; b_3 is not. The R^2 is near 1 and the RMSE has been reduced considerably relative to the standard deviation around the overall mean of the baseline model. Total Sales per year is significant above and beyond Total Sqft and FTE Managers being in the equation. Likewise, Total Sqft is significant above and beyond Total Sales per year and FTE Managers being in the equation. However, FTE Managers is not significant. This multiple regression equation is a better model than when only the overall mean of Y is used.

This seems to be a good equation, but it might not necessarily be the best one. You need to determine what should be done with FTE Managers, the non-significant variable. Should it be removed from the equation? The reduced model based on Total Sales per year and Total Sqft is considered next.

16.3.5 Regression Model Using Total Sales per Year = X_1 and Total Sqft = X_2

A representation of a multiple regression model with two linear X-variables looks like a plane suspended in space. As Figure 16.7 indicates, such a fitted model is characterized by the equation $\hat{Y}_i = b_0 + b_1 X_{1i} + b_2 X_{2i}$. The subscript *i* in this equation refers to a particular observation. In the branch profits example, the subscript ranges from i = 1, ... , 18.

Figure 16.7 Estimated Linear Response Yhat as Function of X_1 and X_2

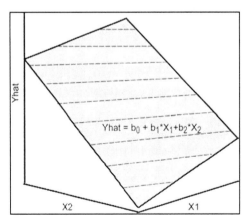

Figure 16.8 shows the Summary of Fit of this model. The R^2 of the full model, containing all three variables, is 0.998724, which is marginally higher than the current reduced model. This also demonstrates one peculiar aspect of the R^2. Whenever a variable is added to the equation, R^2 always increases or, in some rare cases, might remain the same.

The Root Mean Square Error (RMSE) is different. Removing an X-variable from the equation might increase or decrease its value. In the branch profits example, the model with X_1 and X_2 results in a smaller RMSE than the full model. Removing X_3 from the model resulted in a smaller RMSE = 11785.42 compared to the RMSE = 12194.79 of the three-variable model (see Figure 16.5). Models with a significantly smaller RMSE offer a tighter fit compared to models with a higher RMSE.

Figure 16.8 Summary of Fit of Two X-Variable Model

Summary of Fit	
RSquare	0.998723
RSquare Adj	0.998553
Root Mean Square Error	11785.42
Mean of Response	494487.1
Observations (or Sum Wgts)	18

From the Parameter Estimates table in Figure 16.9, observe that the slope coefficients of Total Sales per year and Total Sqft have not changed much after elimination of FTE Managers from the model. Recall (Figure 16.6) that in the full model, the slope coefficient of Total Sales per year is $b_1 = 99.031$ and of Total Sqft is $b_2 = 24.028$. The coefficients of the reduced model of Figure 16.9, $b_1 = 99.047$ and $b_2 = 24.128$, have changed very little.

Figure 16.9 Parameter Estimates Table of Two X-Variable Model

Parameter Estimates				
Term	Estimate	Std Error	t Ratio	Prob>|t|
Intercept	-278183	9870.104	-28.18	<.0001*
Total Sales per year	99.047475	0.915263	108.22	<.0001*
Total Sqft	24.127716	1.602384	15.06	<.0001*

The multiple regression equation of the model with two X-variables is

$$\hat{Y} = -278183.0 + 99.047 \cdot \text{Total Sales per year} + 24.128 \cdot \text{Total Sqft}$$

Conclusions from Two X-Variable Model

The results show that little is lost in terms of statistical precision by eliminating FTE Managers from the equation. The RMSE is almost the same as before. The coefficients did not change much, suggesting that FTE Managers has little or no impact on Y above and beyond Total Sales and Total Sqft. This simpler model seems preferable to the three-variable model, because it is more economical in achieving similar results.

16.4 Evaluating Model Assumptions and Goodness of Fit

An important step in multiple regression analysis is to evaluate model assumptions and how well the model fits the data. As with simple regression analysis, residuals play an important role in this step. Residuals are used in determining such quantities as RMSE and R^2, but they can also be informative in their own right, especially when plotted against other variables. Large residuals are especially noteworthy. Residual patterns can reveal problems with the data or inadequacies in the model.

In this section, we use the model with the three X-variables Total Sales per year, Total Sqft, and FTE Managers, even though the two-variable model (excluding FTE Managers) appears superior. The reason for choosing this model is to demonstrate that some tools actually show that a variable such as FTE Managers is not needed in the model. These tools help with finding a good equation, a topic discussed in Chapter 18.

16.4.1 Residuals e_i

An actual observation can be represented as

$$Y_i = \hat{Y}_i + e_i = b_0 + b_1 X_{1i} + b_2 X_{2i} + b_3 X_{3i} + e_i$$

where the residual e_i is the difference between the actual observation Y_i and the predicted value $\hat{Y}_i$, i.e., $e_i = Y_i - \hat{Y}_i$. Small residuals indicate a close fit.

Residuals and predicted values can be saved to the data table with the **Save Columns** menu shown in Figure 16.10. The residuals are saved to a new column called Residual Branch Profits. A similar operation saves predicted values to a separate column. These saved variables can be used to create plots or tables for further analysis.

For predicted values, there is a choice of saving them as a formula (**Prediction Formula**) or as calculated values (**Predicted Values**). Both operations yield identical results. However, the **Prediction Formula** option stores the values as the result of a formula that can be altered, whereas the **Predicted Values** option stores them as numbers. The JMP data after saving both prediction formula and residuals are in Figure 16.10.

Figure 16.10 Predicted and Residuals of Three X-Variable Regression

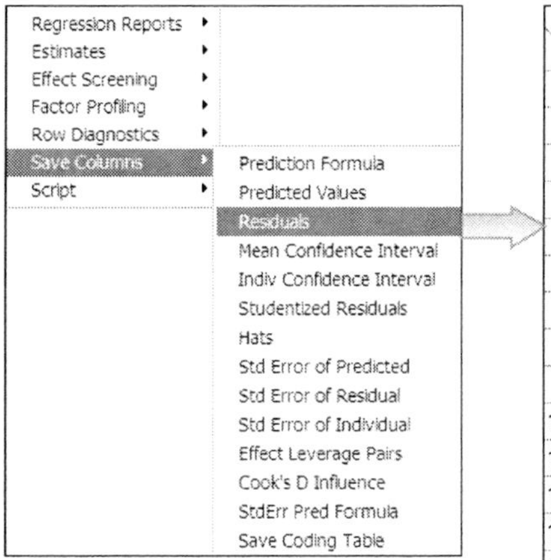

	Total Sales per year	Total Sqft	FTE Managers	Branch Profits	Pred Formula Branch Profits	Residual Branch Profits
1	10591	2474	2	821298	830327.724	-9029.7235
2	9069	4262	3	718488	723264.241	-4776.241
3	7511	7608	3	630752	649370.498	-18618.498
4	1541	3248	2	-40582	-47305.651	6723.65059
5	7569	5095	2	606998	594032.493	12965.5067
6	6014	6455	2	474231	472716.883	1514.11715
7	10506	3628	3	861916	850338.281	11577.7186
8	4588	2818	1.5	246162	243759.859	2402.1411
9	6058	2481	2	386684	381588.272	5095.72781
10	3073	7027	3	187208	195910.722	-8702.7217
11	12021	4052	2.5	1031091	1010207.89	20883.1089
12	7541	1697	2	504632	509613.589	-4981.5888
13	3090	3023	2.5	100223	101037.313	-814.31329
14	7594	3483	3	561795	558475.918	3319.08167
15	10587	5646	2.5	901631	906497.511	-4866.5106
16	1529	5813	3	29199	13837.2201	15361.7799
17	6053	1821	2	346304	365234.852	-18930.852
18	7578	2872	2.5	532738	541860.383	-9122.3834

Menu (left side):
Regression Reports
Estimates
Effect Screening
Factor Profiling
Row Diagnostics
Save Columns
Script

Submenu:
Prediction Formula
Predicted Values
Residuals
Mean Confidence Interval
Indiv Confidence Interval
Studentized Residuals
Hats
Std Error of Predicted
Std Error of Residual
Std Error of Individual
Effect Leverage Pairs
Cook's D Influence
StdErr Pred Formula
Save Coding Table

16.4.2 **Residuals e_i and RMSE**

As seen in ANOVA, the sum of all squared residuals is the SSE. The MSE or Mean Square Error is the SSE divided by the degrees of freedom. The square root of the MSE is the Root Mean Square Error (RMSE) and is the estimated standard deviation around the fitted model. The formula for the RMSE of an equation with (p – 1) X-variables is

$$RMSE = \sqrt{\frac{\sum_{i=1}^{n} e_i^2}{n - p}}$$

For a model with three X-variables, $p = 4$ (p = number of X-variables plus 1 for the intercept term). The divisor for the error degrees of freedom is $n - 4$ (DF_{Error}). The RMSE is available in the Summary of Fit table (see Figure 16.5 and Figure 16.8).

16.4.3 Residuals and R^2

The R^2 measures the proportion of the Total Sum of Squares that is explained by the model. A residual is the unexplained part of an observation. So from this we can calculate

$$R^2 = \frac{SS_{Model}}{SS_{Total}} = 1 - \frac{SS_{Error}}{SS_{Total}} = 1 - \frac{\sum_{i=1}^{n} e_i^2}{\sum_{i=1}^{n}\left(Y_i - \overline{Y}\right)^2}$$

This equation shows that SS_{Error} is the sum of squared residuals. Likewise, the SS_{Total} is a sum of squared residuals, except that the differences are taken between observed values and the overall mean $\overline{Y}$. So the R^2 represents a comparison between the regression and the baseline model.

16.4.4 R^2 and RMSE

One of the disadvantages of the R^2 is that as new X-variables are added to a model, the value of R^2 never decreases, even when other criteria such RMSE show an inferior fit. In the branch profits example, the R^2 of the *full* model containing all three X-variables is 0.998724, which is slightly larger than the R^2 of the model containing only (X_1, X_2). In contrast, the RMSE does not always go down as new variables are added. Conversely, the RMSE might actually decrease (improve), if a variable is eliminated from the model. The criterion is that if the t-Ratio < 1, eliminating that variable from the model will reduce the RMSE. If the t-Ratio > 1, eliminating a variable will increase the RMSE, but it might still improve the model.

16.4.5 R^2_{adj}

The R^2 increases with every X-variable added to the regression equation, regardless of the actual contribution that variable makes in explaining the Y-variable. Consequently, some users prefer the adjusted R^2:

$$R^2_{adj} = 1 - \frac{MSE}{SS_{Total}/(n-1)}$$

The adjusted R^2 follows the same movement as the RMSE by adjusting for the number of variables in the model. It might decrease if non-significant variables are added to the equation.

16.4.6 Plot of Residuals versus Predicted Y (Yhat) or versus Time Order

The plot of residuals versus the predicted values or versus the time variable can be used to examine the residuals for abnormal patterns. Residuals are often examined for constant variance over the range of X (homoscedasticity) and large residual values called outliers, as well as autocorrelation patterns (that is, correlation of residuals over time: the tendency that large residuals are followed by large residuals or large residuals are followed by small ones). Some patterns can be evaluated by special tests; others are often obvious when presented in a graph. Examples of residual patterns that violate the standard regression assumptions are shown in Figure 16.11. The left pattern shows that residuals are increasing as Yhat increases. The middle pattern shows a single outlier residual. This residual might unduly influence the regression estimates. The pattern on the right shows that positive and negative residuals alternate indicating correlations among the observations.

Figure 16.11 Residual Patterns Violating Standard Regression Assumptions

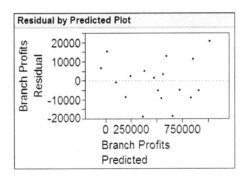

The plot of residuals versus Predicted Ys (Yhats) of the branch profits example, shown in Figure 16.12, reveals no particular pattern. This indicates that the regression assumptions appear to be met.

Figure 16.12 Residual versus Predicted Y Plot

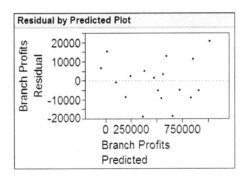

16.4.7 **Durbin-Watson Test for the Independence of Residuals**

More sophisticated tools can be applied to the analysis of residuals. For example, autocorrelation is often a hidden problem in economic data. Assuming that the observations are in time order, the Durbin-Watson test examines correlation between consecutive observations. The Durbin-Watson test does not apply to the branch profits example, because all profits refer to the same time period. However, first-order autocorrelation is sufficiently important to warrant an additional simple example.

Example: Dow Jones 30 Industrials Adjusted Close

The data file *DowClose2009.jmp* contains the 102 (adjusted) closing values from March 2, 2009 to July 24, 2009. Specifically, the file contains a Date column and a Time column. The Time column labels the time order of the dates from 1 to 102. The column Dow 30 Close contains the adjusted closing values for each date, while the column Diff Dow 30 Close contains the differences between consecutive days. As a result of taking differences, there are only 101 values.

In order to demonstrate the Durbin-Watson test, we run two simple regressions with the Fit Model platform. Both regressions have the same X-variable Time. The two Y-variables are Dow 30 Close and Diff Dow Close. For each regression, enter the appropriate variables and select **Run Model**. The residual versus predicted plot is part of the default output. Then ask for the **Durbin-Watson Test** from the **Row Diagnostics** menu (Figure 16.13). This results in a window with the Durbin-Watson statistic, the number of observations used, and the estimated autocorrelation.

Figure 16.13 Selecting the Durbin-Watson Test

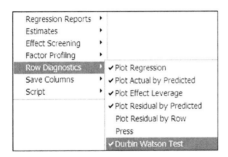

Next, click the red triangle of the Durbin-Watson window to obtain the associated *p*-value to test whether the autocorrelation is significantly different from 0 (no autocorrelation). Figure 16.14 shows complete results. On the left is the plot of Dow 30 Close Residual versus Dow 30 Close Predicted with a pronounced curvilinear pattern. The Durbin-Watson test window shows The Durbin-Watson statistic = 0.2294821 with a *p*-value < 0.0000. The estimated first order autocorrelation is 0.8716. The assumption of the independence of residuals is not satisfied. The conclusion is that in their present form, these data should not be analyzed with standard least squares regression.

Now consider the results on the right of the daily differences Diff Dow 30 Close $=$ $[(\text{Dow 30 Close})_t - (\text{Dow 30 Close})_{t-1}]$, where the subscript $t = 2, 102$. The plot of residuals versus predicted values shows a random pattern. The Durbin-Watson statistic (2.2287757) indicates no significant first-order autocorrelation with p = 0.8541. The autocorrelation coefficient is –0.1156. This suggests that during the indicated time, the daily differences of the Dow Jones 30 Industrials were random and independent, but that the actual Dow Jones 30 Industrials were not. This pattern has been observed frequently and has financial implications. However, here it is used merely to demonstrate the Durbin-Watson test.

Figure 16.14 Durbin-Watson Test of Residuals for Dow Jones 30 Industrials and Daily Differences

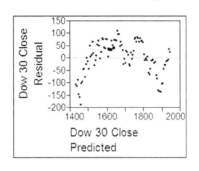 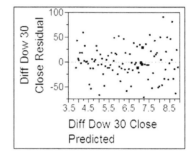

Durbin-Watson	Number of Obs.	AutoCorrelation	Prob<DW
0.2294621	102	0.8716	0.0000*

Durbin-Watson	Number of Obs.	AutoCorrelation	Prob<DW
2.2287757	101	-0.1156	0.8541

Other diagnostics, some quite sophisticated, are available on the **Save Columns** menu, including Leverages and Cook's D statistic for outliers. This book will not cover these tools.

16.5 Model Interpretation

Multiple regression models are more difficult to interpret than simple regression models. Parameter estimates are often correlated, sometimes to a degree that makes interpretation of individual coefficients dangerous at best. The parameter estimates might have different units so that a straightforward comparison of, say, b_1 and b_2 does not make sense.

In the following sections, we discuss leverage plots as a graphical means to evaluate the importance of an X-variable in explaining the response variable. Next we discuss standardized Beta coefficients, widely used to assess the importance of individual variables. Lastly we look at variance inflation factors (VIFs) to determine whether X-variables are correlated amongst themselves, a situation called multi-collinearity.

16.5.1 Leverage Plots and the Importance of X_i

Fit Model produces leverage plots for each X-variable. Leverage plots are useful with continuous variables. They show graphically how each variable contributes in multiple regression. The lines can be viewed as visualized t-Ratios. The slope of the line is the parameter estimate b_i. The vertical axis is in Y-units and JMP keeps an identical Y-range across different ranges in the X-variables.

Leverage plots can be used in two ways:

1. The absolute magnitude of the slope of the fitted line of the leverage plot of each predictor variable is a measure of its relative importance.

2. Data points should be scattered randomly around the line. If they deviate from a random pattern, assume that the model has certain deficiencies. For example, if the data follow a curved pattern along the line, this may be an indication that a quadratic term should be added to the model (see Figure 16.17).

Figure 16.15 is a schematic of a leverage plot. It shows the strength of the effect each X-variable has on the Y-variable. The difference of an observation from the horizontal line is the residual of that observation if the X-variable is excluded from the model. The difference of an observation point from the fitted line is the residual of that observation if the X-variable is included in the model.

The horizontal line is the fitted model (Yhat) without the X_i-variable included. The sloped line represents the fitted model with X_i included. Ideally, observations should be randomly scattered around the horizontal line or the sloped line. When observations hug the sloped line, including X_i results in smaller regression residuals and in a better fitting regression model. When observations are scattered around the horizontal line, including X_i will not help in explaining the data. So X_i does not improve the model above and beyond what the other X-variables already explain.

Figure 16.15 Leverage Plots of Individual X-Variables

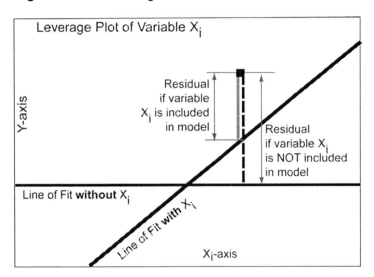

The slope of the leverage plot is a good indication in absolute terms of how important a variable is. A steeper slope points to a more important variable. A relatively steep slope indicates that the X-variable can explain a relatively large amount of the variation in the Y-values, assuming other X-variables are already in the equation.

The three leverage plots of X_1, X_2, and X_3 explain the story in the branch profits example (Figure 16.16). Total Sales per year (in the left panel) is much stronger than the Total Sqft, because the slope of the Total Sales per year leverage plot is steeper than that of the Total Sqft leverage plot in the middle panel. In a model without Total Sales per year, the residuals would be considerably larger than in a model without the Total Sqft variable. Lastly, the leverage plot of FTE managers (right panel) shows a near zero slope, which is consistent with the statistical insignificance of this variable.

Figure 16.16 Leverage Plots for Total Sales per Year, Total Sqft, and FTE Managers

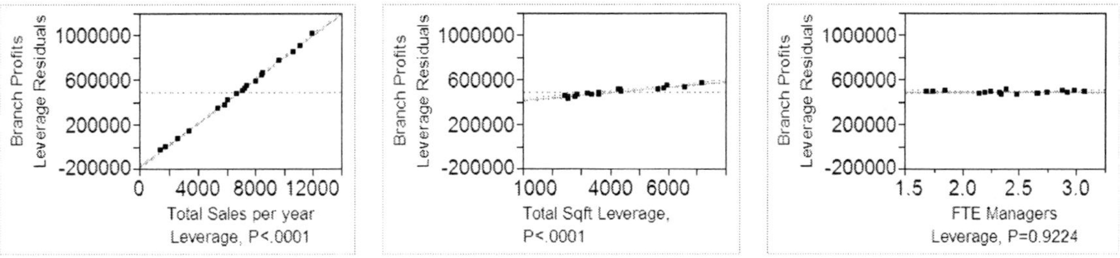

Leverage plots also give important graphical information regarding the adequacy of the model. The pattern of the scatter might also suggest outlier observations. Often one or two outlying observations can incorrectly render an X-variable significant, as shown in the left panel of Figure 16.17. There are two outlier points that are responsible for this variable having a significant *p*-value. This variable is kept in the equation only because of these two outliers. If they were absent, the *p*-value would not be significant.

Depending on patterns of the data plotted, leverage plots may suggest transformations of the X-variables or the Y-variable. The right panel of Figure 16.17 shows that the leverage residuals form a curved pattern (as indicated by the dashed line) around the straight line. In such situations, a simple transformation such as adding a quadratic X-term might linearize the pattern of the leverage residuals to be random around the line.

Figure 16.17 Two Leverage Plot Patterns

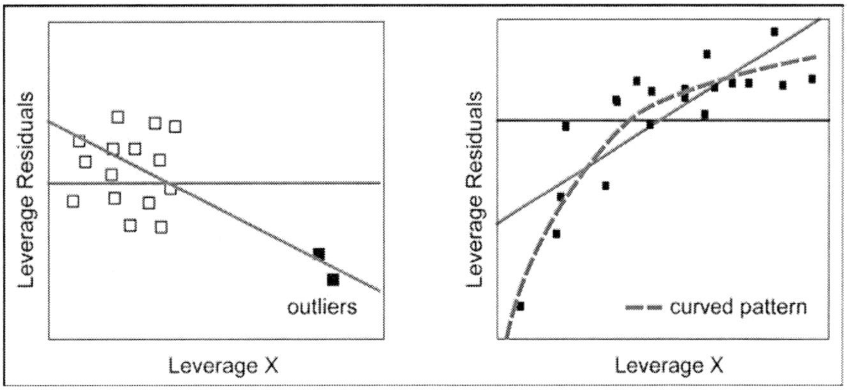

16.5.2 **Standardized Beta and the Importance of X$_i$**

Either standardized Betas or leverage plots can be used to determine the relative importance of variables in the model. The relative size of the absolute value of the standardized Beta of X$_i$ measures its importance in the model. Variables with relatively large Betas in absolute value are more important than those with relatively smaller ones. Standardized Beta coefficients are needed for the comparison, because multiple regression slopes have different units. In the branch profits example, the units of X$_1$ are Dollars/Total Sales per year, whereas those of X$_2$ are Dollars/Total Sqft.

Standardized Beta coefficients are unitless quantities. Their absolute values allow a numerical comparison of the importance of the X-variables in the model. In multiple regression, the Betas can be larger than +1 and smaller than −1. In JMP, Betas are obtained by placing the cursor over the Parameter Estimates table and right-clicking and selecting **Columns → Std Beta**. This produces an additional column Std Beta in the Parameter Estimates table on the right of Figure 16.18.

Figure 16.18 Parameter Estimates Table with Standardized Betas and VIFs

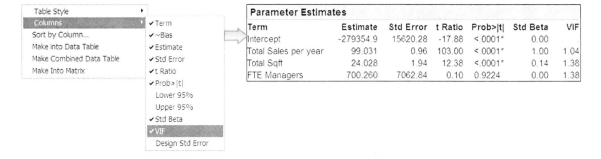

In the branch profits example, Total Sales per year has a standardized Beta B$_1$ = 1.003. This value is considerably higher than the standardized Beta B$_2$ = 0.139 of Total Sqft. Total Sales is the much more important variable because it has a higher standardized Beta. FTE Managers is unimportant with B$_3$ = 0.001. (*Always use absolute values when comparing standardized Betas.*)

The standardized Beta of X$_i$ is

$$B_i = b_i \cdot \frac{S_{x_i}}{S_y}$$

For Beta$_{\text{Total Sales per year}}$ calculate the two standard deviations S$_{\text{Total Sales per year}}$ and S$_{\text{Branch Profit}}$ using the Distribution of Y platform. With S$_{\text{Total Sales per year}}$ = 3138.47 and S$_{\text{Branch Profit}}$ = 309775.14,

$$\text{Beta}_{\text{Total Sales per year}} = b_{\text{Total Sales per year}} \cdot \frac{S_{\text{Total Sales per year}}}{S_{\text{Branch Profit}}} = 99.031 \cdot \frac{3138.47}{309775.14} = 1.003$$

16.5.3 Understanding the Role of X-Variables: Column Diagnostics

The outcome of the equation-finding process should be one or more good candidate equations that can be examined and modified appropriately to suit the problem at hand. You need to examine the properties of a regression equation in more detail. In this section we discuss checking regression slopes and variance inflation factors (VIFs).

Check the Slopes of the Equation

Ideally, regression coefficients should make sense. They should have the right sign and magnitude, and they should also be consistent with expert judgment. We can use leverage plots and standardized Betas for these checks: X-variables with an expected positive effect should have a positive slope, etc. However, there are exceptions to this rule. In the branch profits example, consider the equation with three X-variables:

Yhat = –278183.0 + 99.031*Total Sales per year + 24.028*Total Sqft + 700.260*FTE Managers

The Total Sales per year slope coefficient estimates that for each increase of one unit in Total Sales, Branch Profits will rise on average by an estimated 99.031 dollars. The slope coefficient of Total Sqft shows the expected increase in profits for large branches in terms of square feet. You could draw similar conclusions from the leverage plots. The important part is that both slopes are positive, as expected, and that the relative impact of each variable makes sense.

Check for Multi-Collinearity Variance Inflation Factors (VIFs)

Multi-collinearity occurs when the X-variables are correlated among themselves. This can be discovered by looking at simple correlation coefficients among the X-variables. There are situations when simple (pairwise) correlations do not reveal multi-collinearity. With multi-collinearity in the X-variables, regression coefficients are often quite difficult to interpret. For example, regression coefficients of correlated variables might have the wrong sign because of this multi-collinearity. Dropping one of the correlated X-variables from the equation can produce considerable changes in the sign and magnitude of regression slopes.

We measure multi-collinearity most conveniently with VIFs. VIF_k measures the correlation of X_k with a linear combination of all other X-variables in the equation. Each X-variable has a VIF greater than or equal to 1. You can apply simple rules concerning VIFs to decide whether multi-collinearity exists:

- If a $VIF_k > 5$, be concerned.
- If a $VIF_k > 10$, consider changing the equation, e.g., by eliminating the X_k responsible for the multi-collinearity.

There are exceptions to this rule. Pairs of transformed variables, such as a linear term X and a quadratic term X^2 in a polynomial fit, might show high VIFs, yet you might want to include the transformed as well as the original variable. Related variables such as X and X^2 are then interpreted jointly.

Adding VIFs to the Parameter Estimates table is similar to adding standardized Betas as shown in Figure 16.18. A new column, VIF, will be added to the Parameter Estimates table. In this example, all three VIFs are nearly 1, the ideal value for absence of multi-collinearity. Total Sales per year, Total Sqft, and FTE Managers are close to being statistically independent variables. The VIFs of 1.38 of Total Sqft and FTE Managers are due to the correlation ($r = 0.505$ from Figure 16.2) between these two variables.

What is the variance inflation factor of X_k? Using regression analysis, we can measure how much X_k can be explained in terms of the other X-variables. For finding VIF_k, run the equation in which X_k is the dependent variable (like the Y before). Predict X_k from the remaining predictor variables ($X_1, X_2, ... , X_{k-1}, X_{k+1}, ..., X_{p-1}$). This will result in an equation of the form:

$$X_k = m_0 + m_1 X_1 + \cdots + m_{k-1} X_{k-1} + m_{k+1} X_{k+1} + \cdots + m_{p-1} X_{p-1}$$

(Using JMP, you need not go through this extra regression equation.) The R^2 of this equation is directly related to the VIF_k. The more the X_i are related with X_k, the higher the R^2 of this equation. We define then

$$VIF_k = \frac{1}{1 - R_k^2}$$

As R_k^2 approaches 0, indicating lack of multi-collinearity, VIF_k approaches 1. As R_k^2 approaches 1, indicating near perfect multi-collinearity, VIF_k approaches infinity. The range of variance inflation factors is always $1 \le VIF \le \infty$.

Global Statistical Test in Multiple Regression

Overall F-tests in the ANOVA table of a multiple regression output test the overall null hypothesis $H_0: \beta_1 = \beta_2 = ... = \beta_p = 0$ versus H_A: not all the slope coefficients are equal to zero. This is accomplished with the overall F-Ratio

$$F = \frac{MS_{Regression}}{MS_{Error}}$$

with p and $(n - p - 1)$ degrees of freedom. This F-Ratio is not as useful as individual t-Ratios for X-variables. However, the F-Ratio provides a quick assessment of the overall significance of the model.

16.6 Summary

- Multiple regression models have one dependent variable and several independent variables.

- The dependent variable is always continuous. Independent variables are usually continuous, although JMP allows nominal variables in the model as well (see Chapter 17).

- A multiple regression model should help characterize the relationship between the Xs and Y or allow predictions of Y.

- Multiple regression analysis is different from analyzing simple pairwise relationships between the dependent and independent variables.

- Multiple regression analysis fits the dependent variable to several independent variables.

- The overall fit is evaluated by several statistics, such as R^2, RMSE, and the global F-test.

- The *p*-value of the t-Ratio tests the statistical significance of individual parameters.

- Checks of model adequacy should be part of regression analysis.

- Residual plots, leverage plots, standardized Beta coefficients, variance inflation factors, and others are also used to evaluate multiple regression models.

- Residuals of a fitted model show the closeness of individual observations to the fitted model and are useful for assessing the adequacy of model assumptions.

- Residual plots are tools to find trends and other patterns in the data relative to the model.

- Leverage plots indicate how much an X-variable contributes toward explaining Y, given that all other X-variables remain in the model. The slope of the line is the slope of the variable in the equation.

- Leverage plots help identify outliers or patterns that might lead to changes in the model.

- Standardized Beta coefficients are used to place numerical values on the importance of each X-variable in explaining Y.

- Variance inflation factors (VIFs) are numbers that indicate the strength of relationships between all X-variables in the equation. Each X-variable has a variance inflation factor. A VIF = 1 indicates that this variable is unrelated to the other X-variables. There are situations in which large VIFs cannot be avoided.

16.7 Problems

1. Create a checklist to use for the process of building a multiple regression model.

2. An article in the *Journal of Real Estate Practice and Education* (Benjamin, Guttery, and Sirmans 2004) presents the following multiple regression equation obtained from data on 185 homes for predicting the sales price of a home without a fireplace:

   SalesPrice = 64.46LA – 8.19DOM + 16.10GARAGE – 2555.02AGE

   where LA = square footage of living area, DOM = days on the market, GARAGE = square footage of garage area, and AGE = age of structure (years). All regression coefficients are significant at the 1% level.

   a. Give an interpretation of the regression coefficient of LA.

   b. Give an interpretation of the regression coefficient of AGE.

   c. Do all the regression coefficients make sense? Explain why or why not.

   d. Use the model to predict the sales price for a home that has 2400 sqft of living area, was on the market for 4 months, has a 500 sqft garage, and is 10-years old.

3. The office manager of a medical practice wishes to increase the number of patients seen per day and has collected data to understand which factors significantly affect the total length of an office visit, measured in minutes. Data were collected from 25 randomly selected office visits over a two-week period and include the following:

Length of Visit	Total length of office visit (in minutes)
Admission Time	Time to check in and fill out necessary paperwork (in minutes)
Physician Time	Time the physician spends with the patient (in minutes)
Intake Nurses	Number of nurses doing initial screening of patients.
Patients Waiting	Number of patients who are waiting for initial screening
Checkout Time	Time needed to pay bill, schedule next appointment, etc. after seeing the doctor (in minutes)

3. Develop a multiple linear regression model to predict the total length of an office visit using the data found in *OfficeVisits.jmp*.

   a. Describe the data.

   b. Run the multiple regression model with all X-variables. Examine which Xs are non-significant and eliminate them one at a time. Always select the most non-significant variable (highest *p*-value) to delete from the equation. After all X-variables are significant, stop deleting and write down this equation as your final model.

   c. Give an interpretation for each of the significant regression coefficients.

   d. Is this a good equation? Look at R^2, RMSE, and residual plots, etc., to make this determination.

   e. Do the regression coefficients make sense (e.g., do they have the right sign)?

   f. Based on your analysis, what recommendations would you make to reduce the length of office visits?

4. A college senior is preparing to apply to business schools and is interested in predicting his post-MBA income. The file *Post_MBA_salary.jmp* gives data on some of the best U.S. business schools, as identified on the Forbes Web site.

   a. Evaluate the correlation coefficients of each pair of variables. Based on this analysis, which variables might be promising in predicting post-MBA salary? Does it appear that multi-collinearity might be a problem?

   b. Fit a simple regression for post-MBA salary for each of the independent variables. Which of these simple regressions are significant?

   c. Fit a multiple regression for post-MBA salary using all of the independent variables. Eliminate insignificant variables. Check for the presence of multi-collinearity.

    d. Compare the results of the simple regressions with those of multiple regression.

5. A sports and entertainment complex is considering offering a "family package" with discounted pricing for certain events that are likely to be of interest to families. The family package would include tickets (minimum of 4 and maximum of 12) plus discount coupons for concessions. The file *TicketPurchases.jmp* contains data randomly obtained from surveys mailed to recent patrons who have bought children's priced tickets. The file contains the number of members in the family (including parents), weekly discretionary income ($/week), ticket price of most recently attended event, and number of tickets purchased. Use this data to fit a multiple regression model to predict number of tickets purchased.

    a. Describe the data.

    b. Evaluate the pairwise correlations. Which X-variables are strongly correlated with the dependent variable? Which X-variables are strongly correlated with each other?

    c. Construct simple regression for each of the independent variables. Create a table that summarizes the results.

    d. Construct a multiple regression using all of the available independent variables. Write down the regression equation.

    e. Evaluate the goodness-of-fit of the multiple regression equation. Find the variance inflation factors. What is your assessment?

    f. Which of the three models is best? How much do the regression estimates vary between the three models?

    g. Give the interpretation of each of the regression coefficients in the best model.

6. Using the best equation obtained in Problem 5, answer the following questions:

    a. Which of the independent variables is most important in predicting number of tickets sold? Explain how you arrived at your answer.

    b. How does significance differ from importance? What is used to measure each?

    c. Use the regression model to predict the number of tickets sold for a family of four with discretionary income of $150/week for an event whose ticket price is $30.

    d. Find a 95% confidence interval on the average number of tickets sold for the situation given in part c. How could this interval be useful to management in planning the family discount package?

   e. Find a 95% prediction interval. How could this interval be useful to management in planning the family discount package?

   f. Evaluate the residuals.

   g. Give any suggestions for obtaining an improved model.

7. A company wishes to decide the length of their television commercials and the number of times per day the ads should run. The file *SalesIncrease.jmp* shows the sales increase experienced from previous advertising campaigns. The file contains the sales increase in tens of thousands of dollars, the length of the commercial (15, 30, 45, or 60 seconds), and the number of days the commercial is run. Construct a multiple regression equation with sales increase as the dependent variable.

   a. Examine the plot of residuals vs. the predicted values. Is there cause for concern?

   b. Save the residuals.

   c. Using **Graph → Overlay**, plot the residual by each of the independent variables in the model. Are there any apparent patterns?

   d. Suggest ways in which the model could be improved based on your examination of the residuals.

8. The Forbes Web site has identified 88 U.S. companies from its Forbes Global 2000 list as being high performers. The criteria used to make this determination included manageable debt, balance of long- and short-term return of equity, sales growth, earning growth, and return to shareholders. Use the data contained in the file *USA_High_Performers.jmp* to build a regression model that predicts market value. Identify those variables that are significant predictors of market value and rank them in order of importance. Assess the goodness-of-fit of your model. Suggest other variables that might improve the predictive power of the regression model. Given that these data are not randomly selected, what are the restrictions on using multiple regression for prediction?

16.8 Case Study: Forbes Global 2000 High Performers

The Forbes Web site has designated 125 international companies from its Forbes Global 2000 list as being high performers. The criteria used to make this determination included manageable debt, balance of long- and short-term return of equity, sales growth, earning growth, and return to shareholders. Forbes provides the following data for each of the high performers: sales, profits, assets, and market value, all recorded in billions of dollars. The data for 2005 are contained in the file *Forbes_High_Performers.xls*. This data will be used to build a multiple regression equation that predicts profitability. Only a portion of the observations will be used to build the regression model.

Tasks

Perform the following tasks:

1. Use the **JMP Table** → **Subset** to randomly select 80 observations. These are the observations that will be used to develop the regression model.

2. Use multiple regression analysis to find those factors that contribute significantly to predicting profitability. Give the multiple regression equation and assess its goodness-of-fit. Identify those variables that are most important in determining profitability. Discuss your treatment of outlying observations.

3. Select 4 or 5 companies that were not used to develop the multiple regression model. Use the model to predict profitability for those observations and compare it with the actual profitability. (This is a technique used to evaluate the predictive capability of models.)

4. Write a technical report that summarizes your findings and discusses the scope of applicability of this model and its predictive capabilities.

16.9 References

Benjamin, John D., Randall S. Guttery, and C. F. Sirmans. 2004. "Mass Appraisal: An Introduction to Multiple Regression Analysis for Real Estate Valuation." *Journal of Real Estate Practice and Education.* 7, No. 1.

Chapter 17

Multiple Regression with Nominal Variables

17.1 Introduction 482

17.2 Detail Example: Loan Amount versus Sales Revenues 483

17.3 Difference of Intercepts of Two Parallel Lines 485

    17.3.1 Using JMP for Regression Analysis with a Nominal Variable 485

    17.3.2 Interpretation of Parameter Estimates and Associated Hypothesis Tests 489

17.4 Regression Models Including Nominal Variables with Three or More Levels 490

17.5 Both Intercept and Slope of Two Lines Are Different 495

    17.5.1 Detail Example: Growth versus General Equity Fund 495

    17.5.2 Using JMP to Model Two Slopes and Two Intercepts 496

17.6 Summary 501

17.7 Problems 501

17.8 Case Study: Coffee Sales 505

17.1 Introduction

In the previous chapter, multiple regression analysis used only continuous X-variables. However, by internally transforming nominal variables into continuous ones, JMP accommodates both continuous and nominal variables in the same regression model. This enhances the versatility of multiple regression analysis in business applications. The following list is only a small smattering of examples with both continuous and nominal X-variables.

- Chemical Yield (Y) is a function of agitation (X_1) and supplier (X_2) of chemicals. Agitation is a continuous variable, while supplier is a nominal variable.

- Sales revenue (Y) is a function of advertising expenditure (X_1) and geographic region (X_2). Advertising expenditure is a continuous variable, while geographic region is a nominal variable.

- Strength (Y) of a material is a function of operating temperature (X_1) and surface preparation (X_2). Operating temperature is a continuous variable, while surface preparation is a nominal variable.

This chapter deals with how to set up and interpret regression problems that include nominal X-variables. Nominal variables have a finite number of levels and we are interested in testing mean differences between them with appropriate statistical tests. For a nominal variable with two levels, there are two tests of interest in business applications:

1. Is there a statistically significant constant difference between the levels of the nominal variable as represented by the different lines? The left panel of Figure 17.1 shows the difference between two parallel regression lines. Each line represents a different level of the nominal variable. The hypothesis test involves the differences between the two intercepts of the two lines with equal slopes. The two intercepts help us distinguish means of the A and B levels of the nominal variable. This problem is easily extended to more than two regression lines.

2. Do the regression lines have statistically significant different slopes and intercepts? Determining statistically significant differences involves hypothesis tests for differences in both intercepts and slopes of two regression lines. The right panel of Figure 17.1 shows the two regression lines with a different intercept and slope, each representing a different level of the nominal variable.

Figure 17.1 Test Situations for Different Slopes and Intercepts

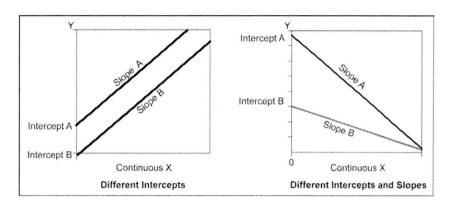

Coding Variables

In its original form, regression analysis required continuous Y- and X-variables. The Fit Model platform does not use nominal X-variables directly. Instead, nominal X-variables are converted into numerical variables. We call these coding variables. For the user, it is important to understand this aspect, because the interpretation of parameter estimates of nominal variables depends on how these variables are coded. JMP automatically codes nominal variables with two levels into a single numeric variable with values +1 and –1. Those with three or more levels use numerical variables with values +1, 0, or –1. The coding method used by JMP is called *effect coding*. The coding variables themselves are present only in the background. However, such variables are used in tests for significance.

17.2 Detail Example: Loan Amount versus Sales Revenues

Situation: SaveTrust is a retail bank with branches operating in several regions of the country. One of its very popular and profitable programs is a loan program to small businesses. Loan amounts are approved based on clearly defined criteria. One such criterion is annual sales revenue of the business applying for the loan. The bank suggests that loan amounts should be related to the annual sales of the business requesting the loan.

Problem: SaveTrust's Senior Vice President for Business Lending is concerned that the criteria for approving loans are applied inconsistently across regions. At first she wants to compare how two regions apply the annual sales revenue criterion. The two regions are Aderly and Purth. The results are to be summarized in a preliminary report to the board of directors.

Figure 17.2 Data File *LoanToSales.jmp* for Loan Amount versus Sales Revenues

LoanToSales

Columns (3/0): Loan Amount in 10000$; Annual Sales in 10000$; Region of Branches

Rows: All rows 13; Selected 0; Excluded 0; Hidden 0; Labelled 0

	Loan Amount in 10000$	Annual Sales in 10000$	Region of Branches
1	13.3	275	Aderly
2	8.9	200	Aderly
3	15.1	320	Aderly
4	10.4	220	Aderly
5	13.1	290	Aderly
6	12.4	265	Aderly
7	13.2	280	Aderly
8	14.8	260	Purth
9	11.5	210	Purth
10	13.9	270	Purth
11	15.4	295	Purth
12	13.1	230	Purth
13	13.8	250	Purth

Data Requirements: The data for this analysis are in *LoanToSales.jmp* (Figure 17.2). There are three columns:

- Column 1 is the continuous Y-variable Loan Amount in 10000$.
- Column 2 is the continuous X-variable Annual Sales in 10000$ representing the criterion used in deciding the loan amount.
- Column 3 is a nominal X-factor Region of Branches: Aderly and Purth.

The data are plotted in Figure 17.3. The dashed line in either panel shows the graphical results when a single line is fitted to the combined Purth and Aderly loans. Observe that Purth observations are above the fitted line, while Aderly observations are below the line. This suggests that the two levels do not follow the same line, supporting the suspicion that the two regions do not apply the criteria equally. This model can be improved by including a term that distinguishes observations by Region of Branches.

In Figure 17.3, as in all subsequent graphs, observations from Purth are represented by black empty squares, while observations from Aderly are represented by black dots. With these data, you can model Loan Amount as a function of Annual Sales, a continuous variable, assuming that businesses with higher annual sales also receive higher loan amounts. In addition, it can be determined whether there are differences in Loan Amount based on Region of Branches, a nominal variable. JMP will code Region of Branches to test whether two parallel regression lines have different intercepts. In practical terms, we test for the mean difference in Loan Amount between two regions, correcting for differences in Annual Sales values.

17.3 Difference of Intercepts of Two Parallel Lines

In this section, we test whether the intercepts of two parallel lines differ. The regression model is $Y_i = \beta_0 + \beta_1 \cdot X_1 + \beta_2 \cdot X_2 + \varepsilon_i$.

In the loan amount example, the dependent variable Y is Loan Amount in 10000$. X_1 is the continuous variable Annual Sales in 10000$. X_2 is Region of Branches, the coded background variable created by effect coding. In the example, this variable contains 1 in rows with Aderly and −1 in rows with Purth observations. This model represents two distinct regression lines. For Aderly, the model equation is $Y_i = (\beta_0 + \beta_2) + \beta_1 \cdot X_1 + \varepsilon_i$. For Purth, the model equation is $Y_i = (\beta_0 - \beta_2) + \beta_1 \cdot X_1 + \varepsilon_i$.

Thus the coefficient β_2 represents the ± average deviation from a line based on a regression model $Y_i = \beta_0' + \beta_1' \cdot X_1 + \varepsilon_i$, where X_1 represents Annual Sales in 10000$.

As with all regression models, the default null hypothesis is H_0: $\beta_i = 0$ against the alternative hypothesis H_A: $\beta_i \neq 0$. If H_0: $\beta_2 = 0$ is accepted, assume that the two lines do not differ. If H_0: $\beta_2 = 0$ is rejected, then assume that the two lines are different but parallel.

17.3.1 Using JMP for Regression Analysis with a Nominal Variable

Use the Fit Model platform with Loan Amount in 10000$ as the Y-variable. The two model effects are Annual Sales in 10000$ and Region of Branches. Figure 17.3 shows that two separate parallel lines produce a much closer fit than can be attained with the single X_1-variable. The dashed centerline is a line from which the two level lines deviate. The centerline is specified by the intercept b_0 and slope b_1 of the continuous X-variable. The deviation for each level is $\pm b_2$. In this example, the estimate b_2 of Region of Branches is negative. Therefore, the line for Aderly, although containing an intercept $(b_0 + b_2)$ is lower than the line for Purth, with an intercept $(b_0 - b_2)$.

Figure 17.3 Two Parallel Lines Fitted To Each Region with $b_2 < 0$

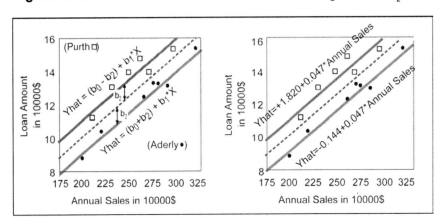

The Parameter Estimates table for this example is shown in Figure 17.4. The estimates of the regression coefficients are in the first numeric column. The column Prob>|t| contains the *p*-values for the significance test of each parameter estimate that the actual regression parameter is zero. The estimate of Region of Branches[Aderly] is –0.9820. It indicates that the intercept of the line for Aderly is estimated to be different by –0.9820. The *p*-value in the Prob>|t| column is <.0001, which is less than the $\alpha = 0.05$ threshold. The parameter b_2 is therefore significantly different from zero. The estimate for Purth is the negative of the estimate for Aderly. Thus the estimate for Region of Branches[Purth] = 0.9820. It is not listed in the Parameter Estimates table, but it is implied, because the estimates of the levels of a nominal variable need to sum to zero.

Figure 17.4 Parameter Estimates Table for Regression Model with Nominal Variable

Parameter Estimates				
Term	Estimate	Std Error	t Ratio	Prob>|t|
Intercept	0.8379	0.9504	0.88	0.3987
Annual Sales in 10000$	0.0472	0.0036	12.95	<.0001*
Region of Branches[Aderly]	-0.9820	0.1249	-7.86	<.0001*

From this table, construct the regression equations (with coefficients rounded to three decimals):

$$Yhat_{Aderly} = (0.838 - 0.982) + 0.047*Annual\ Sales\ in\ 10000\$$$
$$= -0.144 + 0.047*Annual\ Sales\ in\ 10000\$$$

$$Yhat_{Purth} = (0.838+0.982) + 0.047*Annual\ Sales\ in\ 10000\$$$
$$= 1.820 + 0.047*Annual\ Sales\ in\ 10000\$$$

Figure 17.3 shows the two lines with equal slope but different intercepts. The only difference between the equations for Purth and Aderly loans is the value of the intercept. The two fitted lines are parallel and separated by a vertical distance of $2*b_2 = 2*|-0.982|$ = 1.964. In terms of the intercepts of the two equations, the distance is $|1.820 - (-.144)|$ = 1.964.

Residual plots show that the model that includes the nominal variable has smaller residuals. Figure 17.5 shows two residual plots. The left panel shows that residuals of the model without the nominal variable, i.e., $Y_i = \beta_0 + \beta_1 \cdot X_1 + \varepsilon_i$, range between −2 and +2. As observed, the model without the nominal variable has all Purth residuals positive and all Aderly residuals negative. The residuals in the right panel of Figure 17.5 are from the model that includes both Annual Sales in 10000$ and Region of Branches. The model is $Y_i = \beta_0 + \beta_1 \cdot X_1 + \beta_2 \cdot X_2 + \varepsilon_i$. The residuals of the two groups appear randomly mixed. Their values are in the range from −1 to +1. Smaller residuals around the fitted model are a sign of a better fit.

Figure 17.5 Residual Plots for Regressions Without and With Nominal Variable

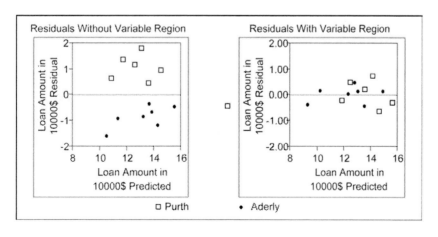

In JMP, the formula (Figure 17.7) for the two equations can be retrieved by selecting **Prediction Formula** from the **Save Columns** submenu as shown in Figure 17.6. This newly created column in the data table is appropriately called Prediction Formula Loan Amount in 10000$.

Figure 17.6 Menu to Save Prediction Formula

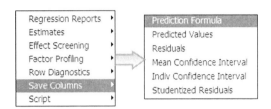

The formula itself can be viewed in the formula window of that column. The coefficients are given unrounded with up to 14 decimal places. We use only three decimal places in the discussion. The first entry is the coefficient $b_0 = 0.838$. The term 0.047*Annual Sales in 10000$ represents the estimated average change in Loan Amount as a linear function of Annual Sales. The coefficient $b_1 = 0.047$ is the common slope of the two regression lines. The Match function shows how the intercept is computed. If for a particular row the variable Region of Branches matches the word "Aderly", then subtract 0.982 from $b_0 = 0.838$. If the variable matches "Purth", then add 0.982 to $b_0 = 0.838$. If it does not match either of these two words, then do not calculate the predicted value. The coefficient $b_2 = -0.982$ belongs to Region of Branches[Aderly] of the estimated equation.

Figure 17.7 Prediction Formula in Formula Window after Save Columns

			"Aderly" = -0.9819958722836
0.83793856986767 +	0.04724778438752 * *Annual Sales in 10000$* + Match⌈*Region of Branches*⌉		"Purth" = 0.9819958722836
			else = .

The default Parameter Estimates table shown in Figure 17.4 contains only the coefficient b_2. This coefficient is associated directly with Aderly, because it was coded as +1. Selecting **Expanded Estimates** from the **Estimates** menu (Figure 17.8) produces an Expanded Estimates table that shows the coefficients associated with each level of the nominal variable.

Figure 17.8 Menu to Obtain Expanded Estimates

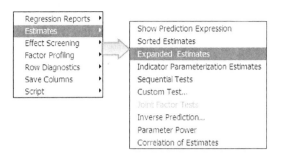

The Expanded Estimates table in Figure 17.9 is a convenience. It should be noted that the coefficients of the nominal variable always sum to zero, e.g., +0.9820 − 0.9820 = 0. This is true even when there are more than two levels of the nominal variable. The *p*-value of both the Aderly and Purth coefficients is the same. However, the signs of the coefficients for the two levels are negatives of each other.

Figure 17.9 Expanded Estimates Table Showing Parameters for All Levels of Region of Branches

Expanded Estimates

Nominal factors expanded to all levels

Term	Estimate	Std Error	t Ratio	Prob>|t|
Intercept	0.8379	0.9504	0.88	0.3987
Annual Sales in 10000$	0.0472	0.0036	12.95	<.0001*
Region of Branches[Aderly]	-0.9820	0.1249	-7.86	<.0001*
Region of Branches[Purth]	0.9820	0.1249	7.86	<.0001*

17.3.2 Interpretation of Parameter Estimates and Associated Hypothesis Tests

The hypothesis test for the regression coefficient of the nominal variable Region of Branches is used to determine a statistically significant average difference in Loan Amount between the two regions. In the example, the null hypothesis H_0: $\beta_2 = 0$ can be rejected, because the *p*-value < 0.0001. Conclude that Aderly loans have significantly lower loan amounts than Purth loans after adjustment for differences in annual sales between the two regions. The estimated average difference between Purth and Aderly loan amounts is 2*0.9820 = 1.964 in units of $10000, or $19640. The estimated slope coefficient b_1 = 0.0472 of Annual Sales is also significantly different from zero with p < 0.0001. This indicates that businesses with larger annual sales on average receive larger loan amounts. The estimated increase in loan amount per $10000 in Annual Sales is on average 0.047*$10000 = $470.

The coefficient b_0 represents the midpoint between the intercept of the Aderly and the Purth lines. It is used to calculate the intercepts of the two lines as $b_0 \pm b_2$.

The two X-variable model has a higher $R^2 = 0.952$ and a smaller RMSE = 0.442 compared to the model containing only Annual Sales with $R^2 = 0.659$ and RMSE = 1.130. Thus by including Region of Branches, the RMSE has been reduced by 61%! The reduction in RMSE is reflected in the residuals of the two models. The left panel of Figure 17.5 shows the residuals of the straight line without Region of Branches. These residuals are used in calculating the RMSE = 1.130. The right panel of Figure 17.5 shows the residuals of the regression equation with Region of Branches. These residuals were used in calculating the RMSE = 0.442. Another aspect of the model with the nominal variable is that the residuals follow a random pattern indicating that the deficiencies of the smaller model have been remedied. This analysis shows that there are significant differences in Annual Sales by Region of Branches. This confirms the original concern of the Vice President for Business Lending. Lastly, a normal quantile plot of the residuals would confirm that you can assume normality.

17.4 Regression Models Including Nominal Variables with Three or More Levels

Regression methods for nominal variables with two levels can be easily extended to nominal variables with three or more levels. As a rule, a nominal variable with k levels requires (k –1) coding variables. In JMP, each coding variable will then have (1, 0,–1) values. In the loan amount to sales revenue example, suppose that Region of Branches is a nominal variable with k = 3 levels. A third region called SouthWells is added. In this case, because there are three branch regions, i.e., k = 3 levels, we need (k – 1) = 2 coding variables. The three levels will be identified by two coding variables called Region of Branches [Aderly] and Region of Branches [Purth] with the internal coding shown in Table 17.1.

Table 17.1 Effect Coding of Variable with Three Levels

Level of Region of Branches	Region of Branches [Aderly]	Region of Branches [Purth]
Aderly	1	0
Purth	0	1
SouthWells	–1	–1

This is a standard coding for three levels. JMP assigns the values alphabetically. Thus, Aderly is assigned the value 1 in the first coding column, and Purth in the second. SouthWells is alphabetically third, and therefore receives the values –1 in both columns. This assignment allows a unique identification of each of the three levels. The alphabetical order can be overridden by list checking the order of the levels in the Column Info window.

The JMP data table *LoanToSales3Regions.jmp* with the additional region SouthWells is shown in Figure 17.10.

Figure 17.10 JMP Data File for Loan Amount versus Annual Sales for Three Regions

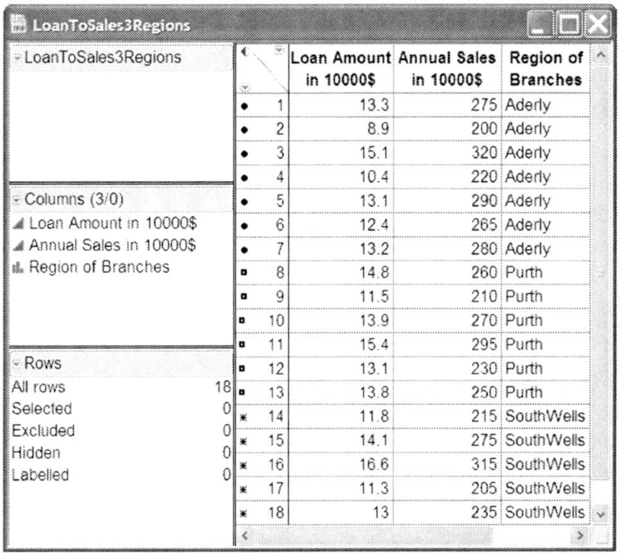
LoanToSales3Regions

LoanToSales3Regions

Columns (3/0)
Loan Amount in 10000$
Annual Sales in 10000$
Region of Branches

Rows
All rows 18
Selected 0
Excluded 0
Hidden 0
Labelled 0

	Loan Amount in 10000$	Annual Sales in 10000$	Region of Branches
1	13.3	275	Aderly
2	8.9	200	Aderly
3	15.1	320	Aderly
4	10.4	220	Aderly
5	13.1	290	Aderly
6	12.4	265	Aderly
7	13.2	280	Aderly
8	14.8	260	Purth
9	11.5	210	Purth
10	13.9	270	Purth
11	15.4	295	Purth
12	13.1	230	Purth
13	13.8	250	Purth
14	11.8	215	SouthWells
15	14.1	275	SouthWells
16	16.6	315	SouthWells
17	11.3	205	SouthWells
18	13	235	SouthWells

The Fit Model platform is used to generate all relevant output. The regression plot in Figure 17.11 shows the data including three parallel lines, one for each level. (Note that lines fitted to each level individually using simple regression generally would not be parallel.) There appears to be little if any difference between Purth and SouthWells, but Purth and SouthWells are different from Aderly. Hypothesis tests confirm these intuitive conclusions.

Figure 17.11 Regression Plot of Loan Amounts versus Annual Sales for Three Regions

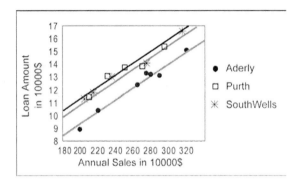

The results of the model with standard coding are shown next. Figure 17.12 shows $R^2 = 0.96$ and a RMSE = 0.40, compared to $R^2 = 0.72$ and RMSE = 1.02 for the simple regression model without Region of Branches. The lower RMSE and higher R^2 suggest that including the nominal variable in the model reduces the residuals considerably.

Figure 17.12 Summary of Fit for Loan Amount Example with Three Regions

Summary of Fit	
RSquare	0.9614
RSquare Adj	0.953128
Root Mean Square Error	0.405132
Mean of Response	13.09444
Observations (or Sum Wgts)	18

The Estimate column of Figure 17.13 gives the estimated regression coefficients. The last column contains the p-values for the significance test of each parameter estimate. The underlying null hypothesis is that the unknown regression parameter is zero. The two-sided alternative hypothesis claims that it is different from zero. In Figure 17.13, the p-values of both coding variables are less than the $\alpha = 0.05$ threshold and are therefore significantly different from zero.

Figure 17.13 Parameter Estimates and *p*-Values for Loan Amount Example with Three Regions

Parameter Estimates				
Term	Estimate	Std Error	t Ratio	Prob>|t|
Intercept	1.2157	0.6883	1.77	0.0991
Annual Sales in 10000$	0.0468	0.0027	17.51	<.0001*
Region of Branches[Aderly]	-1.2300	0.1330	-9.25	<.0001*
Region of Branches[Purth]	0.7282	0.1359	5.36	0.0001*

The regression coefficient of Annual Sales is 0.0468 with a *p*-value < 0.0001. This shows that there is a statistically significant linear relationship between Loan Amount and Annual Sales both in 10000$. The term labeled Intercept is the intercept of a center equation from which the three level equations deviate and to which they are parallel. The intercepts of the three parallel equations can be calculated as shown in Table 17.2. For example, the calculated intercept for the Purth equation is 1.9439 = 1.2157 + 0.7282. This and other intercepts of this example represent extrapolations beyond the range of the data. They should be considered fitting constants, useful in comparing the average difference between loan amounts by region.

Table 17.2 Intercepts and Equations by Region

State	Intercept = $b_0 + b_i$	Equation by Region
Aderly	1.2157 − 1.2300 = −0.0143	$(b_0 + b_{Aderly}) + b_1$*Annual Sales = −0.0143 + 0.0468*Annual Sales
Purth	1.2157 + 0.7282 = 1.9439	$(b_0 + b_{Purth}) + b_1$*Annual Sales = 1.9439 + 0.0468*Annual Sales
SouthWells	1.2157 + 0.5018 = 1.7175	$(b_0 + b_{SouthWells}) + b_1$*Annual Sales = 1.7175 + 0.0468*Annual Sales

For the intercept of SouthWells, consult the Expanded Estimates table, or note that the three level estimates sum to zero, i.e., Estimate[Aderly] + Estimate[Purth] + Estimate[SouthWells] = 0. This means that Estimate[SouthWells] = − (Estimate[Aderly] + Estimate[Purth]) = − (−1.2300 + 0.7282) = 0.5018.

To summarize, the regression coefficient of Aderly is −1.2300 (*p*-value < 0.0001), that for Purth is 0.7282 (*p*-value < 0.0001), and that for SouthWells is 0.5018 (*p*-value = 0.0035). The *p*-values in Figure 17.13 indicate that all three-regression coefficients are significantly different from zero. The three lines share the same slope of Annual Sales, $b_1 = 0.0468$. Therefore, because the three lines are parallel, they differ by the same amount everywhere in the range of the X_1-variable Annual Sales in 10000$.

Differences Between Lines: However, the real question is whether the regression lines generated by these coefficients are significantly different from each other? The simplest approach to test which if any of the equations are different from each other is by a least squares means comparison. This approach is used in ANOVA. In this example, the LSMeans Differences Student's t can be used because the *p*-values of the nominal variables are less than 0.05. Instead of comparing the three intercepts, use LSMeans Differences to compare the Region of Branches levels. Although this comparison is based on the three LSMeans, it is perfectly permissible to use, because the three lines are parallel. Their LSMeans differ by the same amount as the three intercepts. Least squares means are adjusted to permit this kind of comparison. Figure 17.14 shows the menus to obtain the Connecting Letters Report. The Connecting Letters Report has already been discussed in connection with ANOVA. Purth and SouthWells have the higher means, but are not significantly different from each other, because they share the same letter "A". Aderly has the letter "B" and does not share it with any other level of Region of Branches. It is significantly different from the other two levels. This conclusion is based on LSMeans, but it also applies to the intercepts and the three parallel lines.

Figure 17.14 Connecting Letters Report for Comparison of Level Means

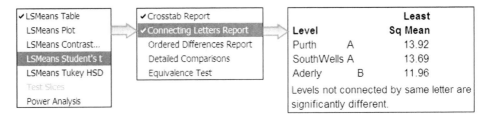

Checking Assumptions: The plot of residuals versus predicted values (not shown here) shows no deviation from randomness and fairly similar variation within each region. To firm up visual impressions, save the residuals with the **Save Columns** menu (Figure 17.6). Then test for unequal residual variances using the UnEqual Variances option in Fit Y by X, with residuals as Y and Region of Branches as X. Then check for normality of residuals with the Distribution platform. A normal quantile plot of the residuals would support the assumption of normality.

17.5 Both Intercept and Slope of Two Lines Are Different

What if the data indicate that the lines will not be parallel? We want to test whether the slopes of the two lines are different and/or whether the intercepts are different. This has an application in finance discussed in the next example.

17.5.1 Detail Example: Growth versus General Equity Fund

Situation: A group of advisors to a retirement plan looks at the yield of various types of funds. They assume that the yield of equity stocks depends on the prevailing long-term interest rate, but are not quite sure of the yield of specific types of funds with different investment objectives. One of the trustees suggests growth funds for consideration as part of the overall allocation. Others argue that the fiduciary responsibility limits the choices to low-risk funds. The trustees decide to examine the performance of growth and general equity funds.

Problem: Are there differences in yield versus interest rates when considering the two fund types? The trustees pick the annual yields of funds spread over a number of years for which different average annual long term interests prevailed. Twenty different funds are selected; 10 are from the growth fund sector and 10 are from the general equity sector.

Data Requirements: Figure 17.15 contains the three columns of *GrowthGenEquity.jmp*. Column 1 contains the type of fund as either Growth or GenEquity. Column 2 shows the annual interest rate. Column 3 is the Y-variable Yield in %.

Figure 17.15 Mutual Fund Yields by Type and As a Function of Interest Rates

		Fund Type	Interest Rate in %	Yield in %
□	1	Growth	9.8	7.65
□	2	Growth	9.1	8.12
□	3	Growth	6.6	10.7
□	4	Growth	8.9	7.95
□	5	Growth	8	10.28
□	6	Growth	8.1	8.41
□	7	Growth	8.5	8.86
□	8	Growth	8	9.13
□	9	Growth	8.4	8.4
□	10	Growth	7.4	10.1
•	11	GenEquit	9.4	7.24
•	12	GenEquit	8.7	7.26
•	13	GenEquit	8.8	7.57
•	14	GenEquit	7.2	8.58
•	15	GenEquit	6.7	8.75
•	16	GenEquit	9.1	7.83
•	17	GenEquit	9	7.4
•	18	GenEquit	6.9	8.17
•	19	GenEquit	7.9	7.98
•	20	GenEquit	6.7	8.01

GrowthGenEquity

Columns (3/0)
- Fund Type
- Interest Rate in %
- Yield in %

Rows
All rows	20
Selected	0
Excluded	0
Hidden	0
Labelled	0

17.5.2 Using JMP to Model Two Slopes and Two Intercepts

Interaction Term to Test Equality of Slopes

To model this situation, two lines with different slopes are needed. This requires an additional variable representing the interaction between continuous Interest Rate in % (X_1) with nominal Fund Type (X_2). This interaction term adjusts the slopes of the two equations. The model is $Y = \beta_0 + \beta_1 X_1 + \beta_2 X_2 + \beta_3 (X_1 \cdot X_2) + \varepsilon$.

The estimated equation is $Yhat = b_0 + b_1 \cdot X_1 + b_2 \cdot X_2 + b_3 \cdot (X_1 \cdot X_2)$. The coefficients b_2 and b_3 depend on the specific level of the nominal variable for which Yhat is to be estimated. The difference in the change of yield is tested with the slope coefficient of the interaction term. The null hypothesis assumes that both funds have the same slope (change in yield) as a function of interest rate. If b_3 is significantly different from zero, the two slopes will differ, because the slopes of the two lines require a significant + and – correction. The test for the difference in two slopes turns into a test based on b_3 with the hypotheses H_0: $\beta_3 = 0$ versus H_A: $\beta_3 \neq 0$. To reject or fail to reject H_0, look for the p-value of the t-Ratio associated with b_3. Figure 17.16 shows two regression lines with different slopes and intercepts. The two lines are not parallel and therefore have different slopes. The test for the difference in two intercepts turns into a test based on b_2 with the hypotheses H_0: $\beta_2 = 0$ versus H_A: $\beta_2 \neq 0$. Use the p-value of the t-Ratio associated with b_2 to reject or fail to reject H_0.

Figure 17.16 plots the data using different symbols for each fund type.

Figure 17.16 Regression Plot from Fit Model

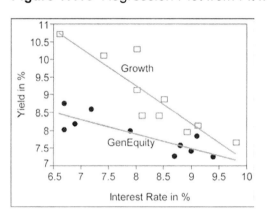

The regression model needs to be specified in the Fit Model window. The Y-variable is Yield in %. The model effects can be entered by simultaneously highlighting **Fund Type** and **Interest Rate in %** and then by choosing **Factorial to degree** (default degree is 2) from the **Macros** menu. This action will enter Fund Type, Interest Rate in %, and the interaction Fund Type*Interest Rate in % in the **Construct Model Effects** field (see Figure 17.17).

Figure 17.17 Model Specification for Two Equations with Different Slopes and Intercepts

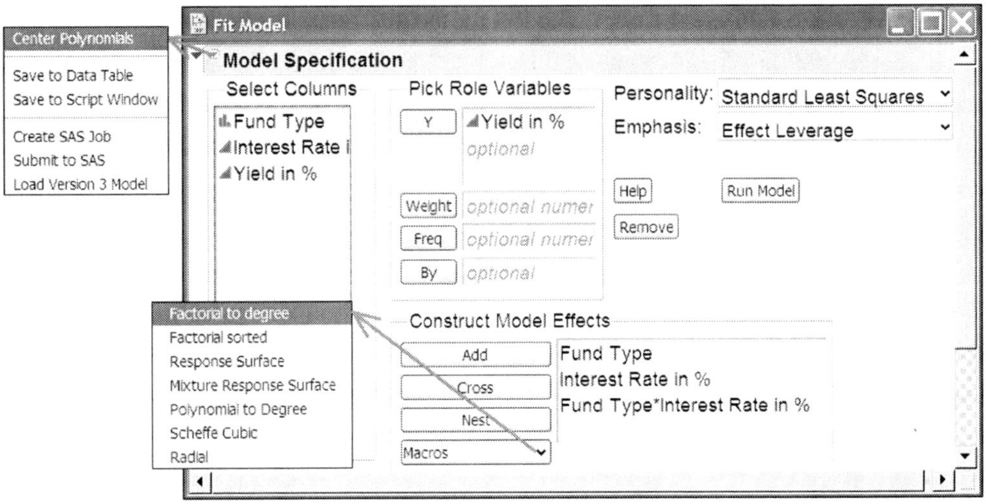

JMP by default centers polynomial equations. Although the conclusions are not affected, it is sometimes more convenient to view uncentered polynomials. This can be done in the Fit Model window (Figure 17.17) by clicking the red triangle to the left of **Model Specification** and deselecting **Center Polynomials**. This action must be completed before clicking **Run Model**.

The regression plot of Figure 17.16 suggests that the two lines have both a different slope and intercept. However, are these differences statistically significant? The *p*-values of the Parameter Estimates table in Figure 17.18 provide the answer.

Figure 17.18 Uncentered Parameter Estimates with Interaction Term

Parameter Estimates				
Term	Estimate	Std Error	t Ratio	Prob>|t|
Intercept	14.4459	0.8326	17.35	<.0001*
Fund Type[GenEquity]	-3.2806	0.8326	-3.94	0.0012*
Interest Rate in %	-0.7338	0.1011	-7.26	<.0001*
Fund Type[GenEquity]*Interest Rate in %	0.3250	0.1011	3.21	0.0054*

This table shows that both the intercepts are different because the *p*-value of the estimate of Fund Type[GenEquity] is $0.0012 < \alpha = 0.05$. In addition, the slopes are significantly different, because the *p*-value of the interaction term Fund Type[GenEquity]*Interest Rate in % is $0.0054 < \alpha = 0.05$. The Expanded Estimates table of Figure 17.19 gives a more complete picture of all the terms involved in the equation. However, note that Fund

Type[GenEquity] and Fund Type[Growth] have the same *p*-value. Their estimates are negatives of each other, because they need to sum to zero. The same is true for the two estimates of the interaction of Fund Type with Interest Rate in %.

Figure 17.19 Expanded Estimates with Interaction Terms

Expanded Estimates				
Nominal factors expanded to all levels				
Term	Estimate	Std Error	t Ratio	Prob>|t|
Intercept	14.4459	0.8326	17.35	<.0001*
Fund Type[GenEquity]	-3.2806	0.8326	-3.94	0.0012*
Fund Type[Growth]	3.2806	0.8326	3.94	0.0012*
Interest Rate in %	-0.7338	0.1011	-7.26	<.0001*
Fund Type[GenEquity]*Interest Rate in %	0.3250	0.1011	3.21	0.0054*
Fund Type[Growth]*Interest Rate in %	-0.3250	0.1011	-3.21	0.0054*

The terms Fund Type[GenEquity] and Fund Type[Growth] are level-specific corrections to the Intercept estimate. Likewise, the interaction estimates for Fund Type[GenEquity]*Interest Rate in % and Fund Type[Growth]*Interest Rate in % are corrections to the overall slope.

Use the **Save Columns** menu to save the **Prediction Formula**. Then examine the estimated equation in the formula window (Figure 17.20) in more detail.

Figure 17.20 Formula for Predicted Values

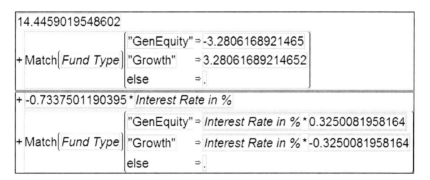

Actually, there is a separate equation for each level of the nominal variable. Imagine a centerline, consisting of the Intercept = 14.4459 and a slope estimate for Interest Rate in % = –0.7338. The coefficients ± –3.2806 represent the corrections to the two intercepts. The coefficients ± 0.3250 associated with the two levels of the interaction term represent the corrections to calculate the slopes. The calculations of the slope and intercept of each line are shown in Table 17.3.

Table 17.3 Calculation of Intercepts and Slopes by Fund Type

Fund Type	Intercept	Slope
GenEquity	(14.4459 − 3.2806) = 11.1653	(−0.7338 + 0.3250) = −.4088
Growth	(14.4459 + 3.2806) = 17.7265	(−0.7338 + −0.3250) = −1.0588

The equation for the estimated average Yield for GenEquity is

$$\text{Yhat}_{\text{GenEquity}} = (11.1653) + (-0.4088) \cdot \text{Interest Rate in \%}$$

The coefficient (−.4088) says that on average, the yield of general equity funds is estimated to decrease by −0.4088 % for every 1% increase in interest rates. The equation for the estimated average Yield for Growth is

$$\text{Yhat}_{\text{Growth}} = (17.7265) + (-1.0588) \cdot \text{Interest Rate in \%}$$

The coefficient (−1.0588) says that on average, the yield of growth funds is estimated to decrease by −1.0588% for every 1% increase in interest rates.

Based on the fitted model, you can conclude the following: The two types of funds have yields that are negatively related to interest rates. As interest rates increase, fund yields tend to decrease. In addition, there is a significant difference in the rate of change in yield (slopes) by fund type. For the values within the range of the data, growth funds have a higher average yield than general equity funds.

The residuals follow a nearly straight line in a normal quantile plot in support of the assumption of normally distributed residuals. The plot of residuals versus predicted values (not shown here) shows one larger residual in the Growth group. However, this is not enough to suggest differences in variances or other deviation from randomness within each region.

17.6 Summary

- Nominal X-variables can be included in a multiple regression model.
- JMP uses effect coding to change a nominal variable with k levels into the (k − 1) coding variables.
- The levels are expressed by coding variables containing +1, 0, and −1.
- There is more than one way to code nominal variables. As a default, JMP assigns the values +1, 0, and −1 alphabetically to a specific coding variable. Use List Check to change the default order of the values.
- The coded variables and interactions of coded variables with other variables are treated as regular (continuous) X-variables.
- A nominal variable with two levels requires a single (+1, −1) coding variable.
- The ANOVA test for the difference between the level means can be applied to test differences in intercepts.
- Using an interaction term between the nominal variable and a continuous X-variable, you can test whether the response slope differs between levels of the nominal variable.

17.7 Problems

1. A real estate agency has developed a regression model for setting sale prices for new residential home listings.

   SalesPrice = 23839 + 91.17SquareFootage + 3650Fireplace

   where Fireplace = −1 if the house does not have a fireplace
   +1 if the house does have a fireplace

   a. Use the regression equation to predict the sales price for a 2500 square foot house without a fireplace.

   b. Use the regression equation to predict the sales price for a 2500 square foot house with a fireplace.

   c. What is the estimate of the average difference between the price of a house with a fireplace and a house without a fireplace?

2. A property tax assessor is concerned with the value that an in-ground swimming pool adds to a home. A multiple regression analysis based on a random sample he collected is given as follows:

HousePrice = 47648 + 125.6SquareFootage + 2826Pool +
    6.2(SquareFootage*Pool)

where Pool = −1 if the house does not have a swimming pool
    +1 if the house does have a swimming pool

a. Write down the regression equation for a house with a swimming pool.
b. Use the equation to predict sales prices for a 2500 square foot house with a pool.
c. Write down the equation for a house without a swimming pool.
d. Use the equation to predict sales prices for a 2500 square foot house without a pool.

3. A real estate agent has created a regression model to help in understanding the value that a garage adds to a house. The capacity of the garage is categorized as No Garage, 1 Car Garage, or >2 Car Garage.

SalesPrice = 52765 + 127.8SquareFootage + 3209[1 car] + 5190[>2 car]

where the coded variables are as follows:

	1 car	2 car
1 Car Garage	+1	0
>2 Car Garage	0	+1
No Garage	−1	−1

a. What is the effect on sales price for a house with a one-car garage?
b. What is the effect on sales price for a house with a three-car garage?
c. What is the effect on sales price for a house without a garage?

4. A municipality wishes to compare the response time (in minutes) between the public and private ambulance service. JMP output from a multiple regression relating distance to the call and the nominal variable, type of ambulance service, to response time is shown below:

a. Write down the multiple regression equation.
b. Is there a significant difference in response time between the two ambulance services? Set up and conduct an appropriate test of hypothesis.

Parameter Estimates

Term	Estimate	Std Error	t Ratio	Prob>|t|
Intercept	2.454023	0.885174	2.77	0.0217
Distance (miles)	2.2068966	0.171884	12.84	<.0001
Ambulance Service[Private]	−1.821839	0.516248	−3.53	0.0064

5. A producer and seller of organically grown turkeys wants to evaluate the productivity of farms located in different regions of the country. Turkeys are raised for periods from 20 to 35 weeks and their weight is measured in kilograms. Of interest is whether there is a consistent difference in weight between turkeys raised in the North and in the South. Data on 13 turkeys are contained in the file *turkey_weights.jmp*. Construct a multiple regression to predict the weights of turkeys by age and region of origin.

   a. In the Fit Model platform, enter the column "Weight (kg)" as the Y and the columns "Age (weeks)" and "Origin" as the model effects.

   b. Write down the multiple regression equation.

   c. Assess the goodness-of-fit of the regression model.

   d. Is there a significant difference in the weights of turkeys based on region of origin? Explain!

6. An environmental advocacy organization wishes to investigate the claim that automobile fuel consumption is greater for front-wheel drive cars as compared to rear-wheel drive cars. The file *FuelConsumption_DriveSystem.jmp* contains the fuel efficiency (in miles/gal) and fuel consumption (in gal/mile) as reported by the Environmental Protection Agency for 18 randomly selected two-wheel drive cars. Also included are the car class, manufacturer, curb weight, and drive system (front- or rear-wheel). Curb weight is defined as the weight of a vehicle, without cargo, driver, and passengers, but including the maximum amounts of fuel, oil, coolant and standard equipment (e.g., the spare tire).

   a. Describe the data.

   b. Construct a regression model for highway fuel consumption using curb weight and drive system as the independent variables. What conclusion can be drawn from the model with regard to drive system and highway fuel consumption?

7. A corporate ethics office is investigating a claim that year-end bonuses are inequitably distributed between two sales divisions—residential and commercial. *Sales_Bonus.jmp* contains a random sample with three variables pertinent to the analysis for each sampled person: Experience (years), the Sales Division (Commercial, Residential), and the size of the Bonus ($). Use this data to construct a regression model that relates bonuses to sales division and experience. Is the claim of inequity supported by the data? Write a memo that summarizes the data, analysis, and findings.

8. A credit card company is evaluating factors that relate to the percentage of outstanding credit card balances paid. The company has obtained the FICO credit scores for 150 randomly selected cardholders in three age groups. FICO credit scores range from 300 for the poorest credit risk to 850 for the best credit risk. The age groups are defined as follows:

   Young Adult = ages 21–35 years
   Middle = 36–60 years
   Senior = over 60 years

   The data are contained in *CreditCardPayments.jmp*.

   a. Construct a multiple regression equation that relates percentage of outstanding credit card balance to the independent variables age group and FICO score.

   b. Use the Connecting Letters Report to describe differences between the three age groups.

   c. Check the residuals for normality. Could you use the regression equation to predict the percentage of outstanding credit card balance for a Young Adult with a FICO score of 700? Explain!

9. Collegiate Gifts sells customized novelty items such as sweatshirts, T-shirts, and baseball hats to schools, civic organizations, and small businesses. The items are imprinted with an organization's logo or slogan. Typically, the lot sizes are relatively small, less than 100. The file, *CollegiateGifts.jmp*, contains information pertaining to Collegiate's profits in dollars per lot size for both sweatshirts and T-shirts.

   a. Write down the multiple regression model to fit two lines with different slopes based on a nominal X-variable. Fit a multiple regression that accounts for different intercepts and slopes between lot size as a continuous X-variable and sweatshirts and T-shirts as levels of a nominal X-variable. (Use uncentered polynomials by clicking on the red triangle in the Fit Model window and unchecking Centered Polynomial.)

   b. Check the R^2 and RMSE, as well as residual and leverage plots, to make sure this is a good equation. Are the slopes of the sweatshirts and T-shirt lines significantly different? (Explain why you can conclude that!)

   c. Write down the fitted multiple regression model. Explain the coefficients. (Hint for simplified analysis: Click on the red triangle of the Fit Model output.

Go to Save Columns and Save Prediction Formula. Click on the heading of the newly saved column and bring up the Formula window. The Formula of the regression equation is right there.)

d. Could you have obtained the same results by fitting two individual simple regression lines? (Do this with the Fit Y by X platform by alternately excluding rows for sweatshirts and T-shirts.)

17.8 Case Study: Coffee Sales

A retailer is considering running a special next week on one-pound bags of coffee and must decide whether to put the generic brand or a national brand on sale. In addition, he must set the price and place an appropriately sized order to meet the anticipated demand. Historical data are available for both national and generic brands as given in the following tables.

National Brand No. of bags sold/week	Sales Price (Euros)
2025	5.89
1481	6.99
625	8.99
1947	5.49
1514	6.88
1226	7.29
573	8.99
3200	3.19

Generic Brand No. of bags sold/week	Sales Price (Euros)
212	7.19
4050	2.99
1556	4.89
1769	4.99
195	6.99
1186	5.89

1. Create a single data file for both brands.

2. Create a single model that can be used to predict sales for both brands. Summarize the model and goodness-of-fit in a table. Are there significant differences between the two brands? Explain how you arrived at your answer.

3. Use your model to determine the brand to put on sale, price, and amount to order. Summarize your recommendations and justification for your choices in a memo format.

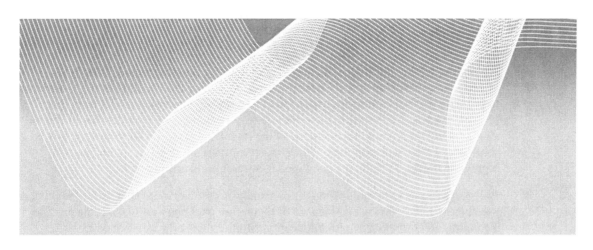

Chapter 18

Finding a Good Multiple Regression Model

18.1 Introduction 508

18.2 Detail Example: Profit of Bank Branches 510

18.3 All Possible Regression Models 512

    18.3.1 R^2 Criterion 515

    18.3.2 RMSE Criterion 515

18.4 Stepwise Regression 516

    18.4.1 Stepwise Regression Algorithms 516

    18.4.2 Stepwise Regression in JMP 517

18.5 Candidate Models 523

18.6 Model Recommendation 530

    18.6.1 Recommended Model for the Bank Branch Profits Example 531

    18.6.2 Other Criteria for Including or Excluding X-Variables 532

18.7 Summary 533

18.8 Problems 534

18.9 Case Studies 539

    18.9.1 Case Study 1: Real Estate Appraisal 539

    18.9.2 Case Study 2: Discrimination in Compensation? 539

18.1 Introduction

An analyst, having the choice of many X-variables for building a model, needs to identify those variables that are most useful in explaining the variation of the Y-variable. In multiple regression analysis, you need to find a good regression model from many possible models. According to the *principle of parsimony*, such a model should contain all the important X-variables and none of the unimportant ones. Unfortunately, this is sometimes easier said than done. This chapter outlines a process of finding a good regression model. In finding a good model, several problems may arise.

Building a model with multiple regression analysis requires more than merely looking at *p*-values to determine the significance of regression parameter estimates. One problem is a condition referred to as multi-collinearity. Multi-collinearity is indicated by high variance inflation factors and exists because two or more X-variables are highly correlated with each other. In such a case, selection of the best model might be ambiguous. Any one of these collinear X-variables (or any small subset of them) might usefully improve the model, but together as a group they might become statistically non-significant. Recall that the t-Ratios measure the significance of each X-variable as if it were added to the model last. Suppose three variables with multi-collinearity are in the model. As long as all three are in the model, each might be non-significant, because the other two X-variables explain most of what the third could explain. The third variable is not needed.

Another frequently encountered situation is that single observations are responsible for turning a particular X-variable significant. Deleting such an observation from the data turns a significant variable non-significant. However, problematic observations should not be removed without a sound justification. Sometimes there are reasonable explanations for strange observations, such as an error in recording the observation or an inadequate filter used in acquiring the data. Other times, these strange observations point to phenomena that have not yet been incorporated into the model.

For example, in many applications the same phenomenon can be measured by several different but related variables. An investment manager needs to consider a reliable measure of business activity in a region. There are several available. Some business activity measurements are available weekly; others become available monthly or quarterly. Some are detailed measures of business activity; others are more superficial. Some are readily available and inexpensive to acquire; others are difficult to obtain and only available through expensive subscriptions. Because they all measure the same phenomenon—business activity—they might be highly correlated. If all are simultaneously incorporated into the model, they will be collinear. Since it is desirable to keep the number of variables as small as needed, one or two variables might suffice. But what subset of the candidate X-variables should be used? In selecting relevant measures of business activity, the subject matter expert needs to compromise between cost of data, precision and validity of the measures, frequency of their availability, and ease of obtaining or recording each variable. Variables that are available on a timelier basis, can be recorded more precisely, or are less costly to obtain might be preferred.

The question, What is a *good* model? does not always have a simple or clear answer. In finding a good regression model, the principle of parsimony leads us to the least complicated model that serves its purpose. It says that a good model should contain as many variables as are needed to explain the Y-variable for the purpose for which the model is built. It should never have more than that number. The principle of parsimony advises frugality with the number of X-variables, but not to the point where the model loses vital characteristics in solving the problem.

Finding a good regression model requires statistical and non-statistical judgment. In finding a good regression model, the significance threshold for individual variables is often applied with more flexibility than in other hypothesis testing. In deciding to include or exclude an X-variable, the traditional 0.05 significance threshold is often relaxed to values as high as 0.25. You need to determine whether including a variable makes sense.

Finding a good model requires three steps. The **first step** generates one or more candidate models that are modified and refined in a second step. The approach starts with asking, What if we were faced with a data set containing many X-variables and need to come up with a good model in 15 minutes? We would need to apply a method that uses automatic variable inclusion and exclusion rules. One such method generates all possible regression models. This method can produce a long list of candidate models. Another method, stepwise regression, is criticized because it suspends subject knowledge in favor of automatic selection rules of X-variables. Its appeal is that it produces a single candidate model quickly. Both all possible regressions and stepwise regression are discussed in this chapter.

The **second step** analyzes and modifies candidate models using all the tools available in regression analysis. For example, variance inflation factors are used to evaluate multi-collinearity; leverage plots are used to determine the fit of each variable as if it were added last to the model; residual plots are used to identify outliers, distributional assumptions, or dependencies. According to the insights gained from this analysis, the candidate regression model might be left as is or might be modified. For example, a high degree of multi-collinearity between two variables might lead to the elimination of one variable even though it is significant. A consequence of multi-collinearity is that the regression coefficients of multi-collinear variables are difficult to interpret and often have the wrong sign. Eliminating one variable usually returns the coefficients of the other to an expected and interpretable range.

The **third step** selects a proposed regression model and examines its properties in further detail. The estimated coefficients are once more examined to see that they make sense. The fit of each variable is characterized on the basis of leverage plots, standardized beta coefficients and variance inflation factors. Overall characteristics such as RMSE and R^2 are checked to be within an acceptable range. The residuals are analyzed for violations of the assumptions. At this stage any concerns with the model should be noted. If the concerns are overwhelming, the three-step process should be repeated, possibly using additional data. In some situations, a validation step can be used to include data that have not yet been used in the model building process.

18.2 Detail Example: Profit of Bank Branches

The various approaches to finding a good regression will be illustrated with an example on the profitability of bank branches. This problem is similar to the one in Chapter 16 on multiple regression analysis. However, it uses different variables and observations.

Situation: A regional bank is planning to expand its network of branches. The goal is to increase total profitability of the bank. The bank is not interested in adding branches that are money losers or stodgy performers. That is why the bank executive committee wants to understand the drivers of bank branch profitability and incorporate the findings into their decision making.

Branches vary in size. Some have large facilities, others much smaller ones. Depending on the size of the branch, some have many employees, while other branches have just a few. Branches also focus on different aspects of the banking business. Some seek to increase their deposits, others try to attract as many customers as possible, yet others focus on loan or insurance sales. In the executive committee's opinion, the bank research department should be engaged to study the factors that contribute to branch profitability.

Problem: Find a model that accounts for the variations in bank branch profitability. The model should be as simple as possible, yet explain a large amount of the variation. It should make sense and should be easy to explain to the ultimate users. Since it is expected that not all variables will be included in the model, it is important to find understandable reasons why variables were omitted from the model.

Table 18.1 Variable List of Bank Branch Profits

Variable	Explanation
Branch Profit (per year in $)	Y-variable management is trying to understand.
July Deposits (in $100,000)	Deposits at the bank branch on the first business day of July.
Useable Square Foot	Space available for customer services. Not included in this number is space used for employee accommodations and entrance foyers.
Total Staff (in FTE)	Average staff size in full-time equivalent engaged in customer services, including branch management, during the month of July.
Customer Transactions	Number of transactions (approval of a loan, the issuance of a credit card, etc.) concluded at a branch during the month. One customer can have more than one transaction.
Insurance Sales	Number of life insurance policies (involving savings and investment programs) sold by the branch to its retail customers during the month.
Loan Sales	Number of loans written by each branch during the month, although this includes different loans with different credit structure.

Data Requirements: The executive committee gives the research department little time to perform the study. Consequently, the research will have to be performed using existing data. The research department comes up with a list of available variables that appear to have some relationship to branch profitability. The complete list of variables is described in Table 18.1. Data from the month of July are available on 31 branches.

Figure 18.1 Excerpt of *BranchProfitsFindEquation.jmp*

BranchProfitsFindEquation

BranchProfitsFindEquatio

Columns (8/0)
- Branch #
- Branch Profit
- July Deposits
- Useable Square Foot
- Total Staff
- Customer Transactions
- Insurance Sales
- Loan Sales

Rows

All rows	31
Selected	1
Excluded	0
Hidden	0
Labelled	0

	Branch #	Branch Profit	July Deposits	Useable Square Foot	Total Staff	Customer Transactions	Insurance Sales	Loan Sales
1	1	671558	169	511	37	43776	112	3994
2	2	424577	321	253	36	29490	118	2601
3	3	910263	524	144	46	35612	135	2601
4	4	628273	651	101	42	13163	77	1316
5	5	-37556	169	315	16	17245	59	1744
6	6	338007	600	292	35	6020	35	780
7	7	293449	499	95	25	13163	35	1101
8	8	53470	321	367	29	23367	65	2173
9	9	551887	372	275	41	5000	71	459
10	10	661373	372	315	35	23367	6	2601
11	11	475501	296	186	32	21327	88	2173
12	12	754309	245	182	33	25408	106	2387
13	13	306179	423	259	23	15204	59	1744
14	14	302360	473	410	27	15204	100	1744
15	15	368561	296	311	32	5000	53	459
16	16	346282	296	275	39	27449	118	2816

Each branch represents an observation. Each observation needs a row in the data table, and each variable needs a column. Not including the Branch #, there are seven variables, one Y-variable and six X-variables. Branch # represents coded identifiers of branches. This column will not be used in building the regression model. An excerpt of the data table *BranchProfitsFindEquation.jmp* is shown in Figure 18.1.

18.3 All Possible Regression Models

Examining all possible equations is practical as long as there are only a relatively small number of X-variables. For example, with six X-variables, there are $2^6 - 1 = 63$ equations to examine, plus one baseline model with no X-variables. When the number of X-variables is only moderately large, the task can be daunting. For a problem with 10 X-variables, there are $2^{10}-1$ or 1023 different possible equations.

Figure 18.2 Model Specification Window for Stepwise Regression

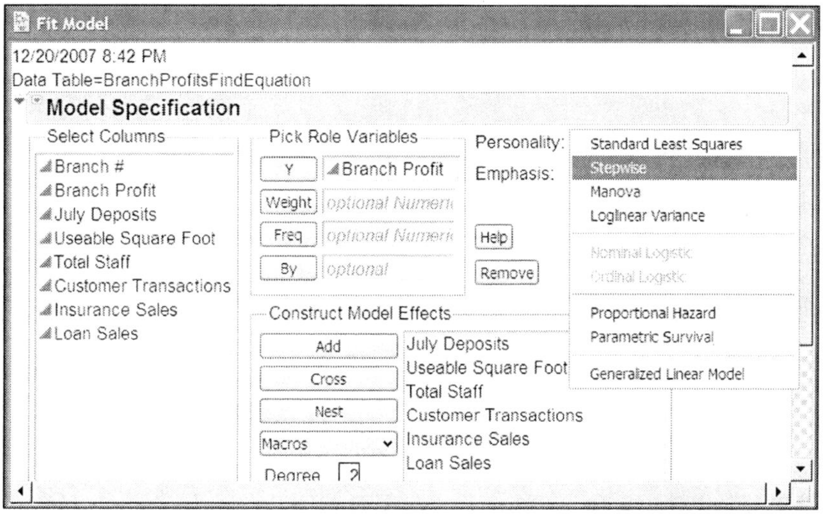

Using the Fit Model platform, you can analyze all possible regressions by selecting the **Stepwise** personality in the model specification window. Figure 18.2 shows how to specify the variables and how to select the Stepwise regression approach.

Next click **Run Model** and select **All Possible Models** (Figure 18.3) from the red triangle of the resulting Stepwise Fit bar (see Figure 18.5).

Figure 18.3 Selecting All Possible Regression Models in JMP Stepwise

The outputs for All Possible Models in Figure 18.4 are limited to the Number of variables in the equation, RSquare, RMSE and AICc (not used in this text). The highlighted models indicate the best models in terms of RSquare and RMSE from among those with the same number of X-variables.

Figure 18.4 Excerpted Results for All 63 Possible Models with Model Selection

All Possible Models				
Model	Number	RSquare	RMSE	AICc
July Deposits,Useable Square Foot,Customer Transactions,Loan Sales	4	0.4883	260529	871.191
Useable Square Foot,Customer Transactions,Insurance Sales,Loan Sales	4	0.4185	277725	875.154
July Deposits,Useable Square Foot,Insurance Sales,Loan Sales	4	0.4176	277957	875.205
July Deposits,Useable Square Foot,Total Staff,Customer Transactions,Loan Sales	5	0.8432	147078	837.896
July Deposits,Total Staff,Customer Transactions,Insurance Sales,Loan Sales	5	0.8326	151952	839.917
July Deposits,Useable Square Foot,Total Staff,Customer Transactions,Insurance Sales	5	0.8143	160057	843.139
Useable Square Foot,Total Staff,Customer Transactions,Insurance Sales,Loan Sales	5	0.8092	162250	843.983
July Deposits,Useable Square Foot,Total Staff,Insurance Sales,Loan Sales	5	0.7872	171340	847.362
July Deposits,Useable Square Foot,Customer Transactions,Insurance Sales,Loan Sales	5	0.5681	244080	869.301
July Deposits,Useable Square Foot,Total Staff,Customer Transactions,Insurance Sales,Loan Sales	6	0.8439	149760	841.427

The six X-variables can be combined to construct 63 different regression models. This yields 63 different lines of output. Figure 18.4 shows an excerpt of the All Possible Models output with two highlighted models. Altogether there are six highlighted models, one for each number of X-variables. They are summarized in Table 18.2. The RSquare column shows that the range of R^2 values is from 0.7389 to 0.8439. The RMSE ranges from 147078 dollars for the five-variable equation to 176210 dollars for the equation with one variable. The Std Dev[RMSE] = 339058 dollars for the base model without X-variables. This is almost twice as large as the largest of the other regression RMSEs listed in the table. It points out that any of these regression models is a better model than just fitting a mean to the branch profits.

Table 18.2 The Best Models for Each Number of Variables

Model	Number	RSquare	RMSE ($)
July Deposits, Useable Square Foot, Total Staff, Customer Transactions, Insurance Sales, Loan Sales	6	0.8439	149760
July Deposits, Useable Square Foot, Total Staff, Customer Transactions, Loan Sales	5	0.8432	147078
July Deposits, Total Staff, Customer Transactions, Loan Sales	4	0.8324	149100
July Deposits, Total Staff, Customer Transactions	3	0.8080	156592
July Deposits, Total Staff	2	0.7623	171113
Total Staff	1	0.7389	176210
None	0	0	339058

In Table 18.2 and Figure 18.4, two important criteria are displayed, both derived from residuals:

- R^2 measures the proportion of the SS_{Total} explained by the fitted model.
- RMSE or Root Mean Squared Error is the estimated standard deviation around the fitted model.

18.3.1 R^2 Criterion

Recall that R^2 measures the proportion of the Total Sum of Squares explained by the statistical model with $0 \leq R^2 \leq 1$. The R^2 criterion looks for a model with a reasonably large R^2. The problem with this criterion is that some models never achieve a large R^2. For example, models that seek to determine the value of firms based on accounting and market data might not achieve R^2 larger than 0.20, whereas in engineering, models with R^2 below 0.70 might indicate a poor fit. The question of what is a high or low R^2 is problem specific and can only be answered in comparison with R^2 from other similar models.

In the bank branch profits example, note that the models with three or more variables have R^2 greater than 0.80. However, even the models with one or two variables have reasonably high R^2. In other regression examples, the decrease might not always be as gradual as it is here. Sometimes decreases in R^2 are fairly abrupt, depending on the inclusion or exclusion of an important variable.

As pointed out in Chapter 14, "Simple Regression," R^2 can be influenced by one or a few observations. Therefore the R^2 criterion is of limited use, but it offers a quick look at the strength of the model in terms of explained proportion of the SS_{Total}. R^2 never decreases when an additional variable is added to the model. For this reason, the R^2 criterion is not "bigger is better," because following that rule, you would always select the model with all variables in the model. Instead, look for a reasonably large value of R^2 with an acceptable number of variables.

18.3.2 RMSE Criterion

The RMSE is the estimated standard deviation of the observations around the fitted model. It is calculated from the residuals and is therefore a measure of how tightly observations are distributed around the model. Relatively small RMSEs are more desirable, because they indicate a tighter fit than relatively large RMSEs. The RMSE criterion looks for models with an RMSE near the minimum achievable RMSE. Alternatively, look for an RMSE that is small enough to serve the original purpose of the model. It is not necessary to look for the model with the smallest RMSE, since this might lead to models with more variables than are needed to explain the Y-variable adequately.

Table 18.2 gives the RMSEs for the bank branch profits example. RMSE = Std Dev = 339058 dollars for the model that does not use any X-variables. The model with five X-variables has the lowest RMSE (147078 dollars). However, the models with four and six variables have RMSEs very close to the smallest value. Eliminating Useable Square Foot from the model with five X-variables increases the RMSE by a relatively small amount, from 147078 to 149100 dollars. In fact, a good candidate for some applications might be the model with only one X-variable (Total Staff), which has an RMSE = 176210 dollars, approximately 20 percent larger than 147078. Given the simplicity of a one-variable model and depending on the purpose and the required precision, this simple regression model might be adequate for predicting profitability.

18.4 Stepwise Regression

Stepwise regression is an automatic regression algorithm that enters or removes X-variables into the regression model, one X-variable at a time. The X-variables are entered or removed based on statistical criteria, usually F-Ratios and their corresponding *p*-values. The stepwise process is stopped when either all significant X-variables have been entered, or all non-significant X-variables have been removed.

Stepwise regression does not necessarily produce a "best" model, nor do the various stepwise algorithms necessarily produce the same model. The final stepwise regression model also depends on the strictness of inclusion or exclusion criteria. In JMP, these are specified as Prob to Enter and Prob to Leave. They are used as thresholds similar to the α significance level. For example, smaller Prob to Enter or larger Prob to Leave result in a model with fewer X-variables.

Stepwise regression has been criticized because the automatic selection of X-variables relies solely on formal statistical criteria. It does not involve expert knowledge and judgment that is so important in building valid and useful models. Here, stepwise regression is used as a start in building good models, particularly when there are a large number of possible X-variables or when analysis time is limited.

18.4.1 Stepwise Regression Algorithms

Three stepwise regression algorithms are available in JMP: Forward, Backward, and Mixed. These algorithms are outlined next.

Forward stepwise regression starts with no X-variables in the model and adds X-variables one at a time. At each step, the stepwise algorithm selects the X-variable that best explains the remaining unexplained SS_{Error} when added to the current model. An X-variable is added to a model only if it will be statistically significant in the new model, i.e., it has a *p*-value less than Prob to Enter. Forward stepwise regression is stopped when all X-variables not in the model have *p*-values larger than Prob to Enter.

Backward stepwise regression starts with all X-variables in the model and eliminates them one at a time. At each step, only non-significant X-variables are considered for elimination. At each step, the X-variable with the largest *p*-value is eliminated from the model, as long as its *p*-value is greater than the Prob to Leave. Backward stepwise regression is stopped when all X-variables remaining in the model have a *p*-value less than Prob to Leave.

The problem with the Forward algorithm is that after a variable has been entered into the model, the algorithm does not allow its removal at a subsequent step, no matter how insignificant that variable turns out to be at later steps. Likewise, the Backward algorithm does not allow a variable that has been removed at a previous step to be reentered into the model, no matter how significant that variable might have become at subsequent steps. The Mixed stepwise algorithm avoids these concerns.

The **Mixed stepwise regression** algorithm can start either with all X-variables in the model or with no X-variables in the model. At each step, the Mixed stepwise algorithm will either delete a non-significant X-variable from the model or enter a significant X-variable that is not already in the model. The algorithm is stopped when all variables in the model are significant and those not in the model are non-significant according to Prob to Enter for inclusion and Prob to Leave for elimination.

As mentioned before, the three methods might or might not lead to the same model. Because running stepwise regression is so simple, it might be useful to try all three versions of the algorithm. You can compare the set of X-variables contained in the resulting regression models. These regression models serve as starting points for a modeling approach that uses statistical and expert knowledge to improve a model's meaning and usefulness.

18.4.2 Stepwise Regression in JMP

Figure 18.2 shows the Fit Model platform where you select Y and assign the X-variables to Model Effects. Choose **Stepwise** regression from the **Personality** pull-down menu in the upper right corner (see Figure 18.2). Click **Run Model** to produce a window to control the stepwise algorithm. Figure 18.5 shows that Stepwise Regression Control window. Use this window to choose either **Forward**, **Backward**, or **Mixed** stepwise algorithms. Depending on this choice, start with no X-variables or with all X-variables by clicking **Remove All** (for Forward or Mixed) or **Enter All** (for Backward or Mixed) respectively. The defaults for **Prob to Enter** and **Prob to Leave** can be changed by typing over the existing values.

Figure 18.5 Stepwise Regression Control Window

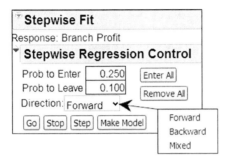

Choice of Prob to Enter and Prob to Leave

Prob to Enter is the inclusion threshold. When all *p*-values of the X-variables not in the model are greater than Prob to Enter, the Forward algorithm is stopped. Any variable with a *p*-value smaller than Prob > F is a candidate to be included at the next step. Prob to Leave is the removal threshold. Any variable with a *p*-value larger than Prob > F is a candidate to be removed from the model.

In JMP, the default value for Prob to Leave is 0.10 and for Prob to Enter is 0.25, except for the mixed algorithm, where both are 0.25. The effect of choosing larger error probabilities is that often more X-variables are included in the model than for the traditional 0.05 threshold. Smaller Prob to Leave or Prob to Enter tend to under-fit the model, i.e., tend to select fewer X-variables. In modeling, with the principle of parsimony in mind, you want to consider all variables that improve the model. If desired, the Prob to Enter and Prob to Leave values can be adjusted to suit the analyst. Indeed, it is a good exercise to change these probabilities to see by how much the resulting model changes. In this example, changing Prob to Enter from the default 0.25 to 0.10 with the Forward algorithm results in a model consisting of the single variable Total Staff.

Forward Stepwise Regression

Forward stepwise regression starts with no X-variables in the model. It adds one X-variable at a time, selecting the one that is conditionally most significant at that step. The Forward regression algorithm is stopped when all X-variables not in the model have conditional *p*-values larger than the Prob to Enter.

Figure 18.6 shows the initial Current Estimates window for Forward stepwise regression. The JMP default places no X-variables in the model. The window shows those parameters and statistics that make sense for a model with no X-variables. This book focuses on MSE (or the RMSE) and the RSquare and ignores the other statistics. Slope coefficients of X-variables not in the model are assigned the value 0. However, the *p*-values for each of these X-variables measure their significance if the variable were added next to the current model.

The Current Estimates window for the model with no X-variables shows that for this model the intercept is the overall mean bank branch profits (469464.194 dollars). The *p*-values in the "Prob>F" column give the significance of each X-variable in the simple regression model. July Deposits, Total Staff, Customer Transactions, and Insurance Sales have *p*-values less than Prob to Enter = 0.25. The lowest *p*-value is for Total Staff, marking this variable to be included next in the Forward regression algorithm. Rsquare = 0.0, because there are no X-variables in this model. The MSE = 1.15e + 11 represents the square of Std Dev[RMSE] = 339058 dollars.

Figure 18.6 Starting Window of Forward Algorithm

To begin the stepwise algorithm, click the **Go** button (shown in Figure 18.6). The stepwise algorithm can be performed one step at a time by clicking the **Step** button each time another step is desired. The final results are shown in Figure 18.7. They include the updated Current Estimates of the completed Forward stepwise regression model and the Step History.

Figure 18.7 Current Estimates and Step History of Completed Forward Algorithm

Current Estimates

		SSE	DFE	MSE	RSquare	RSquare Adj	Cp	AIC
		5.408e+11	25	2.163e+10	0.8432	0.8118	5.1127601	743.0523

Lock	Entered	Parameter	Estimate	nDF	SS	"F Ratio"	"Prob>F"
☑	☑	Intercept	-661756.05	1	0	0.000	1.0000
☐	☑	July Deposits	556.986961	1	1.18e+11	5.447	0.0279
☐	☑	Useable Square Foot	329.815045	1	3.72e+10	1.720	0.2017
☐	☑	Total Staff	21677.6983	1	1.22e+12	56.581	0.0000
☐	☑	Customer Transactions	23.1805008	1	1.99e+11	9.204	0.0056
☐	☐	Insurance Sales	0	1	2.529e+9	0.113	0.7399
☐	☑	Loan Sales	-182.44651	1	1.02e+11	4.719	0.0395

Step History

Step	Parameter	Action	"Sig Prob"	Seq SS	RSquare	Cp	p
1	Total Staff	Entered	0.0000	2.55e+12	0.7389	13.148	2
2	July Deposits	Entered	0.1082	8.06e+10	0.7623	11.554	3
3	Customer Transactions	Entered	0.0173	1.58e+11	0.8080	6.5195	4
4	Loan Sales	Entered	0.0627	8.41e+10	0.8324	4.7714	5
5	Useable Square Foot	Entered	0.2017	3.72e+10	0.8432	5.1128	6

In the bank branch profits example, all but Insurance Sales are checked. This model is the best of the five-variable models generated by the All Possible Models approach. Insurance Sales is not checked because it cannot be entered with a Prob > F = 0.7399. Useable Square Foot has Prob > F = 0.2017. This variable was entered because the Prob to Enter was chosen by default to be 0.25 (see Figure 18.5). A value of Prob to Enter = 0.10 would not have included that variable. The stepwise algorithm would have stopped earlier.

JMP Stepwise has a Lock feature that forces variables into the model. Simply select the **Lock** check box for the desired column. Columns can be locked for non-statistical reasons. Note that the intercept is automatically locked into the model. It should not be removed except in very special cases.

The Current Estimates table gives the coefficients of the stepwise regression model. In the bank branch profits example, the model is Yhat = –661756.05 + 556.986961*July Deposits + 329.815045*Useable Square Foot + 21677.6983*Total Staff + 23.1805008*Customer Transactions + –182.44651*Loan Sales.

The RSquare = 0.8432, with the MSE = 2.163e + 10 (RMSE = 147078). (In this book, Cp and AIC value are not used.)

In stepwise regression, it is traditional to measure the significance of each X-variable by the F-Ratio, which here is simply the square of the t-Ratio of the regression parameter estimates. The interpretation of the *p*-value Prob > F is the same as that of the t-Ratio *p*-values. The *p*-values in the "Prob>F" column of Figure 18.7 show that July Deposits, Total Staff, Customer Transactions, and Loan Sales are highly significant, but that Useable Square Foot has a *p*-value = 0.2017. Insurance Sales has a p = 0.7399, much larger than the Prob to Enter = 0.25, and is therefore not included in the stepwise regression model.

The Step History shows the results of each step. Each row shows the changes to the model at that step. The column Action shows that all variables have been Entered, because this is the Forward algorithm. A Backward algorithm would show only Removed variables, while the Mixed algorithm might show both. The Step History also gives for each step the RSquare and Cp, a statistic that is related to the MSE. The column SeqSS is not used in this book.

Clicking the **Make Model** button transfers the variables for analysis to the model by Standard Least Squares. Under Standard Least Squares, you can use leverage plots and other important tools discussed previously to validate assumptions and identify potential outliers.

Backward Stepwise Regression

Backward stepwise regression starts with all X-variables in the model. It deletes one X-variable at a time, selecting the one that is conditionally the least significant at that step. The Backward regression algorithm is stopped when all X-variables in the model have conditional *p*-values smaller than the Prob to Leave.

Figure 18.8 shows the initial Current Estimates window for Backward stepwise regression. This algorithm requires that all X-variables have to be entered at the start. This will be indicated by a check mark in the Entered column of each variable. In JMP, click the **Enter All** button to place all X-variables into the model. The window also shows the parameters of the model with all (six) variables, its R^2 and MSE. These are the same values shown in Figure 18.4 and Table 18.2 for the single model containing all six variables. The *p*-value of each X-variable measures its significance as if this variable were added last to the current model.

Figure 18.8 Starting Window for Backward Algorithm

Stepwise Regression Control							
Prob to Enter	0.250	Enter All					
Prob to Leave	0.100	Remove All					
Direction: Backward ▾							
Go Stop Step Make Model							

Current Estimates							
SSE	DFE	MSE	RSquare	RSquare Adj		Cp	AIC
5.383e+11	24	2.243e+10	0.8439	0.8049		7	744.907

Lock	Entered	Parameter	Estimate	nDF	SS	"F Ratio"	"Prob>F"
☑	☑	Intercept	-673795.74	1	0	0.000	1.0000
☐	☑	July Deposits	574.38065	1	1.2e+11	5.344	0.0297
☐	☑	Useable Square Foot	339.822546	1	3.9e+10	1.737	0.1999
☐	☑	Total Staff	21203.8536	1	9.51e+11	42.407	0.0000
☐	☑	Customer Transactions	23.0219277	1	1.96e+11	8.724	0.0069
☐	☑	Insurance Sales	243.694777	1	2.529e+9	0.113	0.7399
☐	☑	Loan Sales	-182.53517	1	1.02e+11	4.556	0.0432

Figure 18.9 shows the Current Estimates window for the final model of the Backward stepwise regression algorithm. Two variables are removed from the six X-variable model. Step 1 removed Insurance Sales, while step 2 removed Useable Square Foot. The resulting model is identical to the best four-variable model of the All Possible Models algorithm shown in Figure 18.4 or Table 18.2.

The Forward and Backward models differ because, among other possible reasons, Prob to Enter (0.25) and Prob to Leave (0.10) are different. The variable Useable Square Foot has a p-value = 0.2017. This is small enough to include it into the model at step 5 of the Forward algorithm. However, it is sufficiently large to be eliminated at step 2 of the Backward algorithm. The differences between Prob to Enter and Prob to Leave represent one reason why Forward and Backward models might differ.

The Current Estimates table gives the coefficients of the Backward model. In the bank branch profits example, the model is Yhat = –582918.98 + 499.653703*July Deposits + 22428.1971*Total Staff + 21.6864947*Customer Transactions + –162.94547*Loan Sales.

The RSquare = 0.8324 with the MSE =2.223e+10 (RMSE = 149100).

Figure 18.9 Current Estimates and Step History of Backward Algorithm

Current Estimates

		SSE	DFE	MSE	RSquare	RSquare Adj	Cp	AIC
		5.78e+11	26	2.223e+10	0.8324	0.8066	4.7714081	743.1145

Lock	Entered	Parameter	Estimate	nDF	SS	"F Ratio"	"Prob>F"
☑	☑	Intercept	-582918.98	1	0	0.000	1.0000
☐	☑	July Deposits	499.653703	1	9.81e+10	4.413	0.0455
☐	☐	Useable Square Foot	0	1	3.72e+10	1.720	0.2017
☐	☑	Total Staff	22428.1971	1	1.36e+12	61.355	0.0000
☐	☑	Customer Transactions	21.6864947	1	1.78e+11	8.017	0.0088
☐	☐	Insurance Sales	0	1	7.649e+8	0.033	0.8570
☐	☑	Loan Sales	-162.94547	1	8.41e+10	3.781	0.0627

Step History

Step	Parameter	Action	"Sig Prob"	Seq SS	RSquare	Cp	p
1	Insurance Sales	Removed	0.7399	2.529e+9	0.8432	5.1128	6
2	Useable Square Foot	Removed	0.2017	3.72e+10	0.8324	4.7714	5

Mixed Stepwise Regression

The Mixed stepwise regression algorithm avoids potential problems of the Forward and Backward algorithms. The Mixed stepwise algorithm allows that variables be entered or removed depending on the best course of action at any step. The values of Prob to Enter and Prob to Leave are the same by default. In JMP, you could select different values, but this might lead to convergence problems.

In this example, the models of the Mixed stepwise are either that of the Forward, provided you start either with no variables and Prob to Enter = Prob to Leave = 0.25; or that of the Backward, provided you start either with all variables and Prob to Enter = Prob to Leave = 0.10.

18.5 Candidate Models

Candidate models are chosen on the basis of certain overall characteristics, such as the relative magnitude of the RMSE, the type of variables included in the model, or the desired complexity or precision that the model is to provide. Candidate models will undergo a more detailed scrutiny so that a final model can be recommended. Consider the bank branch profits models with five, four, and three variables listed in Table 18.2 as the three candidate models. These three models share three variables (July Deposit, Total Staff, and Customer Transactions). The four-variable model also contains Loan Sales, while the five-variable model includes Loan Sales and Useable Square Foot. The reasons for their inclusion among the candidate models are as follows:

- Forward stepwise regression with Prob to Enter = 0.25 results in a regression model with five variables: July Deposits, Useable Square Foot, Total Staff, Customer Transactions, and Loan Sales. One of these variables (Useable Square Foot) has a rather large p-value = 0.2017.

- Backward Stepwise Regression with Prob to Leave = 0.10 results in a regression model with four variables: July Deposits, Total Staff, Customer Transactions, and Loan Sales. These four variables are in the Forward model. However, the Backward model excludes Useable Square Foot.

- The all possible models approach points to the same four- and five-variable models produced by stepwise regression. However, it also points to a three-variable model with the variables July Deposits, Total Staff, and Customer Transactions (Table 18.2).

Evaluating Candidate Models

Each of the candidate models needs to be analyzed in detail until a final model can be recommended. In JMP, the detailed analysis is performed with the Fit Model platform by choosing the default **Standard Least Squares** personality in the model specification window (instead of **Stepwise**). Each model is analyzed using statistical and graphical tools. Non-statistical considerations can enter the evaluation as well. Some of the more useful criteria in evaluating a model are as follows:

The *p*-value "Prob>|t|" of the parameter estimate table can be used to test the significance of each parameter estimate. (The intercept term can be safely excluded from consideration in many applications.) The null hypothesis is that the parameter is zero; the two-sided alternative hypothesis is that it is different from zero. However, in model building you might be willing, at least up to a point, to relax the traditional significance threshold.

The **sign of the slope** is another criterion. X-variables that are expected to be positively related to Y should have a positive slope and vice versa. Models containing variables with counterintuitive slopes do not necessarily make sense. Check the data for consistency or outliers. Reexamine the original interpretation of the coefficients.

Standardized Beta coefficients can determine the importance of each X-variable in explaining Y. Always use the absolute value of the Betas.

Variance Inflation Factors (VIFs) measure multi-collinearity between X-variables. Expect multi-collinearity between X-variables that are transformations of each other, such as X and X^2. In such cases, both variables are intended to be included in the model. Individual parameter estimates of such correlated variables should be interpreted together. However, multi-collinearity also arises between other variables. This is a more troublesome case, because the parameter estimates of such variables need to be interpreted, but might not make sense, especially when they have the wrong sign. A remedy might be to consider a model from which one of these variables has been eliminated (despite statistical significance).

Leverage plots provide a view of how each variable fits as it is added last to the model. Look for unusual patterns, such as outliers or a curvilinear pattern in the residuals. Leverage plots have been discussed in the chapter on multiple regression analysis.

Residual plots show how close the existing data fit the model. It should be noted that residuals are from the data that were used in fitting the model. They often tend to fit closer than when new data are compared to the model. Plotting residuals versus predicted values, other X-variables, or the sequence of data collection gives insight into the model fit. Normal quantile plots of residuals check some aspects of the regression assumptions.

In the following sections, we discuss the three candidate models, using for each only those criteria that give especially good insight.

Model with Five Variables from Forward Stepwise Regression

The model with five variables is the result of Forward stepwise regression with Prob to Enter = 0.25 and was identified by the All Possible Models option as a candidate model. Some of its characteristics are shown in Figure 18.10.

- It is the model with the lowest RMSE (147078 dollars).
- It has five variables and thus is more complicated than the other candidate models.
- The *p*-value of Useable Square Foot is 0.2017, a relatively large value. This variable has the lowest standardized Beta (0.11).
- The VIFs of Customer Transactions and Loan Sales are nearly 10. Since all other VIFs are near 1, assume that these two variables explain similar variation of the branch profits.

Figure 18.10 Parameter Estimates of Five-Variable Model

Parameter Estimates						
Term	Estimate	Std Error	t Ratio	Prob>|t|	Std Beta	VIF
Intercept	-661756	150241.1	-4.40	0.0002*	0.00	.
July Deposits	556.98696	238.6612	2.33	0.0279*	0.22	1.41
Useable Square Foot	329.81504	251.5051	1.31	0.2017	0.11	1.14
Total Staff	21677.698	2881.895	7.52	<.0001*	0.68	1.32
Customer Transactions	23.180501	7.640857	3.03	0.0056*	0.72	8.86
Loan Sales	-182.4465	83.98602	-2.17	0.0395*	-0.52	9.10

It would be expected that increased Loan Sales increases Branch Profit. However, the parameter estimate of Loan Sales is –182.4465, suggesting the opposite. The reason this coefficient does not make sense is because of the collinearity with Customer Transactions (see VIFs). The leverage plots of Customer Transactions and Loan Sales (Figure 18.11) show that observation 21 seems to influence the significance of either of these variables. All other observations are randomly scattered, not suggesting a linear or other relationship.

Figure 18.11 Selected Leverage Plots of Five-Variable Model

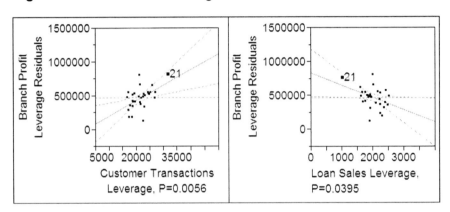

This model has the lowest RMSE among the models considered by the All Possible Models option. However, it contains one term of questionable significance (Useable Square Foot) as well as two terms (Customer Transactions and Loan Sales) of which one might be redundant as demonstrated by VIFs or the leverage plots. The parameter estimate of Loan Sales has the wrong sign, making Loan Sales a candidate for elimination from the model.

Model with Four Variables from Backward Stepwise Regression

The model with four variables is the result of Backward stepwise regression with Prob to Leave = 0.10 and was identified by the All Possible Models option as a candidate model. Some of its characteristics are shown in Figure 18.12.

- It is the model with the second-lowest RMSE (149100 dollars).
- The *p*-value of Loan Sales is 0.0627; it is larger than the common threshold of $\alpha = 0.05$. Loan Sales has a relatively large standardized Beta (–0.46).
- As with the five-variable model, the VIFs of Customer Transactions and Loan Sales are above 5. The other two VIFs are near 1. Assume that these two variables explain similar variation of the bank branch profits.

In a normal business environment, you would expect that increased Loan Sales increase Branch Profit. However, the parameter estimate of Loan Sales is –162.9455, suggesting the opposite. This parameter estimate might have the wrong sign because of collinearity between Loan Sales and Customer Transactions.

Figure 18.12 Parameter Estimates of Four-Variable Model

Parameter Estimates						
Term	Estimate	Std Error	t Ratio	Prob>|t|	Std Beta	VIF
Intercept	-582919	139581.3	-4.18	0.0003*	0.00	.
July Deposits	499.6537	237.8475	2.10	0.0455*	0.20	1.36
Total Staff	22428.197	2863.322	7.83	<.0001*	0.71	1.27
Customer Transactions	21.686495	7.659298	2.83	0.0088*	0.67	8.66
Loan Sales	-162.9455	83.7952	-1.94	0.0627	-0.46	8.82

Similar to the model with five variables, the leverage plots of Customer Transactions and Loan Sales (Figure 18.13) show that observation 21 seems to influence the significance of either of these variables.

Figure 18.13 Selected Leverage Plots of Four-Variable Model

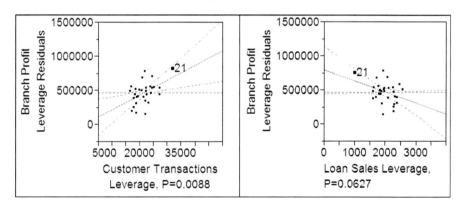

This model has the second-lowest RMSE among the models considered in Table 18.2. All terms have *p*-values below 0.10. However, like the five-variable candidate model, it contains two terms (Customer Transactions and Loan Sales) with high VIFs and a suspect pattern in the leverage plots, while the parameter estimate of Loan Sales again has the wrong sign.

Model with Three Variables from All Possible Models

The model with three variables was identified by the All Possible Models option as a candidate model. Some of its characteristics are shown in Figure 18.14.

- The RMSE (156692 dollars) is 6.5% higher than the five-variable model and 5% higher than the four-variable model. Thus the residuals are not appreciably increased.
- It is the simplest of the three candidate models.
- All *p*-values of the parameter estimates are below $\alpha = 0.05$.
- The VIFs are below 5.

The parameter estimates make sense. Higher deposits and more customer transactions result in profits. Total Staff seems to be a variable that measures the size of a branch rather than the (profit reducing) cost associated with human resources. The parameter estimate of this variable indicates that larger branches have higher profits.

Figure 18.14 Parameter Estimates of Three-Variable Model

Parameter Estimates						
Term	Estimate	Std Error	t Ratio	Prob>|t|	Std Beta	VIF
Intercept	-693577.2	133858.1	-5.18	<.0001*	0.00	.
July Deposits	638.63709	238.2529	2.68	0.0124*	0.25	1.24
Total Staff	23291.829	2970.793	7.84	<.0001*	0.73	1.24
Customer Transactions	7.978202	3.145295	2.54	0.0173*	0.25	1.32

The leverage plots of Customer Transactions and Loan Sales (Figure 18.15) show a linear relationship between Customer Transactions and Branch Profits, given the other two variables are in the model. Observation 21 does not unduly influence the fit of this variable as it did with the four- and five-variable models.

Figure 18.15 Leverage Plot of Customer Transactions in Three-Variable
Model

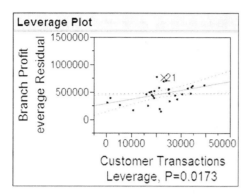

The plot of residuals versus predicted values shows no particular pattern. The residual of
observation 21 is large, but it is not the largest residual. The plot as shown in Figure
18.16 is a default output of the Standard Least Squares personality.

Figure 18.16 Residuals versus Predicted Plot in Three-Variable Model

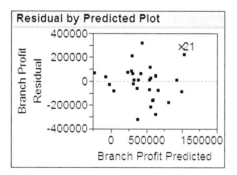

The Normal quantile plot of residuals (Figure 18.18) is produced using the Distribution
platform. This plot reveals that residuals follow a Normal distribution. It requires that the
residuals be saved to the data table using the **Save Columns** option shown in Figure
18.17.

Figure 18.17 Fit Model Menu to Save Residuals

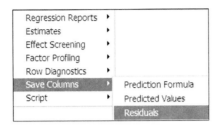

Figure 18.18 Normal Quantile Plot of Residuals of Three-Variable Model

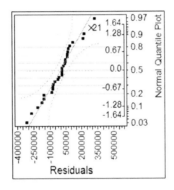

18.6 Model Recommendation

In certain complex applications, it is possible and often desirable to use several competing models. Examples are models of climate change or models to simulate the second-to-second behavior of a financial derivatives market. These models are sufficiently complex to be difficult to understand and to validate. They need constant review and updating. In such situations, competing models taking different approaches to the same problem are useful in providing insight and understanding.

Many business problems are of a less sweeping scope and can be represented by simpler models. In such cases, a single model is recommended from the set of candidate models. The choice of the final model is a step involving both statistical and non-statistical criteria. It requires an evaluation of several dimensions:

Step 1 chooses candidate models because they meet certain basic criteria. They have relatively low RMSE and relatively high R^2. They may contain a reasonable mix of variables.

Step 2 analyzes candidate models in more detail using additional criteria. Based on the analysis, it might modify certain candidate models by eliminating variables or adding transformed variables, e.g., logarithmic or squares terms of an X-variable.

Step 3 selects the recommended model on the basis of the analysis in step 1 and step 2. It might add additional considerations, such as the following questions:

- Is the model understandable and explainable?

- Can variables in the model be acquired and updated with sufficient speed?

- Can the model variables be measured with sufficient precision or are they subject to measurement errors?

- If the purpose of the model is to draw general conclusions about the problem, do the parameter estimates make sense?

- If the purpose of the model is to predict future outcomes, is the precision of the model adequate for the purpose? The precision can be measured directly by the size of the residuals and more indirectly by the RMSE and other measures of fit.

- Does the recommended model meet relevant model assumptions?

18.6.1 Recommended Model for the Bank Branch Profits Example

The problem statement of the bank branch profits example included the following: "Find a model that accounts for the variations in bank branch profitability. The model should be as simple as possible, yet explain a large amount of the variation. It should make sense and should be easy to explain to the executive committee."

It is easy to argue that the three-variable model best meets the objectives outlined in the problem statement. Although it has a higher RMSE than the other candidate models, it is the simplest model. All terms have p-values below the 0.05 threshold. All VIFs are near the minimum 1, indicating that the parameter estimates of the model have a straightforward interpretation. Leverage plots show no suspect patterns. The three parameter estimates have the appropriate signs. The residuals meet the assumptions of randomness and normality.

The three-variable model is a parsimonious model, one that contains the fewest number of variables but also just as many terms as are needed to explain the data. There are no redundancies, as shown by the lack of high VIFs. The increase in RMSE compared to other equations is negligible. The three-variable model should be easy to explain.

Is this the best model? It is difficult to make such a categorical statement. Model recommendations require a degree of humility. Another data set from the same source might yield different results. Additional observations might change the mix of variables in the model. Adding new variables to the data set might change the model as well. A different purpose might lead to a different recommendation. We conclude that this is a good model for now.

18.6.2 Other Criteria for Including or Excluding X-Variables

For many problems, a strict application of statistical criteria in formulating the best model might be insufficient. Non-statistical considerations enter into the variable selection. There are good reasons to override statistical criteria in the inclusion or exclusion of certain X-variables.

Non-Statistical Reasons

The following are some of the non-statistical reasons for excluding certain X-variables:

- Other X-variables are available that capture similar dimensions and are easier, cheaper, more timely to obtain, or more precise.

- Data for what is considered an important X-variable are available only for a very narrow range of values.

The following are some of the reasons for including statistically non-significant X-variables (to be applied with caution):

- The X-variable is part of a theoretical model and should not be ignored.
- The X-variable is deemed important by the user.
- The X-variable is required by a regulatory agency.

Overriding Statistical Criteria

A significant X-variable might be excluded from the model in the following situations:

- It is collinear (see VIFs) with other X-variables in the model and including it in the model results in distortions of the regression coefficients (e.g., signs of b_i).

- It is significant only because of a relatively small number of observations.

Conversely, under certain conditions it may be useful to include statistically non-significant variables when in a small data set they improve the RMSE. In regression, including any X-variable with a t-Ratio (or F-Ratio) greater than 1 will reduce the RMSE, even though the variables might not be significant using common *p*-value criteria of hypothesis testing. Yet another practical rule is to consider any X-variable with a t-Ratio greater than 1.4 as a good candidate for the model. This is taken as a compromise between traditional significance and the need to reduce the RMSE.

Finding a good regression model requires more than formal statistical rules. Subject knowledge and a clear purpose for the regression model are essential to successful modeling.

18.7 Summary

- Good multiple regression models are not necessarily unique. Several models might satisfy the criteria of a good model in a similar measure.

- The criteria for judging a good model are statistical and non-statistical. Some statistical criteria are explained in this chapter. Non-statistical criteria usually are based on expert knowledge of the area of application.

- Finding a good model is often complicated by a large number of available X-variables.

- Finding a good model is also complicated by multi-collinearity among the X-variables.

- For applications with a modest number of X-variables, it is possible to find the set of all possible regression models. From the set of candidate models select the recommended model.

- Good model-finding follows the principle of parsimony.

- The simple statistical criteria for judging candidate models are RSquare and RMSE.

- Stepwise regression is a quick way to select a candidate model based on statistical criteria.

- The three types of stepwise model algorithms are Forward, Backward, and Mixed. The three types do not necessarily lead to the same model.

- The Forward method starts with no X-variables in the model. It adds one variable at a time until all variables not in the model have a p-value higher the Prob to Enter criterion.

- The Backward method starts with all X-variables in the model. It removes one variable at a time until all variables in the model have a p-value lower the Prob to Leave criterion.

- The Mixed method may start with no or all X-variables. At any one step, it will add or remove one variable at a time until all variables not in the model have a *p*-value higher than the Prob to Enter criterion and all the variables in the model have a *p*-value lower than the Prob to Leave criterion.

- The selected values for Prob to Enter and Prob to Leave influence the final set of variables.

- Each stepwise algorithm suggests a single candidate model that needs to be further analyzed.

- The fit of the model should be checked against statistical criteria using leverage plots and plots of residuals versus predicted, but also VIF and even autocorrelation among the residuals for data involving time (see the Durbin-Watson test).

- The model should also be judged against non-statistical criteria. Ask, Does the model make sense? Do the coefficients have the proper sign with respect to expert knowledge? Are the residuals sufficiently small to yield a precision that matches the requirements of the problem statement?

18.8 Problems

1. The file *BestBSchools.jmp* contains data on post-MBA salaries for graduates of 60 business schools. The following variables are available for analysis:

Variable Name	Description
5 year salary gain ($thousand)	Total compensation after graduation, less tuition and forgone compensation. The five-year gain is before taxes and adjusted for the time value of money.
Years to payback	Payback on MBA tuition.
Pre-MBA salary	In thousands of dollars.
Post-MBA salary	In thousands of dollars.
Tuition	Total out-of-state tuition.

a. Use the All Possible Models method to suggest candidate regression models that are good predictors of the 5-year salary gain.

b. Analyze the candidate models and select the model that best predicts 5-year salary gain. Write a paragraph providing justification for the model you selected.

2. A college admissions office is seeking to increase the number of students that remain at the school after their freshman year. The admissions files and transcripts for thirty randomly selected sophomores have been obtained. Those factors believed to be good predictors of freshman grade point average (GPA) have been entered into the file *freshman_gpa.jmp*. The definitions of the variables are as follows:

Variable Name	Description
High School GPA	High school grade point average on a four-point scale.
SAT Critical Reading	Score on the Critical Reading component of the SAT Reasoning Test. Scores can range from 200 to 800.
SAT Math	Score on the Math component of the SAT Reasoning Test. Scores can range from 200 to 800.
SAT Writing	Score on the Writing component of the SAT Reasoning Test. Scores can range from 200 to 800.
Motivation	Rated on a scale from 0–100 using a questionnaire.
School Type	Public or private school.
Freshman GPA	Grade point average on a four-point scale for the freshman year of college.

Use the data to build a multiple regression model that predicts freshman grade point average. Give the regression equation and an assessment of the quality of the model. Discuss the method used to create the model (Stepwise, All Possible Models) and why you chose that method.

3. A regional economic development commission is investigating the effect of professional sports teams on local economies. The file *professional_sports.jmp* contains information on cities that have both Major League Baseball (MLB) and National Football League (NFL) franchises. You will find the variable definitions in the Notes option found in the Column Properties menu in the Column Information dialog box.

a. Use stepwise regression to find candidate models to predict Major League Baseball attendance. Compare the Forward, Backward, and Mixed stepwise regression results.

b. Select a final model for predicting MLB attendance.

c. Use stepwise regression to find candidate models to predict National Football League attendance. Compare the Forward, Backward, and Mixed stepwise regression results.

d. Select a final model for predicting NFL attendance.

e. Discuss the differences in the final models to predict MLB and NFL attendance.

4. The file *Canton_homes.jmp* contains realtor-supplied selling prices and 10 other characteristics of homes, such as square footage, lot size, and so on, for a sample of 28 homes in the Canton, Ohio area. Use this data to develop a multiple regression model.

a. Transform any of the continuous variables as needed (e.g., Year Built).

b. Compare the models from Forward, Backward, and Stepwise regression.

c. Change the Prob to Leave and Prob to Enter values and note the effect.

d. How do the stepwise regression models compare with those suggested by the All Possible Models method?

e. Use one of the stepwise models as a starting point to develop a final model. Which variables are most important in predicting house price?

f. Evaluate the precision associated with this model. Is the model sufficiently precise to be of use to a real estate agent or buyer? If not, what additional actions would you take to improve the model?

5. A food scientist is analyzing the nutritional value of breakfast cereals and would like to develop a predictive model for calories. The data on 76 different cereals with calories as the Y-variable and 12 X-variables are in *Cereal_Calories.jmp*.

a. Obtain the pairwise correlations and scatterplots for all continuous variables. Mark or label any outliers you identify on the scatterplots. Are there highly correlated X-variables?

b. Develop a good multiple regression model for predicting calories.

c. Assess the goodness-of-fit.

d. Analyze the residuals. Do any of the outliers identified from the scatterplots appear as outliers in the residual plots? Exclude the outliers and rerun the regression to determine their influence on the model. Should these outliers be removed from the model? Explain.

6. One of the criteria a rental car company uses to select new cars for its fleet is average annual fuel cost. The data in *fuel_economy.jmp* were obtained from the US Department of Energy Web site. You will find the variable definitions in the Notes option found in the Column Properties menu in the Column Information dialog box. Determine those factors that affect the average annual fuel cost by building a multiple regression model. Consider creating interaction terms to improve the model. Assess the goodness-of-fit of the model and the impact of any outliers. What additional factors not in the data set might potentially improve the model? How could the model assist in deciding which model cars the rental company should purchase?

7. An office manager is preparing next year's budget for computer equipment based on employee requests for replacement. Data for different types of computers have been collected from each of two authorized vendors and can be found in *new_computer_purchases.jmp*. The following variables are available for predicting the price of a new computer:

Variable Name	Definition
RAM	Amount of random access memory in gigabytes.
HD Capacity	Amount of hard drive storage in gigabytes.
Processor	One of two models: GX2 or GX3.
Type	Laptop or desktop.
Screen Width	Measured in inches.
Vendor	OfficePlus or Clips.
Price	U.S. dollars.

Build a multiple regression model that can be used to predict the price of a computer. Give the regression equation and associated goodness-of-fit measures. Use your model to answer the following questions.

a. Is one vendor preferable to the other for purchasing new computers?

b. What factor is most important in determining the price of a new computer?

c. Department heads have been discouraging employees from replacing desktop computers with laptops on the premise that laptops are more expensive than desktops. Based on the available data, are the department heads justified in their actions?

d. Should HD Capacity be included in the multiple regression model? Explain why or why not.

8. A two-income family is evaluating childcare options in their area. The file *child_care_centers.jmp* contains data as published in a local business magazine. Note that one of the childcare centers is open for only 3¼ hours; this is an after-school program. The variables have the following definitions:

Variable Name	Definition
Weekly Rate ($)	Cost of childcare for one child for one week, in dollars.
FTE Enrollees	The number of full-time equivalent children enrolled at the childcare center.
Licensed Capacity	The maximum number of children that can be enrolled at a childcare center.
FTE Staff	The number of full-time equivalent staff employed at the childcare center.
Hours open/day	The number of hours the childcare center is open each day.
Summer Program	Indicates whether a center offers special activities (weekly field trips, swimming lessons, etc.) during the summer.

a. Build a multiple regression equation to predict the weekly rate (cost) of childcare. Decide whether to include the after-school program in your analysis.

b. Create new two variables, Hourly Rate and Enrollee to Staff Ratio, by adding two columns with appropriate formulas to your JMP data table.

c. Look for a better regression model using either Hourly Rate or Weekly Rate as the dependent variable. Consider the use of Enrollee to Staff Ratio as an X-variable in your model.

d. Briefly discuss the differences between the various models you constructed.

e. Select a final model. Write a brief summary presenting your final model. Discuss the quality of the model and justification for excluding any outliers. Explain your choice for the dependent variable. Is your final equation easy to interpret? What does this model reveal about the childcare market in this area? Are there additional variables that should be investigated?

18.9 Case Studies

18.9.1 Case Study 1: Real Estate Appraisal

The real estate tax assessment process often uses multiple regression analysis to assist in determining residential property values. This is an efficient means for valuing a large number of properties as compared to other methods such as the market comparison approach. In this case, you will find data on residential properties in a region of your choice and construct a multiple regression model that predicts residential property values.

Tasks

Perform the following tasks:

1. Select approximately 20–30 residential properties that are representative of the geographic area chosen. Create a data file that contains the current value (or selling price) and characteristics of the property such as lot size, square footage, garage size, presence of air conditioning, presence of a swimming pool, etc. This information can be found on the Web sites of local realtors, newspaper on-line classified ads, or from the assessment records of municipalities. (Select adequate ranges for selling price and property characteristics.)

2. Use a multiple regression model to determine those variables that significantly predict property value.

3. Assess the goodness-of-fit of your model.

4. Use your model to predict the value of a specific residential property and find the associated prediction interval.

5. Summarize your findings in a short report that describes the real estate market in the chosen region, the data source, the multiple regression model, and the quality of the model. Illustrate the use of the model for predicting real estate values and discuss the scope of applicability of the model.

18.9.2 Case Study 2: Discrimination in Compensation?

A corporation has been accused of gender discrimination in wages. The CEO has requested a complete review of compensation practices. You are provided with a random sample of 50 employees. The variables are defined as follows:

Variable Name	Definition
Gender	Male or female.
Joblevel	Unskilled is the lowest level, clerical is the next lowest, mid-level is the third level, and managerial is the highest level.
Educlevel	Basic means that the employee did not finish high school, high school means that the employee did finish high school and/or additional limited schooling, and college means that the employee did finish college.
Years	Represents the number of years an employee has been with this company.
Compensation	Represents wages and salaries, but not mandated compensation such as workers' compensation, social security, etc.

The data are contained in the file *DiscriminationCase.jmp*.

1. Develop a regression model relating salary to years, position, education, and gender. Examine some interaction terms for their usefulness. Justify your choice of interactions.

2. Describe the compensation practices. In particular, show whether the data support gender discrimination. If yes, do the following:

   a. Predict the salary of employees assuming gender is not a determinant of wages (eliminate any terms relating to gender) and gender is a determinant of wages (include all terms relating to gender).

   b. What do the predictions suggest about the presence of discrimination? What is the impact of education, job level, and years with the company? How good a model is this?

3. Write a substantial report with a clear structure that summarizes the problem, the analysis, and conclusions.

Chapter 19

Exponential Smoothing Models for Time Series Data

19.1 Introduction 542

19.2 Detail Example: 10-Year Treasury Note Closing Prices 543

19.3 Smoothing Models 550

    19.3.1 Simple Moving Averages 551

    19.3.2 Exponential Smoothing Models 554

    19.3.3 Simple Exponential Smoothing (SES) 555

    19.3.4 Double Exponential Smoothing For Linear Trend (DES) 561

    19.3.5 Winters' Additive Seasonal Method 565

19.4 Summary 569

19.5 Problems 569

19.6 Case Study: Lockheed Martin Stock in Changing Times 572

19.7 References 573

19.1 Introduction

Managerial resource allocations are often based on forecasting demands for products, services, human resources, or material supplies. Many forecasts use time series data. Such data, recorded at regular intervals (minutes, hours, days, weeks, months), represent a historical record. Time series forecasts use historical patterns in the data to predict future outcomes.

The following are characteristics of economic time series data:

- They have been collected over time.
- Observations are not necessarily independent of each other.
- They typically have not been collected as part of an experiment.
- The mean value at any time *t* is a function of time, often including trends, seasonalities, and instabilities.

Managerial forecasts are arbitrarily classified as short-term, medium-term, or long-term. In a stock market environment, a day trader is interested in price changes on a moment-by-moment basis, whereas a pension fund manager has a performance horizon of several years, if not decades. Obviously, day traders and pension fund managers have drastically different perceptions of what is short-, medium-, and long-term. Table 19.1 gives a few short- and medium-term forecasting applications.

Table 19.1 Short- and Medium-Term Forecasting Applications

Application	Forecasting Objective	Business Purpose
Supermarket	Forecast demand for each product sold per week.	Keep out-of-stock to less than 1% of orders. Keep inventory to minimum levels.
Plastics manufacturer	Track performance characteristics of a product.	Keep performance characteristic within specifications by taking corrective action.
Electric utility	Forecast demand for energy sources by type for next day.	Purchase correct amounts of energy short term. Provide sufficient energy for peak periods.
Hospital	Track hospital admissions per day and time of week.	Maintain proper staffing levels.
Car manufacturer	Estimate the number of cars requiring a specific part.	Inform suppliers to deliver proper number of parts at appropriate time.
Finance department	Estimate cash on hand and cash needed.	Determine need for mix of financial instruments.

Government agencies regularly make forecasts using historical data collected at fixed time intervals. Such government forecasts include daily volume of airline travel, monthly housing starts, monthly Consumer Price Index (CPI), and daily high, low, and average temperatures in a city. In the study of the stock market or business cycles, the historical patterns of economic indicators are of interest. The purpose of such forecasts is to be prepared for the future.

Forecasting is at its best when a regular pattern underlies the process that needs forecasting. It is at its worst when one-of-a-kind events, such as catastrophes, interfere with that pattern. Forecasts might be off target. In a good forecasting system, off-target forecasts are investigated and taken as occasions to improve the system.

Data Requirements for Exponential Smoothing

The JMP Times Series platform is designed to analyze time series data. Exponential smoothing is a univariate method, because it has only a single continuous Y-variable. The minimum data for exponential smoothing consist of a single column of continuous Y-values, ordered from the oldest (first row) to the most recent (last row). *One important assumption for exponential smoothing in JMP is that the time intervals between observations are equal and that there are no missing observations in the series.* Columns containing identifiers, such as dates, are typically added to the data file. It is possible to analyze multivariate time series with more than one Y-variable, but this is beyond the scope of this book.

19.2 Detail Example: 10-Year Treasury Note Closing Prices

Situation and Problem: Examine the monthly yields of 10-Year Treasury Notes (10-year T Notes) to assess the volatility of this investment for a period of 10 years.

Data: The data file *10yearTNote_10years.jmp* containing the yields of 10-Year Treasury Notes was obtained from finance.yahoo.com. It contains 120 observations of the first trading day in each month, beginning with September 3, 1996, and ending with August 1, 2006. An excerpt of the data is shown in Figure 19.1. (The column Dif(Adj.Close, 1) is discussed in the following section.)

Figure 19.1 Excerpt of *10yearTNote_10years.jmp* of Monthly Adjusted
Close of 10-Year T Notes

		Date	10 year T Note - Adj. Close*	Dif(Adj.Close,1)
□	1	09/03/1996	6.7	.
□	2	10/01/1996	6.35	-0.35
□	3	11/01/1996	6.04	-0.31
□	4	12/02/1996	6.41	0.37
□	5	01/02/1997	6.5	0.09
□	6	02/03/1997	6.53	0.03
□	7	03/03/1997	6.91	0.38
□	8	04/01/1997	6.7	-0.21
□	9	05/01/1997	6.66	-0.04
□	10	06/02/1997	6.5	-0.16

10yearTNote_10years

Columns (3/0)
Date
10 year T Note - Adj. Close*
Dif(Adj.Close,1)

Rows
All rows 120
Selected 0
Excluded 0
Hidden 0

Describing Time Series Data in JMP

In JMP, time series data are analyzed with the Time Series platform obtained from the
Modeling menu. Figure 19.2 shows the menu to enter the Time Series platform.

Figure 19.2 Menu to Select Time Series Platform

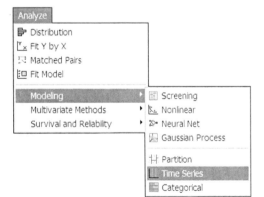

The first window to appear is the column selection window. It is used to specify the data
column that is to be represented as a time series. In Figure 19.3, this is the column 10
year T Note – Adj. Close*. Select **Date** as the optional **X, Time ID** variable for the
horizontal axis. This window also allows specification of the length of autocorrelation
lags to be graphed in the autocorrelation plot and the partial autocorrelation plot. In this

example, 15 lags are specified. The number of periods for which forecasts are desired can also be specified. In this example, 12 forecasting periods are desired.

Figure 19.3 Variable Specification Window for 10-Year T Notes

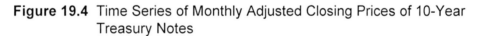
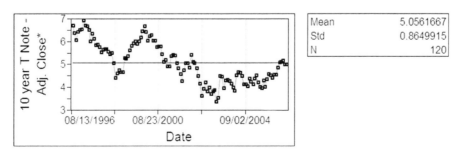

Clicking **OK** results in the time series plot shown in Figure 19.4. The vertical axis is the yield of 10-Year Treasury Notes. The horizontal axis is labeled by the date. The horizontal line through the plot represents the overall mean = 5.0561667%. The standard deviation around the mean = 0.8649915%. This standard deviation can be compared against the RMSEs of other models.

Figure 19.4 Time Series of Monthly Adjusted Closing Prices of 10-Year Treasury Notes

The 10-year Treasury Note yields are typical time series data. They are collected every trading day as part of the data collection by the exchanges. However, the data as presented here only show the first trading day of each month. The observations are clearly not independent. The graph of monthly data suggests strong dependencies over months. The mean value of the 10-year Treasury Note yield seems to change over time. The graph suggests lower yields for more recent trades than for earlier ones.

The presentation of the time series plot needs to be enhanced by adjusting the default labels of the horizontal axis. In order to obtain labels on the horizontal axis that mark the beginning of calendar years, the Date column needs to be assigned a proper Time Frequency property as shown in Figure 19.5. ① Open the **Column Properties** facility by double clicking the variable name **Date**. ② Then open the **Column Properties** menu and select **Time Frequency**. ③ Toggle the Time Frequency to **Monthly**, because the data were recorded on the first trading day of each month. Exit the window by clicking **Apply** and then **OK**.

Figure 19.5 Selecting the Time Frequency Property for Monthly Data

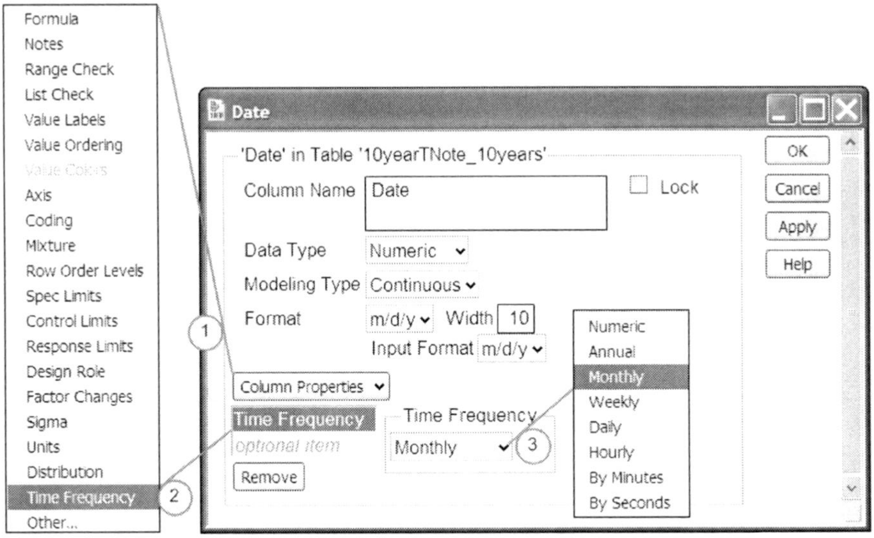

Next, redo the steps outlined in Figure 19.2 and Figure 19.3 to create a new time series plot. Doubling-clicking with the cursor over the horizontal axis produces the X Axis Specification window. Figure 19.6 shows this window with modified entries marked. The minimum and maximum dates are 01/01/1996 and 01/01/2007, respectively. Both dates are outside the data range to give a neater appearance. Then, in the **Tick marks and Grid Lines** section, select **Major** and **Minor** for tick marks and their gridlines. Selecting **Gridline** produces vertical lines for each major and minor tick. Also, toggle **Tick Label Orientation** to **Vertical**.

Figure 19.6 Axis Specification Window for Horizontal Axis

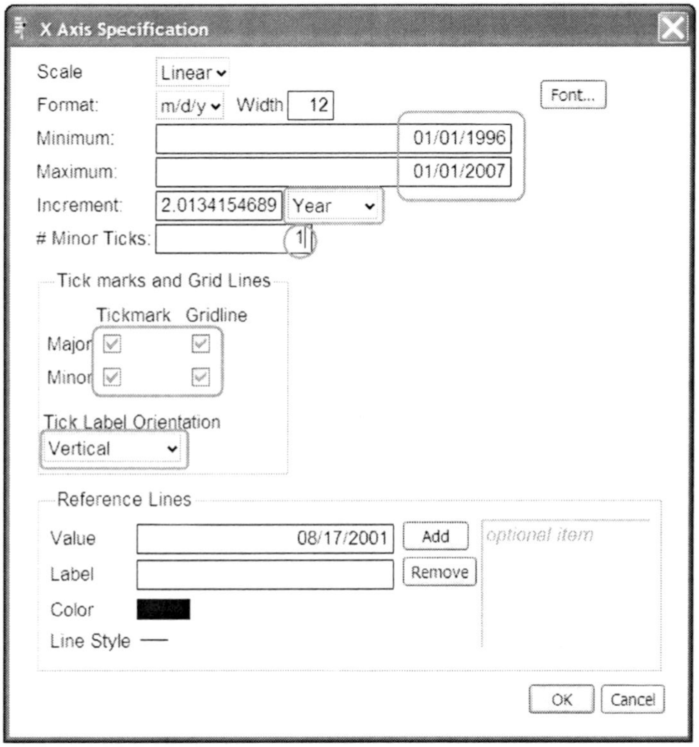

The resulting time series plot (Figure 19.7) shows vertical Date labels starting with 01/01/1996 in two-year increments corresponding to the major ticks. Minor tick marks at in-between years can be seen on the axis.

Figure 19.7 Revised Time Series Plot

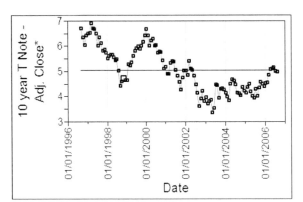

Other Time Series Methods

The default output also shows the autocorrelation and partial autocorrelation functions that are used for model identification in Box-Jenkins (ARIMA) time series models. The ARIMA approach is available in JMP, but is beyond the scope of this book (see *Time Series Analysis Forecasting and Control*, by G. E. P Box, G. Jenkins, and G. C. Reinsel (2008)). Still other methods include econometric methods. In this chapter we cover only smoothing methods, because of their simplicity.

Differences

Economic data sometimes require the analysis of differences between consecutive observations, rather than the original data. These differences between consecutive row values of a column are placed in a new column by using the Dif(Variable, 1) function from the formula editor shown in Figure 19.8. In this example, consecutive differences of 10 year T Note – Adj. Close* are formed. (The parameter 1 could be changed to form differences of 2 or more rows apart.)

Figure 19.8 Dif Function in Formula Editor

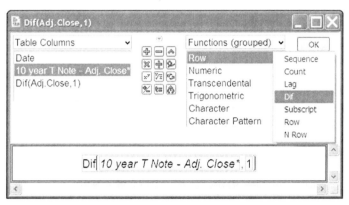

The time series plot in Figure 19.9 shows these differences plotted over time. The pattern appears entirely random. This suggests that changes in 10-Year Treasury Notes are random from month to month. The cumulative effects of random changes cause considerable fluctuations in actual values.

Figure 19.9 Time Series Plot of Consecutive Differences of Monthly 10-Year Treasury Notes

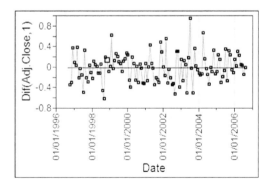

19.3 Smoothing Models

Smoothing models are used to analyze time series data and to perform (1) estimates of the current average of the time series, and (2) simple forecasts of future values. Smoothing models attempt to remove short-term fluctuations from the data in order to observe longer-term data patterns. They are used in many business and stock market applications. These methods can be performed by simple programs, but do have methodological disadvantages over more sophisticated methods such as Box-Jenkins (in JMP referred to as ARIMA) methods. Smoothing models consist of several components. Different methods incorporate different components. The following are the key components:

- Mean component: A stationary time series is in equilibrium around a constant mean. In a non-stationary time series, the mean component might undergo changes.

- Trend component: A time series has a trend when the series increases and decreases over a selected period of time. Trends can be linear or curvilinear. Trends can reverse their pattern from increasing to non-increasing and vice versa.

- Seasonal component: A time series has a seasonal component when the mean value changes in a recurring pattern of a fixed length. Demand for electricity has a seasonal component due to fluctuating energy requirements from season to season.

Table 19.2 Model Terms of General Smoothing Model

Model Term	Explanation	Comments
μ_t	Time-varying mean of time series at time t.	Stationary series exhibit a constant mean.
β_t	Time-varying slope term for trend at time t.	$\beta_t = 0$ indicates that there is no trend.
$s(t)$	Seasonal term. Depending on series, there might be more than one seasonal term.	$s(t) = 0$ indicates no seasonality.
ε_t	Random shocks.	Random shocks are often assumed to be normally distributed for prediction purposes.

A general smoothing model can be written as $Y_t = \mu_t + \beta_t + s(t) + \varepsilon_t$. The model terms are explained in Table 19.2. Smoothing models estimate the current mean as a linearly weighted combination of past observations. We will examine four smoothing models:

- simple moving averages without trend or seasonality
- simple exponentially weighted moving averages without trend or seasonality
- double exponential smoothing for linear trend
- Winters' additive seasonal model for linear trend and seasonality

19.3.1 Simple Moving Averages

Simple moving averages are used over short time periods to estimate the mean of recent time series values. However, they are not a sophisticated forecasting method. They nevertheless play an important role in day-to-day business analysis. Many business reports contain simple moving averages.

Simple Moving Average Calculation

A simple moving average (MA) of length k (also called uniformly weighted moving average or UWMA) gives equal weight to the most recent k observations. The data consist of a series of equally spaced observations $y_t, y_{t-1}, y_{t-2}, \ldots, y_{t-k+1}, y_{t-k}, y_{t-k-1}$, and so on, where y_t is the most recent observation. The spacings are the times between consecutive observations. They could be in seconds, minutes, hours, days, etc.

The k-period simple moving average is

$$MA(k) = \frac{y_t + y_{t-1} + \ldots + y_{t-k+1}}{k}$$

The MA is updated from one period to the next by dropping the oldest and adding the most recent observation to the average:

$$MA(k)_{updated} = \frac{y_{t+1} + y_t + y_{t-1} + \ldots + y_{t-k+2}}{k}$$

Weights of Observations in a MA(k)

In a simple moving average MA(k), the k most recent observations, y_i with $(i = t, t - 1, \ldots, t - k + 1)$, receive a weight of $1/k$. All others receive a weight of zero. An observation receiving a weight of 0 is ignored. In a MA (3), the weight is $1/3$ for the three most recent observations and 0 for earlier ones.

Average Age of Observations in a MA(k)

The average age of data in a MA(k) is (k – 1)/2 time periods. Accordingly, the moving average MA(3) of the three most recent monthly observations has an average age of (3 – 1)/2 = 1 month (the average of 0, 1, and 2). Similarly, an MA(4) has an average age of 1.5, and a MA(7) has an average age of 3. Applications with rapidly changing means require shorter MAs, whereas those with stable means can use a longer history and estimate the mean with higher precision. In stock market applications, three-month and six-month averages of stock values and indices are common.

Example of MA(3): EDO Stock Price

The EDO Corporation is a maker of airborne electronic warfare systems and mine counter-measure systems. The stock price is monitored using the average of prices taken on the first trading day of each month. The moving average is taken over three months. To assess long-term trends in the stock price, five years of monthly data are used. An excerpt of the data contained in *EDOstockUWMA.jmp* is shown in Figure 19.10.

Figure 19.10 Excerpt of Monthly Adjusted Close Prices of EDO Stock with Three-Month UWMA

		Date	EDO adj. close	UMWA(3)	Lag(EDO,1)	Lag(EDO,2)
▪	1	05/09/2001	28.07	·	·	·
▪	2	01/10/2001	26.31	·	28.07	·
▪	3	01/11/2001	23.76	26.05	26.31	28.07
▪	4	03/12/2001	25.86	25.31	23.76	26.31
▪	5	02/01/2002	25.46	25.03	25.86	23.76
▪	6	01/02/2002	25.96	25.76	25.46	25.86
▪	7	01/03/2002	26.43	25.95	25.96	25.46
▪	8	01/04/2002	30.29	27.56	26.43	25.96
▪	9	01/05/2002	27.05	27.92	30.29	26.43

EDOstockUWMA

Columns (6/0)
- Date
- EDO adj. close
- UMWA(3)
- Lag(EDO,1)
- Lag(EDO,2)

Rows
All rows   60

Figure 19.12 shows two time series relating to the EDO Corporation stock price. These plots were generated by using similar steps to those used for Figure 19.4. EDO adj. close represents observations of the EDO stock price. UWMA(3) is the three-month moving average calculated from three consecutive EDO adj. close observations.

The formula in Figure 19.11 is a one-pass evaluation of UWMA(3), rounded to two decimal places. Consecutive values are entered into the formula with the Lag function. The Lag function is taken from the **Row** submenu, and reaches back as specified by the Lag parameter (here 1 and 2). The Round function is part of the **Numeric** submenu. The Lag parameters and rounding decimals are circled. To show the function of the Lag operator, Lag(EDO,1) and Lag(EDO,2) are given as two separate columns of *EDOstockUWMA.jmp*.

Figure 19.11 Formula Window for UWMA(3)

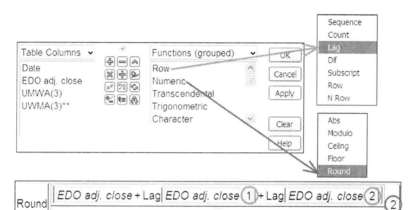

The left plot in Figure 19.12 shows the adjusted closing prices of the first day of trading in each month. The right plot shows the three-month moving averages. The smoothing effect of the MA is clearly visible. The raw data are more scattered than the smoother MA. The difference between the raw data and the moving averages is comparatively small, because only the three most recent months are averaged.

A longer averaging period increases smoothing. The forecast value of a simple MA is its most recent mean value. UWMA models allow no trend or seasonality, although moving averages could be extended to take them into account. The actual stock price of 1/05/2002 is 27.05. The three-month moving average is 27.92. On that date, this is the forecast for the EDO stock price for future time periods.

Figure 19.12 EDO Corporation Stock Price and Three-Month Moving Average

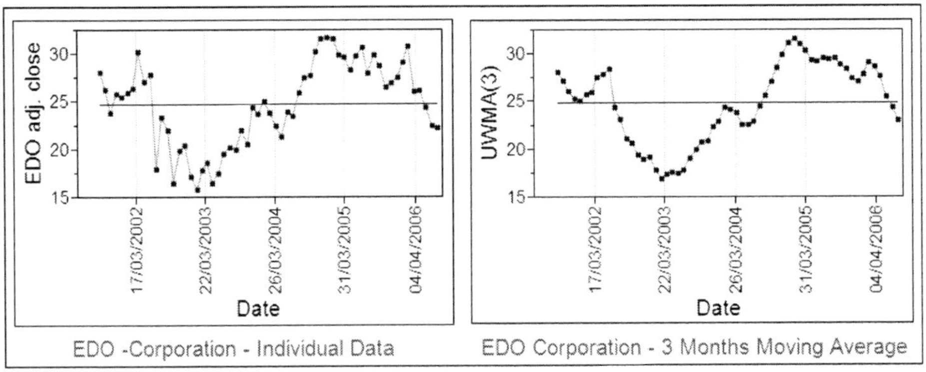

19.3.2 **Exponential Smoothing Models**

Exponential smoothing models are forecasting models with several variants to suit particular applications. In this section we discuss the following:

- simple exponential smoothing for applications with no trend and seasonality
- double exponential smoothing for applications with a linear trend but no seasonality
- Winters' additive model for applications with trend and seasonality

Estimates and Smoothing Weights

Recall the general smoothing model $Y_t = \mu_t + \beta_t + s(t) + \varepsilon_t$. Table 19.3 contains a summary of estimators and smoothing weights. Their meaning will become apparent as each smoothing method is discussed.

Table 19.3 Estimated Quantities in Exponential Smoothing

Estimator Symbol	What Is Estimated	Smoothing Weight	Description of Smoothing Weight
L_t	The **smoothed level L_t** estimates the mean μ_t.	α	Mean level smoothing weight
T_t	The **smoothed trend T_t** estimates slope coefficient β_t.	γ	Trend smoothing weight
$S_{t\text{-}j}$	For $j = 0, ..., s - 1$, the **smoothed seasonality factor $S_{t\text{-}j}$** estimates the seasonality s(t).	δ	Seasonality smoothing weight

19.3.3 Simple Exponential Smoothing (SES)

Use SES when the mean is not expected to change within a reasonable time frame, or if there are no obvious trends or seasonality in the data. The SES model $y_t = \mu_t + \varepsilon_t$ represents an observation at time t by the mean at t plus some random error. The forecast into the future (for any time period) will be the current mean estimate.

The smoothing equation is $L_t = \alpha \cdot y_t + (1 - \alpha) \cdot L_{t-1}$ with $0 < \alpha < 1$. This equation shows that the updated (smoothed) level estimate consists of the new observation y_t multiplied by the smoothing constant α, plus the previous smoothed level estimate multiplied by $(1 - \alpha)$. A more telling expression for the estimated level at time t is

$$L_t = \alpha y_t + \alpha(1 - \alpha)y_{t-1} + \alpha(1 - \alpha)^2 y_{t-2} + \alpha(1 - \alpha)^3 y_{t-3} + \alpha(1 - \alpha)^4 y_{t-4} + \cdots$$

This expression shows that the level estimate L_t is a linear combination of past observations, weighted by exponentially decaying coefficients.

SES Weights

The sum of smoothing weights is always equal to 1. More recent observations are weighted heavier than more distant observations. Figure 19.13 shows the decay of weights for smoothing constants $\alpha = 0.5$ and 0.1. Smaller smoothing constants have weights that decay more slowly. This means that smaller smoothing constants, e.g., $\alpha = 0.1$, smooth the data more than larger smoothing constants, e.g., $\alpha = 0.5$. Smaller constants give more weight to the distant history. For $\alpha = 0.1$, weights are slow to decay and even have noticeable values past 40 observations. For $\alpha = 0.5$ the weights are practically 0 when k = 10.

Figure 19.13 Decaying Weights in SES for $\alpha = 0.5$ and $\alpha = 0.1$

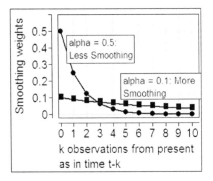

Simple Exponential Smoothing in JMP

For simple exponential smoothing, we return to the 10-Year Treasury Note closing prices example. In order to perform exponential smoothing, choose the Time Series platform (Figure 19.2). Next designate the Y-variable in the context window and click **OK** (Figure 19.3). Select **Smoothing Model** and **Simple Exponential Smoothing** from the variable menu bar as shown in Figure 19.14.

Figure 19.14 Selection of Smoothing Models

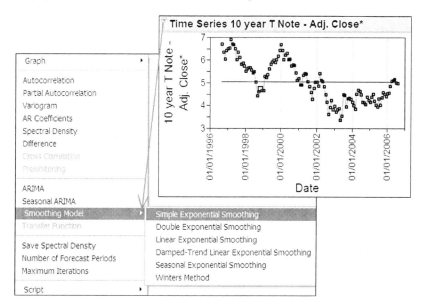

Choosing **Simple Exponential Smoothing** brings up the window shown in Figure 19.15.

Figure 19.15 Default Specification for SES

Click the **Constraints** drop-down menu and change the **Zero to One** default constraint to **Custom**, as shown in Figure 19.16. Selecting **Custom** changes the window with additional selections.

Figure 19.16 Default Window for Custom

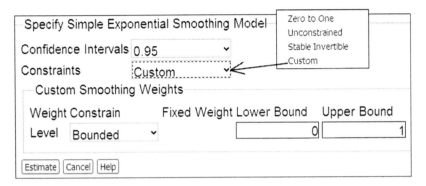

Click the **Level** drop-down menu, and choose **Fixed**, as shown in Figure 19.17. The Fixed field is accompanied by a single field in which you should enter your choice of α. Having changed the **Level** field to **Fixed**, enter the smoothing constant, here α = 0.3. This value will appear in the output in the Parameter Estimates table. A reasonable range for many applications using this method is $0.1 \leq \alpha \leq 0.3$. Lastly, click **Estimate** to obtain the output.

Figure 19.17 Model Specification of SES with α=0.3

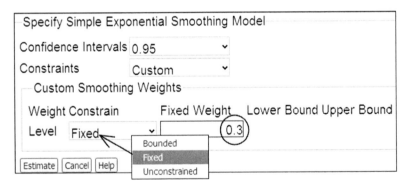

The important output, the Model Summary and Forecast plot, are shown in Figure 19.18 and Figure 19.19, respectively. The well-known criteria for evaluating a model (SSE, standard deviation = RMSE, RSquare) are in the Model Summary. The standard deviation measures the variation around the model. The RSquare measures how much of the original variability is explained. Other criteria are useful in more complicated time series applications. The Forecast graph shows the results of smoothing and the forecast for the number of periods specified in the context window.

Figure 19.18 SES Model Summary for 10-Year Treasury Note

Model Summary			
DF	119	Stable	Yes
Sum of Squared Errors	14.5829979	Invertible	Yes
Variance Estimate	0.1225462		
Standard Deviation	0.350066		
Akaike's 'A' Information Criterion	88.5679113		
Schwarz's Bayesian Criterion	88.5679113		
RSquare	0.83163818		
RSquare Adj	0.83305299		
MAPE	5.91918746		
MAE	0.28790974		
-2LogLikelihood	88.5679113		

The standard deviation is 0.35. Is this a small or large value? This question needs to be answered in relation to other models. However, the RSquare = 0.83 is high by most standards. Thus the model seems adequate, at least numerically. The Detail Plot (Figure 19.19) shows, however, that during periods of an up or down trend, the solid line of the average lags behind the data. During the period from A to B, values tend to increase. The SES is consistently below the actual values—it underestimates the mean. Similarly, during the period from B to C, values tend to decrease. In this period the SES is consistently above the actual values—it overestimates the mean. The mean lags behind the actual values.

Figure 19.19 Time Series Plot with 12-Period Forecast for a 10-Year Treasury Note

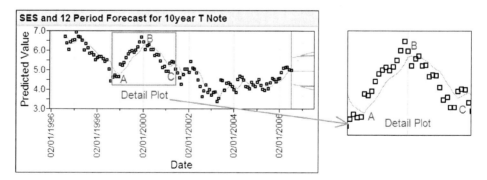

Figure 19.19 shows the raw data as points, the smoothed level L_t as a smoothed line, and 12-period forecasts for individual future values with 95% prediction (confidence) limits as lines. The bold vertical line indicates the end of the data. The middle (horizontal) line is the forecast of the average. The outer two lines are 95% confidence intervals on the forecast estimates. Note the CI gets wider for times farther in the future.

By using the Save Columns option from the contextual menu on the Model bar (red triangle), you can save the values on which the graphical results are based to a new data table. The new data table includes the original data and the time axis values used in this particular run. It also gives the Predicted Value L_t, the standard error of prediction, the residuals, the 95% prediction intervals (labeled CL) for all current observations, and the 12 future periods specified in the context window. The intervals are all prediction intervals, even for past observations. They can be used to examine how well the model would have predicted past values. Figure 19.20 shows the Save Column output for the last 15 rows of the 10-year Treasury Note data. This data table contains the last three original observations (Rows 118–120) and 12 additional rows for the forecast periods (Rows 121–132). For predictions of these future outcomes, no residuals are calculated, because residuals require actual data. These values can be used in plots and other presentations.

Figure 19.20 Save Column Output Table

	Actual 10 year T Note	Date	Predicted 10 year T Note - Adj. Close*	Std Err Pred 10 year T Note - Adj. Close*	Residual 10 year T Note - Adj. Close*	Upper CL (0.95) 10 year T Note - Adj....	Lower CL (0.95) 10 year T Note - Adj....
118	5.14	06/01/2006	4.83945913	0.350066	0.30054087	5.52557588	4.15334239
119	4.99	07/03/2006	4.92962139	0.350066	0.06037861	5.61573814	4.24350465
120	4.98	08/01/2006	4.94773497	0.350066	0.03226503	5.63385172	4.26161823
121	•	08/30/2006	4.95741448	0.350066	•	5.64353123	4.27129774
122	•	09/28/2006	4.95741448	0.36547963	•	5.67374139	4.24108757
123	•	10/27/2006	4.95741448	0.38026901	•	5.70272804	4.21210093
124	•	11/25/2006	4.95741448	0.39450434	•	5.73062878	4.18420018
125	•	12/24/2006	4.95741448	0.4082436	•	5.75755723	4.15727174
126	•	01/22/2007	4.95741448	0.42153528	•	5.78360845	4.13122052
127	•	02/20/2007	4.95741448	0.43442048	•	5.80886297	4.105966
128	•	03/21/2007	4.95741448	0.44693434	•	5.8333897	4.08143927
129	•	04/19/2007	4.95741448	0.45910725	•	5.85724815	4.05758081
130	•	05/18/2007	4.95741448	0.47096563	•	5.88049015	4.03433881
131	•	06/16/2007	4.95741448	0.48253267	•	5.90316115	4.01166782
132	•	07/15/2007	4.95741448	0.49382866	•	5.92530126	3.98952771

In simple exponential smoothing, the forecast estimate k periods ahead is the most recent estimate L_t of the average. The 95% prediction interval increases in width around a constant predicted value. (Observe the forecasted values of the 12 periods.)

19.3.4 Double Exponential Smoothing (DES) For Linear Trend

Double exponential smoothing (DES) is used in situations in which the mean changes in a linear trend up or down. The DES model is basically a straight-line equation with an error term, i.e., $y_t = \mu_t + \beta_t \cdot t + \varepsilon_t$. For DES, two smoothing equations are required. One equation estimates the mean level L_t, and the other equation smoothes the estimate of the trend T_t. (In regression analysis, the equivalent concept to trend is the regression slope. In DES, the trend is updated each time an observation becomes available.)

The DES smoothing equation for the level L_t includes a correction with the most recent trend estimate. The DES equation for the estimated level L_t at time t is

$$L_t = \alpha \cdot y_t + (1 - \alpha) \cdot (L_{t-1} - T_{t-1})$$

The trend is estimated by smoothing differences between level estimates at consecutive times. The smoothed trend at time t is

$$T_t = \gamma \cdot (L_t - L_{t-1}) + (1 - \gamma) \cdot T_{t-1}$$

where γ is the smoothing constant for the trend component. Finally, because this is a linear trend, the k-periods-ahead forecast adjusts the current mean level estimate L_t by adding k times the current trend estimate T_t:

$$L_{t+k} = L_t + k \cdot T_t$$

In JMP, when the same smoothing constant is used for the mean level and the trend, i.e., $\alpha = \gamma$, DES is called Brown's exponential smoothing. When two different smoothing constants are used, i.e., $\alpha \neq \gamma$, DES is called linear exponential smoothing (LES) or Holt exponential smoothing. Here, DES is used with smoothing constants in the range from $\alpha = 0.10$ to 0.30.

DES Example: ABC Stock Price

The following are 60 consecutive weekly closing prices of ABC stock in dollars: 29, 10, 37, 22, 32, 30, 29, 23, 40, 38, 50, 69, 25, 37, 33, 52, 41, 63, 57, 62, 70, 60, 57, 54, 56, 89, 68, 57, 78, 82, 78, 83, 86, 94, 78, 89, 77, 88, 93, 89, 92, 95, 116, 100, 114, 126, 133, 129, 130, 100, 135, 126, 132, 124, 121, 136, 129, 147, 143, 131. These data are contained in *ABC Stock Price.jmp.*

These data are analyzed as a time series using DES (Brown) and SES, both with smoothing constant $\alpha = 0.15$. To perform a double exponential smoothing analysis in JMP do the following:

1. Select the **Time Series** platform from the **Analyze** menu (Figure 19.2).

2. Identify the column of Y-observations as in Figure 19.21. Also specify the desired number of autocorrelation lags (if desired, but not used in smoothing) and the number of periods to be forecast. Click **OK** to produce the basic output.

3. Click the red triangle on the Y-variable menu bar called Time Series ABC Stock Price. A menu appears. Select **Smoothing Models**, and then select **Double Exponential** smoothing from the submenu (see Figure 19.14).

Figure 19.21 Variable Selection for ABC Stock Price Example

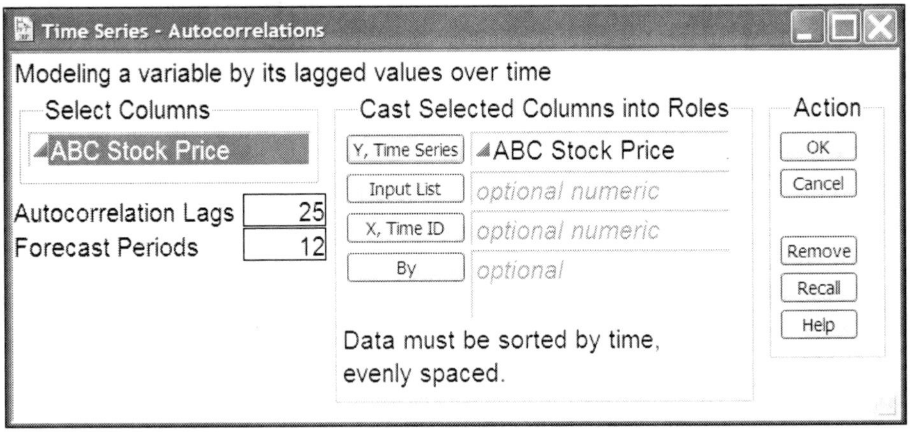

4. A window appears (Figure 19.22) that requires input from the user. Edit the window by selecting Custom as the value for the Constraints field. Then change the **Weight Constrain** (sic) field to **Fixed.** Double exponential smoothing applies the same smoothing constant to both the level and trend components. Enter 0.15 as specified in Figure 19.22. Click **Estimate** to obtain the final output.

Figure 19.22 Model Specification Window for DES

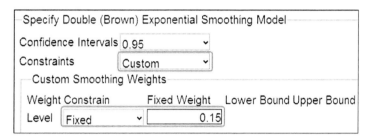

Figure 19.23 shows a comparison of DES with SES based on the same data. Of the two smoothing methods, DES appears to be a better model compared to SES. The following criteria are better for DES: Sum of Squared Errors, Standard Deviation around Model, and RSquare. The graph shows that SES level L_t lags behind (is lower) for these data. The DES forecast predicts the continuation of the linear trend. The SES forecast is flat, using as its value the most recent level estimate.

Side-by-side comparisons are a good method to decide which of the smoothing models will give better results as long as the fundamental behavior of the process continues.

Figure 19.23 Comparison of DES with SES Results for ABC Stock Price

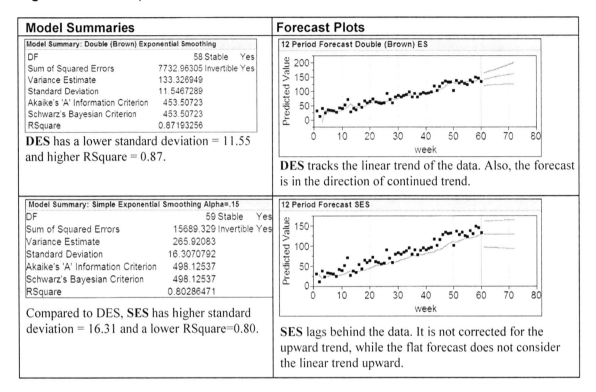

Model Summaries	Forecast Plots
Model Summary: Double (Brown) Exponential Smoothing DF — 58 Stable Yes Sum of Squared Errors — 7732.96305 Invertible Yes Variance Estimate — 133.326949 Standard Deviation — 11.5467289 Akaike's 'A' Information Criterion — 453.50723 Schwarz's Bayesian Criterion — 453.50723 RSquare — 0.87193256 **DES** has a lower standard deviation = 11.55 and higher RSquare = 0.87.	12 Period Forecast Double (Brown) ES **DES** tracks the linear trend of the data. Also, the forecast is in the direction of continued trend.
Model Summary: Simple Exponential Smoothing Alpha=.15 DF — 59 Stable Yes Sum of Squared Errors — 15689.329 Invertible Yes Variance Estimate — 265.92083 Standard Deviation — 16.3070792 Akaike's 'A' Information Criterion — 498.12537 Schwarz's Bayesian Criterion — 498.12537 RSquare — 0.80286471 Compared to DES, **SES** has higher standard deviation = 16.31 and a lower RSquare=0.80.	12 Period Forecast SES **SES** lags behind the data. It is not corrected for the upward trend, while the flat forecast does not consider the linear trend upward.

SES and DES Behavior When the Level Changes

Let us look at the difference in smoothing for two artificial data inputs. The first input shows a variable with a perfect linear trend; the second is a variable in which the Y-variable at first does not change but then changes by a constant value, a so-called step function input.

The comparison when the input is a linear trend shows that SES L_t lags behind the linearly increasing trend in the data. The systematic difference between data and model is called bias (see the left panel of Figure 19.24). Brown's exponential smoothing exactly tracks a linearly increasing trend in the data. There is no bias in the right panel of Figure 19.24.

Figure 19.24 SES and DES Smoothing Response to Perfect Linear Trend

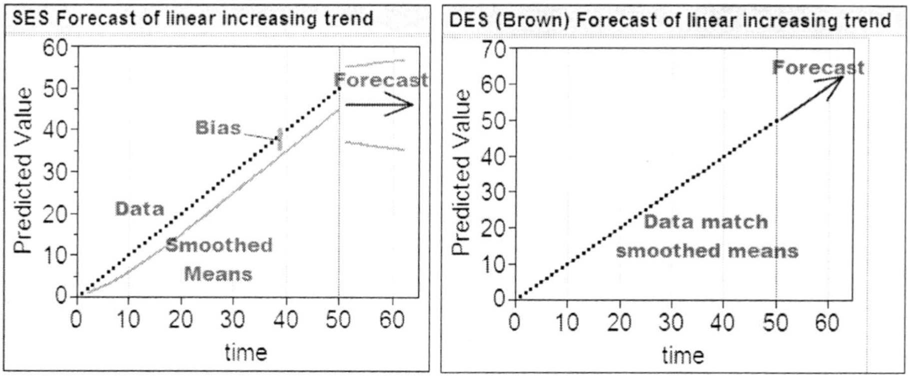

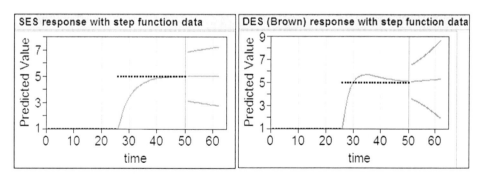

The second comparison between SES and DES shows the level changes by a fixed constant (step function). The response for SES is shown in the left panel of Figure 19.25. SES slowly adjusts to the new level. It never overshoots the new level, but approaches it from below. Such a response is called damping. The forecast continues to be flat. The response to DES is shown in the right panel of Figure 19.25. The DES forecast has a slight overshoot of the new level, because the trend average T_t is not yet 0.

Figure 19.25 SES Response to Step Change

19.3.5 Winters' Additive Seasonal Method

Seasonalities occur when data follow repeating patterns over time. Many economic phenomena follow seasonal patterns: car buying, gas consumption, clothing purchases, travel purchases, etc. Seasonalities can have different periods. In economic time series, common seasonality lengths are quarterly ($s = 4$) and annual ($s = 12$).

Nevertheless, seasonality is any pattern that repeats itself in constant multiples of the time units by which the phenomenon is measured.

The Winters' model is called additive because the level estimates are adjusted by adding a seasonality adjustment to the current estimate. (In multiplicative models the seasonality is introduced by multiplication with a seasonality factor.) The additive model is

$$y_t = \mu_t + \beta_t \cdot t + s(t) + \varepsilon_t$$

Winters' model contains the overall mean level and the adjustment for linear trend, both found in DES. The new additive term, s(t), adjusts for seasonality patterns. This model is accordingly estimated using three equations that are updated each time a new data point is added. One equation is the level equation that is used for prediction. In JMP, the results of this equation are given in the Predicted column. It is similar to the previous level equation but with the seasonality adjustment S_{t-s}:

$$L_t = \alpha\left(y_t - S_{t-s}\right) + (1 - \alpha)(L_{t-1} + T_{t-1})$$

The adjustment for trend is

$$T_t = \gamma\left(L_t - L_{t-1}\right) + \left(1 - \gamma\right)T_{t-1}$$

It is based on estimated level differences between two consecutive levels. The seasonality adjustment adds a constant value periodically to the average. The value of the adjustment depends on the place in the seasonality. The additive seasonality adjustment is

$$S_t = \delta\left(y_t - L_t\right) + (1 - \delta)S_{t-s}$$

Note that the adjustment to the seasonality factor is the observed difference between the current level estimate and the observed value.

Example: Monthly Sales of High End Audio Components

A department store chain needs to analyze sales of high-end audio components with a suspected seasonal variation. The monthly sales data is a variable called Audio Sales in 1000$. In two separate columns, each observation is also identified by year and by month. The data are stored in *AudioSales.jmp*. The Audio Sales in $1000 are as follows in time order:

92.51, 93.71, 93.9, 94.45, 104.99, 94.72, 105.97, 117.53, 129.16, 118.28, 107.91, 108.73, 109.91, 98.94, 99.79, 111.65, 101.09, 112.72, 124.38, 136.42, 125.21, 114.64, 115.71, 113.87, 115.67, 104.61, 105.75, 105.89, 118.15, 106.98, 119.27, 131.38, 144.2, 133.17, 121.14, 122.17, 122.8, 110.65, 110.88, 112.2, 124.68, 112.88, 125.19, 138.49, 152.12,

139.85, 127.77, 128.83, 128.37, 115.83, 116.77, 117.86, 130.7, 118.11, 131.84, 145.75, 159.78, 146.66, 134.52, 134.26

Winters' Additive Method in JMP

An important requirement to use seasonality models is that data are available over multiple seasons so that the seasonality can be estimated. The following steps are used to perform time series analysis with Winters' additive method using three equal smoothing weights = 0.2 and with a 12-period forecast:

1. Select the Time Series platform from the **Analyze** menu.

2. Identify the column of Y-observations as in Figure 19.26.

Figure 19.26 Variable Selection for *AudioSales.jmp*

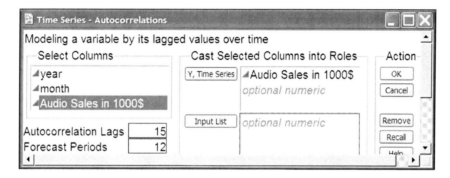

3. The time series plot suggests a seasonality that repeats itself in a 12-month cycle (see Figure 19.28). From the contextual menu of the Y-variable, select **Smoothing Models**, and then **Winters Method**, as shown in Figure 19.14.

4. A model specification window appears (see Figure 19.27). Edit the window as follows: In this example, the seasonality needs to be specified as **Periods Per Season** = 12. Next, change the value in the **Constraints** field to **Custom**. Then change the **Weight Constrain** field to **Fixed**. For each model component (Mean, Trend, Seasonality) enter the desired fixed weight. In this example, 0.2 is used to smooth the three model components. Click **Estimate** to obtain the final output.

Figure 19.27 Model Specification for Winters' Method

Specify Winters Method (Additive) Model

Confidence Intervals	0.95	⌄
Periods Per Season		12
Constraints	Custom	⌄

Custom Smoothing Weights

Weight	Constrain		Fixed Weight	Lower Bound	Upper Bound
Level	Fixed	⌄	0.2		
Trend	Fixed	⌄	0.2		
Seasonal	Fixed	⌄	0.2		

[Estimate] [Cancel] [Help]

The time series plot in Figure 19.28 consists of three phases. The first phase, labeled as Seasonality not yet estimated, shows only raw data. The data in this phase are used to find initial estimates of seasonality. The second phase contains both raw data and a smoothed estimate of the time-varying mean. In this phase, the seasonality estimates are updated. The third phase is the forecast phase and contains no data. Unlike previous methods, the Winters' method gives seasonally changing forecasts.

Figure 19.28 Time Series Plot with Seasonalities of AudioSales.jmp

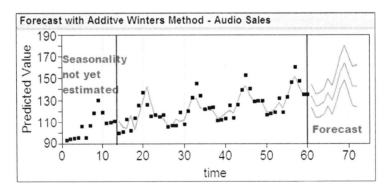

As can be seen from this plot, the data behave in a regular pattern, best seen in the lines of the 12-period forecast. This pattern is based on the most recent estimates and might change over time. Forecasts are adjusted up and down according to the estimated seasonal pattern. Also, the forecast, predicted values, and confidence intervals repeat the seasonality pattern estimated at the most recent available data point.

19.4 Summary

- Time series data require different methods. Smoothing methods are simple and useful. More sophisticated methods are widely used as well.

- Smoothing methods try to eliminate variation from the data to reveal trends and patterns.

- Time series methods include model terms to account for a mean value that changes over time.

- Smoothing models employ smoothing constants to weigh the data in order to estimate the current mean, the trend, and the seasonality coefficients.

- Smoothing weights sum to one for each smoothed component estimate.

- A trend in the mean is a consecutive increasing or decreasing change in the mean. Double exponential smoothing is the recommended model for a linear trend where the changes in the mean are approximately constant.

- A seasonality in the mean suggests a recurring pattern over a fixed period.

- Seasonality models require data on several complete seasons to estimate the seasonality coefficients.

- The forecasts of SES are the current estimate of the mean. The forecast of double exponential smoothing is linearly up or down from the current mean estimated, depending on the current estimate of the trend coefficient. Seasonality forecasts imitate the current pattern estimated from the data.

19.5 Problems

1. Describe a system or process that would have the following time series behavior:

   a. stationarity
   b. trend
   c. seasonality

   Search the Internet to find confirmatory examples.

2. Select a stock of interest. Download five years of monthly and weekly adjusted closing prices.

  a. Create time series plots for each series. For both data sets, display the time series plot with similar spacing and ranges on the two X-axes and the two Y-axes, respectively.
  b. Give numerical statistics that describe each series.
  c. Describe similarities between the two time series.
  d. Describe differences between the two time series. What are the implications for developing smoothing models?

3. A national chain operates two motels in a resort town. One motel is on the beach, about ½ mile from town. The other is in town, conveniently located to restaurants and shopping. The data file *MotelOccupancy.jmp* contains the daily occupancy for each of these motels.

  a. Plot the data using the Time Series platform. Give a statistical summary of the data by motel.
  b. Which exponential smoothing model—simple, double, or Winters' additive— would you recommend? Why?

4. The file *NewEngland_Rainfall.jmp* contains the total monthly rainfall for the six New England states for the years 2001–2006.

  a. Read the file into JMP and use the Time Series platform to obtain a time series plot of the data.
  b. Use the simple moving average method described in Section 19.3.1 to smooth the rainfall time series. Create MA(2), MA(3), and MA(4) plots.
  c. Describe the differences between the three smoothing models and the original time series.

5. In 1925, the New York Cocoa Exchange began trading cocoa futures. In 2004, the New York Board of Trade was formed and is now the marketplace for trading a number of agricultural commodities, including sugar, coffee, cotton, and cocoa. The file *Cocoa_Prices.jmp* contains the closing daily price, in dollars, for a metric ton of cocoa for a two-year period. The data were obtained from the New York Board of Trade Web site.

  a. Using JMP, plot the time series.
  b. Fit simple exponential smoothing models for the smoothing constants of 0.1, 0.2, and 0.3.
  c. Compare the goodness-of-fit of each of these models.

6. The Bureau of Labor Statistics operates the Consumer Price Index program, which publishes monthly data on changes in the prices paid for goods and services. There are varieties of general and specialized indices that serve as economic indicators. These indices are widely used as measures of inflation and as a guide in making economic decisions such as adjusting income payments (e.g., Social Security benefits). In addition, the CPI can be used to adjust

historical prices into inflation-free dollars. The Department Store Inventory Price Index is an inventory weighted price index of goods carried by department stores. This index is designed for department stores in preparing income tax returns. The file *MajorAppliances_DSIPI.jmp* contains the Department Store Inventory Price Index for major appliances from 1980 to 2006. Analyze the series using simple exponential smoothing and double exponential smoothing. Compare the two models and recommend the best model for this series.

7. The New York State Department of Transportation has recorded vehicle miles of travel in the state since 1920. Annual vehicle miles of travel (VMT) are based on travel by all types of motor vehicles as determined by traffic counts and other established procedures during a one-year (365-day) period. The data contained in *NY_VMT.xls* were obtained from the New York State Department of Transportation where the VMT is recorded in billions of miles.

a. Obtain a time series plot for this data.
b. Smooth the data using double exponential smoothing. Try several different smoothing constants.
c. Discuss potential causes for departures from the linear trend in terms of historical events.
d. Select a smoothing constant and use the model to forecast VMT for the years 2001–2004. Compare the forecasts to the actual values as given in the following table:

Year	VMT (billions)
2001	130.83
2002	133.06
2003	135.05
2004	137.52

8. An independent bookseller has recorded quarterly sales (in thousands of dollars) for the last nine years. The data are contained in the file *Quarterly_Sales.jmp*.

a. Give a statistical summary of the data by quarter, year, and overall.
b. Analyze the time series plot. Write a paragraph that qualitatively describes the time series.
c. Smooth the data using Winters' additive model.
   i) Evaluate the quality of the fit.
   ii) Use the Winters' model to forecast quarterly sales for the next year (four forecast periods).
d. Analyze the data using a double exponential smoothing model. Evaluate the quality of the fit and compare it to the fit of the Winters' model.

9. Download the daily adjusted closing prices for General Motors stock for the period January 2, 2002 through May 15, 2002 from http://finance.yahoo.com.

   a. Plot the monthly GM stock prices that you obtained. Describe the time series.
   b. Select an appropriate time series model and briefly explain why you chose the model you did.

10. Find a time series in a field of interest and analyze the data. Develop a smoothing model. Use the smoothing model to give forecasts for the next four forecast periods. Write a summary of your analysis, including an explanation for the length of the series selected, smoothing model, and smoothing constants.

19.6 Case Study: Lockheed Martin Stock in Changing Times

Lockheed Martin is one of the largest defense contractors in the world. This case study will examine the performance of Lockheed Martin stock during the period of the Second Gulf War. The Second Gulf War commenced in March 2003 when U.S. and British forces invaded Iraq, causing the collapse of the Saddam Hussein regime.

Business Problem
Analyze a defense industry stock to understand the effect of geopolitical events on historical performance and the limitations and benefits of a smoothing model in making investment decisions.

Strategy
Analyze Lockheed Martin price and volume for a time period of your choice surrounding the build-up to and commencement of the Second Gulf War. The time series may show a change point, where the behavior of the series visibly changes. Consider fitting more than one smoothing model to see which one better accommodates the series. Additionally you might want to examine the stock performance of other defense contractors to see if these stocks behave in a similar fashion.

Tasks

This case will culminate in a substantial paper describing your analysis.

1. Download the weekly adjusted closing price and volume for Lockheed Martin (symbol LMT) for an appropriate time period.

2. Include a description of the data source, dates selected, and a statistical data summary.

3. Select a good smoothing model and give the associated evaluation of model quality.

4. Make four weekly forecasts from your model. How do they compare with the actual stock prices?

5. Discuss the usefulness and shortcomings of using smoothing models for forecasting stock prices.

19.7 References

Box, G. E. P., G. Jenkins, and G. C. Reinsel. 2008. *Time Series Analysis Forecasting and Control.* 4th ed. New York: Wiley.

574

Index

A

acceptance and rejection regions of distributions 198

ACNielsen 6

activity factors of judgment, in semantic differentials 29

Advanced Questionnaire Design (Labaw) 21

advertising, effectiveness of 102–103

All Pairs, Tukey HSD comparison test 261–262

α (significance level) 167, 189–191, 193, 204, 324, 326

α symbol (error probability, risk) 190, 204

alternative hypothesis 186–190

alternative model, in simple regression analysis 376

American Idol 38

American Society for Testing and Materials (ASTM) 71

analysis of variance (ANOVA)

See one-way analysis of variance

See two-way analysis of variance

Apple Computer Co. 441

appraisal, real estate 539

ARIMA approach, in time series models 548, 550

Art of Asking Questions, The (Payne) 24

Ashenfelter, Orley 427–428

asset variability 419–421

association, strength of 93, 121, 149

ASTM (American Society for Testing and Materials) 71

asymmetric distributions 76, 150

autocorrelation patterns 465, 562

axes scales, in graphs 71

Axis Specifications 68–69

B

Backward stepwise regression 517, 521–523, 526–527

bar charts 56

base model, in simple regression analysis 376

bell-shaped curve

See Normal distributions

benchmarking 3, 6

β_0 symbol (regression line intercept) 382, 384, 451

β_1 symbol (regression line slope)

correlation coefficients and 416

multiple regression analysis 451

outlier sensitivity of 397–399

simple regression analysis 382, 384

t-Ratio for significance of 459

β_1 in exponential smoothing 550

β (error probability, risk)

hypothesis testing 190, 204

regression model evaluation and 524

beta, coefficient for stock returns 421–425

beta, standardized (regression) 472–473

Between Levels Sum of Squares 252, 258

binary outcomes 317, 320

binary splits 96

binomial distributions 111, 113–118

Boeing, Inc. 441

Boston Consulting Group 455

bounded distributions 110–111, 133, 146

Box, G. E. P. 548

Box-Jenkins (ARIMA) time series models 97, 548, 550

box plots

continuous data in 55

improvements in 72

overview of 65–66

problems with 74–76

side-by-side 269

C

Capital Asset Pricing Model (CAPM) 421–425, 444
case studies
    Capital Asset Pricing Model (CAPM) 444
    continuous probability distributions 156–157
    data 17–18
    distributions 128–129
    exponential smoothing models 572–573
    hypothesis testing 221–222
    independence of two nominal variables 366–367
    means, comparing two 244–245
    multiple regression analysis 479–480
    multiple regression models 539–540
    multiple regression with nominal variables 505–506
    one-way analysis of variance (ANOVA) 276
    price elasticity of demand 456
    proportions 337
    simple regression analysis 405
    statistical models 102–103
    survey questionnaires 51
    two-way analysis of variance (ANOVA) 312–313
    variables 79–80
categorical models 97
Caterpillar, Inc. 441
CBA loan example 327–332
CDF
    See cumulative distribution function (CDF), $F(y)$
Cell Chi-Square 345–346
cell deviations 333
censored observations 96
census 8
Center for Research in Security Prices (CRSP) 6
central limit theorem (CLT) 146–148, 193
checklists 27

checks of the reasonableness of predictions 98
Chi-Square (χ^2) distribution
    example of 327–332
    independence tested by 345, 351
    introduction to 327
    overview 106, 150–151
    Pearson Chi-Square statistic 331–334, 345–346, 351, 356, 359–360, 362
choice behavior 93
choice models 97
Claritas.com 6
closed responses, in questionnaires
    See open versus closed responses, in questionnaires
closeness of fit 251
cluster sampling 44–45
clustering methods 97
coding, effect 483
column diagnostics 473–475
compensation discrimination 539–540
competition 17, 51
compliance, demonstrating 4
compound questions 25
COMPUSTAT 6
computer simulation models 10
concentration ellipses 122
conditional probability 341–344
confidence intervals 159–183
    Chi-Square distribution ($\chi 2$) for 149
    Each Pair, Student's t 260–263
    example of 169–173
    mean and standard deviation 165–168
    mean comparison 259–264
    p-values versus 192
    point estimates versus 160–165
    prediction 174–178
    proportions 321
    simple regression analysis 374, 388–390, 392, 401
    standard deviation 214–215
    statistical difference from 258–259
    symbol for incorrect (α) 167
    tolerance 174, 178–180

two-sample t-test example 233
two-sided hypothesis testing and 202–203
variance 150–151
Connecting Letters report 289, 291, 301, 303, 305, 494
constants, variables *versus* 11
consumer preferences 353–357
Consumer Price Index (CPI) 543, 570
Consumer Reports 12
context, in questionnaires 25
contingency tables
    Chi-Square test for equality of proportions 330–331
    independence tests 347–348, 351, 355
    nominal or ordinal variables in 92–93
continuous distributions 109
continuous multiple Y-variables with some X-variables models 96
continuous probability distributions 131–157
    See also Normal distributions
    case study on 156–157
    central limit theorem (CLT) 146–148
    characteristics of 132–135
    sampling distributions 148–152
    uniform distribution 135
continuous scale 12
continuous variables
    describing 58–61
    fixed-effect one-way ANOVA 249
    measures of location for 55–56
    models with single Y and no X 90
    simple regression model for 92–93
control variables 90
convenience samples 38
convergence 147–148
correlation analysis
    See also simple correlation
    autocorrelation patterns 465, 562
    correlation coefficient 121–124
    correlation symbol (ρ) 121
    in time series analysis 96–97
    statistical models and 94

court decisions, hypothesis testing and 189–190
covariance
    sample 408–409, 413
    stock portfolio example 123–124
    two discrete random variables 121–122
credit rating example 327–332
critical values 198–199
cross-classified data
    See stratification in cross-classified data
crosshairs tool 68
crosstab reports
    factor level combinations in 301
    Fit Model 262–264
    mean differences 288–291
CRSP (Center for Research in Security Prices) 6
cubic equations 88
cumulative distribution function (CDF), $F(y)$
    binomial distributions and 114
    continuous data in 55
    continuous distributions and 132–133
    descriptive statistics interpreted by 67–69
    probability of events calculated by 111
curvilinear regression 426–438
    See also simple regression analysis
    case study on 445–446
    Fit Special for 430–433
    price elasticity of demand example 433–438
    quadratic regression for 426–430
customer loyalty 17, 51
customer satisfaction surveys 31–33, 35

D

data 1–18
    See also stratification in cross-classified data
    See also survey questionnaires
    See also surveys, sampling for
    See also variables
    case study on 17–18
    collection of 86

data (*continued*)
  continuous 55–56
  correlation coefficients and 415–416
  definition of 2–3
  distributions of 106
  existing versus new 5–10
  extreme values of 66
  need for 4–5
  nominal 56
  ordinal 56
  rounded 75
  scales for 11–14
data mining 89, 93, 96
Davies, O. L. 239
decision making, variation in 4
decision rule, in hypothesis testing 192
degrees of freedom
  Chi-Square distribution (χ2) 150
  F-distribution 151
  Mean Square Error (MSE) and 463
  Root Mean Square Error 386
  Student's t-distribution 149–150
density function, f(y) 132–133, 139, 150
Department of Economic Analysis, U.S.
Department of Commerce, 77
dependence 450
  See also multi-collinearity
dependent variables 90
  See also multiple regression analysis
  See also simple regression analysis
descriptive statistics
  box plots in 65–66
  CDF (cumulative distribution function) plots
    in 67–69
  continuous variables in 58–61
  histograms in 64–65
  Moments list in 70–71
  nominal variables in 62–64
  quantiles in 69–70
Design and Analysis of Industrial Experiments
  (Davies) 239
design of experiments (DOE) 9–10
deterministic models 83

deviation 345–346
dichotomous outcomes 114
direct observation 24
discrete distributions 109, 111–113
discrete random variables 118–124
discriminant analysis 97
distributions 105–129
  See also continuous probability distributions
  See also Normal distributions
  acceptance and rejection regions and 198
  asymmetric 76
  binomial 113–118
  case study on 128–129
  characteristics of 109
  continuous 109
  discrete 109, 111–113
  error 90
  independent events in 107–108
  random variables in 107
  shapes of 110–111
  skewed 55
  symmetric 55, 66
  two discrete random variables in 118–124
double-barreled questions 25
double exponential smoothing (DES) 561–565
Dow Jones Industrial Average (DJIA) 409,
    441, 467
Dow Jones Utility Average (DJUA) 444
Dunnett's procedure, With Control 261
Durbin-Watson test 466–468
dynamic statistical models 96–97

E

Each Pair, Student's t confidence intervals
    260–263
econometric models 97, 548
economic conditions, return on investment and
    118–119
economic time series data
  See exponential smoothing models
EDO Corporation 552–554
effect coding 483
Effect Tests table 282, 286, 306

elastic demand   434
elements   11, 33, 35
empirical distribution function   67
Engel, Ernst   456
Engel's Law   456
error
    See also residuals
    See also Root Mean Square Error (RMSE)
    ε symbol (residual)   382, 451
    least squares method   383
    random   90
    Smallest Extreme Value error distributions
       90
    standard   164–165, 389, 391
    Type I and Type II   203
    unexplainable   252
Error Sum of Squares   252, 258
estimate
    See also Parameter Estimates table
    data for   2
    point   160–165
    simple regression analysis   384–385
    standard error of   164–165
evaluative factors of judgment, in semantic
       differentials   29
expected cell frequency   333
expected frequencies   345
experience curve   455
experiments, designed   9–10
expert knowledge, for model building   85
expert samples   38
explained variables   90
explanatory variables   90
exponential distribution   146–147
exponential smoothing models   97, 541–573
    case study on   572–573
    double exponential smoothing (DES)
       561–565
    example of   543–549
    overview   542–543, 550–551
    simple exponential smoothing (SES)
       555–560
    simple moving averages   551–554

Winter's additive seasonal method
    565–568
extrapolation   10, 399–400

F

F (cumulative distribution function)   111,
    132–133
f (density function)   132–133
F-distribution   149, 151–152
F-Ratio
    description of   106
    Effects Test table   286
    multiple regression analysis   475
    one-way ANOVA significance by
       258–259, 386
factor analysis   94
factor means   287
factors of judgment, in semantic differentials
    29
fair coin toss example   107–108, 112, 342
financial management
    customer satisfaction surveys on   366–367
    regression and   417–425
    simple correlation and   409–410
Fisher's Exact test   356
Fit Model platform
    curvilinear regression   430–433
    means comparison   262–264
    one-way ANOVA   254–258
    regression analysis with nominal variables
       483, 492
    simple regression analysis   377–380
    stepwise regression   517–518
    two-way ANOVA, with equally replicated
       data   298–305
    two-way ANOVA, without replications
       284–292
Fit Special, in Fit Y by X Platform   378
Fit Spline, in Fit Y by X Model Platform   378
fitting constant   384
fixed effect factors   281
fixed-effect one-way analysis of variance
    (ANOVA)   248–249

focus groups 7–8
Forbes Global 2000 list 479–480
forecasting, exponential smoothing for 542
format, questionnaire 22
Forward stepwise regression 516–521, 525–526
frames, sampling units from 33
frequencies
    charts of 56
    expected cell 333
    histograms of 62
    joint 340
    observed versus expected 345
$F(y)$, cumulative distribution function 111

G

Gaussian models 97
gender discrimination case study 539–540
General Electric, Inc. 418–419, 422–424, 441
General Motors, Inc. 571
goodness of fit 373, 462–468
graphs 54–55, 71–72
greater than or equal to symbol ($\geq$) 117
greater than symbol ($>$) 117

H

Hahn, Gerald J. 174
histograms
    descriptive statistics in 64–65
    frequency charts as 62
    improvements for 72–74
    overview 54
    problems with 74–76
hold-back samples 98
Holt exponential smoothing (LES) 561
homoscedasticity 465
Hsu MCB comparison test, With Best 261
hypergeometric distribution 118
hypotheses
    comparing two means 224–225
    independence of two nominal variables
        345–346
    one-way analysis of variance (ANOVA)
        250–251

two-way (ANOVA), with equally replicated
    data 294
two-way analysis of variance (ANOVA),
    without replications 282
hypothesis testing 185–222
    β symbol (error probability, risk) 424–425
    Chi-Square distribution ($\chi 2$) for 149
    confidence intervals and two-sided
        202–203
    critical value in 198–199
    decision rule for 192
    example of 192–197, 200–202
    mean acceptance or rejection 186–190
    mean comparisons as 259
    p-value in 191–192, 199
    proportions 318
    questionnaires for 22
    regression analysis with nominal variables
        489–490
    significance level for 190–191, 198–200
    test statistic for 198–200
hypothesis testing, sample size for 203–216
    calculation of 205–206
    case study of 221–222
    formulas for 208
    in JMP 206–208
    overview 203–205
    power curves for 209–211
    process standard deviation versus 211–214
    standard deviation (σ) 215–216

I

income elasticity of demand 456
incomplete frames 34
independence, tests of
    conditional probability and 341–344
    contingency tables 93
    Durbin-Watson 466–468
    probability distributions and 107–108
independence of two nominal variables
    339–352
    See also stratification in cross-classified
        data

case study on 366–367
conditional probability and 341–344
example of 346–352
hypotheses for 345–346
overview 340–341
independent variables 90
See also multiple regression analysis
See also simple regression analysis
inelastic demand 434
inference, statistical 160
injuries, lost time occupational 405
input variables 449
integrity, of data sources 7
intensity scales 27–28
interaction effect, in two-way ANOVA 295–296
interaction term 496–500
intercept, of regression line
See multiple regression analysis
See multiple regression with nominal variables
See simple regression analysis
Internet, questionnaires on 22–23
interquartile range 70
interval scale 12–14
interviews, questionnaires for 22–23

J

Jenkins, G. 548
joint frequencies 340
joint probabilities 118–122, 343–344
Journal of the American Medical Association (JAMA) 365
judgment samples 9, 38

K

Kryder's Law 101

L

Labaw, Patricia 21
leading questions 25
Least Squares Means 279
least squares method 383, 397

left-sided alternative hypothesis 187, 193–194
less than or equal to symbol ($\leq$) 117
less than symbol ($<$) 117
Levene test for unequal variance 270
leverage plots 99, 469–471, 525, 527, 529
Leibniz, W. 82, 89
Likelihood Ratio test 331, 351, 356
Likert scales 13–14, 28, 40, 316
Lillefors confidence bounds 215, 271–272
linear equations 88
linear exponential smoothing (LES) 561
lines, in graphs 71
loaded questions 25
location, measures of 109, 134–135
Lockheed Martin, Inc. 572–573
logarithmic transformations, for regression analysis 432
lognormal distributions 90, 146
lost time occupational injuries, case study on 405
loyalty, customer 17, 51
LSMeans Student's *t*-test 287–289
lurking variables 352

M

mailed questionnaires 23
marginal probabilities 118–121, 343–344
market share 17, 353
marker symbols, in graphs 71
marketing research 340
Marks, Bruce 427–428
maximum values 56
mean
confidence intervals for 165–168
covariance and 121
general Normal distribution 136
in stock portfolio example 123–124, 128
marginal distributions to calculate 120
measure of location 109, 134
μ symbol for 163, 550
normally distributed outcomes 145–146
numerical summary as 55
of unimodal symmetric distributions 110

mean (*continued*)
 point estimates for   160–165
 probabilistic center of gravity   111
 Standard Error of the Mean   70
 symmetric unimodal distributions   133
 symmetrical distribution   66
mean component, in exponential smoothing
 models   550
Mean Square Error (MSE)   463
means, comparing several
 See one-way analysis of variance (ANOVA)
 See two-way analysis of variance (ANOVA)
means, comparing two   223–245
 case study on   244–245
 examples of   238–240
 hypotheses for   224–225
 paired t-test for   225–227, 233–237
 two sample t-test for   225–233
measurements, data as   2
measures of location   109, 134–135
measures of spread   109, 134–135
median
 box plots to show   269
 measure of location   109, 134
 numerical summary as   55
 outliers affecting   70
 quantile of   144
 symmetric distributions   66
 symmetric unimodal distributions   110, 133
Meeker, William Q.   174
minimum values   56
Mixed stepwise regression   517, 521–523
mode   109–110, 133
model-fitting step   86, 97
Model Sum of Squares   252, 300
models
 See statistical models
Moments list   70–71
monitoring, statistics in   4
Monte Carlo simulations   83
Moore's Law   101

mosaic plots
 Chi-Square test for equality of proportions
  330
 consumer preferences   355
 independence tests   349–350
 nominal variables described by   62–63
 overview   56
moving averages, simple   551–554
multi-collinearity
 model selection and   508–509, 524–525
 multiple regression analysis process   450
 variance inflation factors of   473–474
multi-phase sampling   46
multi-stage sampling   45
multimodal distributions   110, 133
multiple incompatible frames   34
multiple regression analysis   447–480
 See also simple regression analysis
 case study on   479–480
 column diagnostics   473–475
 example of   451–452
 goodness of fit from   462–468
 leverage plots   469–471
 model interpretation   468
 model representation   460–462
 overview   448–451
 scatterplots in   454–455
 simple correlation as preliminary step in
  453–454
 simple regression analysis step in   455–456
 standardized beta   472–473
 t-Ratio for   459
 three independent variables   456–458
multiple regression models   507–540
 all possible   512–516
 case studies on   539–540
 evaluating   524–525
 example of   510–512
 overview   508–510
 recommendations for   530–533
 stepwise   516–523
 with five variables from forward stepwise
  regression   525–526

with four variables from backward stepwise regression 526–527

with three variables from all possible 528–530

multiple regression with nominal variables 481–506

case study on 505–506

example of 483–485

models for 490–494

overview 482–483

two lines with different intercepts and slopes 495–500

two parallel lines with different intercepts 485–490

multiple Y-variables and no X-variables models 94–95

multivariate models 94, 97

multivariate regression 96

μ symbol (mean) 163, 550

See also mean

See also means, comparing two

mutual exclusivity 108

N

neural networks 97

New York State Department of Transportation 571

New York Times 427

Newcomer, Clarence 428

noise variables 449

nominal scales 12–14, 56

nominal variables 62–64, 90–93, 249

See also independence of two nominal variables

See also multiple regression with nominal variables

non-probability sampling 37–38

non-sampling errors 36–37

nonlinear models 97

Normal distributions 136–146

characteristics to 66

convergence to 147

general 136–138

population mean estimate for 161

residuals 396–397

standard 139–146, 148

z-Ratio for 137, 213

normal quantile plots 55, 215–216

null hypothesis 186–190

See also hypothesis testing

accepting or rejecting 186–190

one-way analysis of variance (ANOVA) 250, 258

p-value for 191–192

proportions 322

null proportion 326

numerical summaries of continuous data 55–56

O

observations

average age of 552

censored 96

cumulative proportion of 67

prediction interval for 174–178

studies based on 10, 24

time series analysis of 96–97

weights of 551

observed frequencies 345

occupational injuries, lost time, case study on 405

$1 - \beta$ symbol (power of test) 209

one-sided confidence intervals 171–173

one-way analysis of variance (ANOVA) 247–276

case study of 276

chemical reactor yield example 380

confidence intervals for mean comparison by 259–264

example of 253–254, 267–269

Fit Y by X, in JMP 254–258

fixed-effect 248–249

hypotheses for 250–251

interpretation of 258–259

model 250

residuals in 265–267, 271–272

side-by-side box plots for 269

one-way analysis of variance (ANOVA)
    (*continued*)
    simple regression analysis versus 372
    statistical models and 93
    unequal variance tests for 270–271
    variation sources 252
online surveys 23
open *versus* closed responses, in questionnaires
    26–33
    checklists 27
    intensity scales 27–28
    quantity scales 27
    ranking questions 30
    semantic differentials 28–30
Ordered Differences report 289, 291, 301, 303
ordinal scales 56
ordinal variables 90–93, 316
outcome categories 91
outcome space 133
outliers
    correlation coefficients affected by 415–
        416
    mean-median discrepancies and 70
    multiple regression analysis affected by
        450
    Normal distributions and 66
    residual values as 465–466
    simple regression analysis affected by 373,
        397–399
overlapping frames 34
Overlay Plot platform 60–61

P

p (probability function) 110–111
p-values (significance)
    Effect Tests table 306
    hypothesis test for beta 424
    hypothesis testing 191–192, 199
    one-way analysis of variance (ANOVA)
        257
    regression line slope and intercept 385, 393
    regression model evaluation 524
    two-sample t-test 232

paired *t*-test, comparing two means by
    225–227, 233–237
pairwise correlation 414, 453
parallel lines, in regression analysis 485–490
Parameter Estimates table 379, 458, 461, 486,
    488–490
parameters (θ) 162–164
parsimony, principle of 83, 508, 531
partial least squares 97
partitioning 96
Payne, S. L. 24
Pearson Chi-Square statistic
    consumer preferences testing example 356
    hospital mortality example 362
    independence of two nominal variables
        345–346
    Likelihood Ratio test versus 351
    on-time performance example 359–360
    test equality of proportion 331–334
Pearson correlation coefficient 408
Pepsi Bottling Group 441
performance comparisons 357–361
performance variables 84, 249
personal interviews 22
personalization, in questionnaires 25
pie charts 56, 63–64
pilot tests, of questionnaires 30
point estimates 160–165
point price elasticity 433–435
Poisson distribution 118
political polling 34, 108, 113, 326
polynomial equations 89
polynomial models 378
popularity voting 38
populations, sampling concept of 33, 35
potency factors of judgment, in semantic
differentials 29
power curves 189, 207, 209–211, 324
predicted variables 90
prediction intervals 174–178, 374, 391–392
predictive questionnaires 20, 22
predictor variables 90

price elasticity of demand
    arc 433
    case study on 456
    point 433–435
    regression and 435
    under unitary demand 435–438
principal components analysis 94
probabilistic center of gravity, mean as 111
probability
    See also continuous probability distributions
    See also distributions
    conditional 341–344
    joint 343–344
    marginal 343–344
    of outcome categories 91
    proportional to size sampling 45
probability function, $p(y)$ 110–111
probability sampling 36–38
process standard deviation 211–214
process variables 449
product life statistical models 96
proportions 315–337
    case study on 337
    Chi-Square test for equality of 327–334
    example of 318–322
    overview 316–318
    sample size and 323–327
    tolerance interval 178

Q

quadratic equations 88
quadratic regression 426–430
qualitative information 4, 7, 249
quality control charting 161
quality management 211
quantiles
    critical value derived from 199
    estimating 68–69, 99
    Normal distribution 143–145
    overview 55
    residuals plotted in quantile plot 99, 266,
        397, 494, 529–530

standard deviation hypothesis test 215–216
    uses of 69–70
quantitative information 4, 249
quantity scales 27
quartiles 56
quasi-experimental studies 10
questionnaires 19–33
    See also sampling
    administration of 22–24
    design of 24
    introduction to 20
    layers in 21–22
    open versus closed responses in 26–33
    questions to avoid in 25–26
quota samples 38

R

R^2 (coefficient of determination) 393
    correlation coefficients and 416
    regression analysis with nominal variables
        490
    regression line evaluation by 88, 393–395
    regression model comparison by 514–515
    Root Mean Square Error and 464–465
    simple regression analysis 373–374, 381
    variance inflation factors (VIFs) and
        473–474
random behavior, models of 83
random-effect one-way analysis of variance
        (ANOVA) 248–249
random error 90
random sampling 9, 36, 38–40
random shocks, in exponential smoothing
        models 550
random variables
    discrete 118–124
    distributions of 107
    standardizing 137–139
range 56, 70
ranking questions 30
ratio scale 12–14
real estate tax assessment, case study of 539
reasonableness of predictions, checks of 98

regression analysis
    See Capital Asset Pricing Model (CAPM)
    See curvilinear regression
    See multiple regression analysis
    See multiple regression models
    See simple correlation
    See simple regression analysis
    See stock market returns
Reinsel, G. C. 548
rejection and acceptance regions 198
reliability 24, 156
replacement, sampling with and without 38, 108, 343
representative samples 9, 35
research questions 186
residual (ε) error 281, 382, 451, 550
residuals
    Durbin-Watson test for independence of 466–468
    model validation 98–99
    multiple regression analysis 462–466
    one-way analysis of variance (ANOVA) 265–267, 271–272
    regression analysis with nominal variables 487, 494
    regression model evaluation and 525, 529
    simple regression analysis 395–397
resource allocation, exponential smoothing for 542
response variables 90
return on investment example 118–119, 128
ρ symbol (correlation coefficient) 121
right-sided alternative hypothesis 187, 194–195
risk analysis
    asset variability in 419–421
    Capital Asset Pricing Model (CAPM) for 421
    probability distributions for 106
    stock return example of 123–124, 128
    stratification in cross-classified data for 362–363

Root Mean Square Error (RMSE)
    ANOVA estimated standard deviation from 283
    degrees of freedom of 386
    multiple regression analysis 461, 463–465
    pooled standard deviation as 231, 257
    regression analysis with nominal variables 490
    regression model comparison of 514–516
    simple regression analysis 373, 379, 381–382, 386, 393
    standard deviation as 56
rounded data 75

S

sample allocation 44
sample covariance 408–409, 413
sample size 165, 323–327
    See also hypothesis testing, sample size for
sample surveys 9
sampling 33–46
    See also questionnaires
    cluster 44–45
    concepts of 33–36
    cost of 8
    multi-phase 46
    multi-stage 45
    non-probability 37–38
    population subsets 35–36
    probability 36–38
    probability proportional to size 45
    simple random 38–40
    stratified 43–44
    systematic 40–43
    with replacement 38, 108
    without replacement 38, 108, 343
sampling distributions 106, 148–152
sampling errors 36
sampling frames 34, 36
sampling units 33, 35
satisfaction surveys 31–33

scales 11–14, 71
   See also questionnaires
scatterplots
   See also simple regression analysis
   election outcomes 428
   multiple regression analysis 454–455
   one-way ANOVA 255
   price elasticity of demand 437
   residuals in 99
   simple correlation 412
   stock market returns 423
seasonal component, in exponential smoothing
   models 550
Second Gulf War 572–573
selection errors, in samples 35
self-administered questionnaires 23
semantic differentials 28–30
sensitivity tests 98
side-by-side box plots 269
σ (standard deviation)
   See standard deviation (σ)
Σ (summation) 111, 162, 386
σ^2 (variance)
   See variance (σ^2)
significance
   See also Student's t-test
   hypothesis test for beta coefficient 424
   F-distribution test 149, 151
   F-Ratio test for 258–259
   level (α) of 167, 189–191, 193, 198–200,
     204, 324, 326
   simple correlation 414
   simple regression analysis 385–386
simple correlation 408–416
   case study on 444
   financial indices 409–410
   multiple regression analysis preliminary step
     453–454
   sample correlation coefficient 408–412,
     415–416
   sample covariance 408–409, 413
   significance testing in 414

simple exponential smoothing (SES) 555–560,
   564–565
simple logistic regression model 93
simple moving averages 551–554
simple random sampling 38–40
simple regression analysis 369–405
   See also multiple regression analysis
   See also multiple regression with nominal
     variables
   case study of 405
   confidence intervals for slope and intercept
     401
   degrees of freedom in 149–150
   estimates of equation from 384–385
   example of 374–381
   extrapolation from 399–400
   interpretation of 381–383
   model for 92
   outlier effects on 397–399
   overview 370–372
   results of 373–374, 387–392
   Root Mean Square Error (RMSE) in 386
   t-ratios to test significance in 385–386
   verifying assumptions of 393–397
simulations 10, 83, 156
single Y-variable and multiple X-variables
   models 94
single Y-variable and no X-variables models
   90–91
single Y-variable and single X-variable models
   91–93
skewed distributions 55, 110, 133
slopes, of regression lines
   See exponential smoothing models
   See multiple regression analysis
   See multiple regression with nominal
     variables
   See simple regression analysis
Smallest Extreme Value error distributions 90
spectral analysis 97
spread, measures of 56, 109, 134–135
spreadsheet format 372
s(t) (seasonal term) 550

Standard & Poor's 500 Index 6, 410, 418–419, 441
standard deviation (σ)
    See also variance (σ²)
    ANOVA estimated 283
    confidence intervals for 165–168
    correlation coefficient calculated from 121
    definition of 56
    factor level combination 281
    general Normal distribution 136
    hypothesis testing of 215–216
    measure of spread 109, 134
    model evaluation by 88
    point estimates for 160–165
    pooled 231, 257
    process 211–214
    sample size and 204
    σ symbol for 163
    Standard Error of the Mean 70
    stock portfolio example 123–124, 128
    variance and 112
standard error 389, 391
standard error of estimate 164–165
Standard Error of the Mean 70
Standard Least Squares personality 524
standard Normal distributions 139–146
standardized beta 472–473
standardized cell deviations 333
statistical estimators 106
Statistical Intervals: A Guide for Practitioners
        (Hahn and Meeker) 174
statistical models 81–103
    building 83–87
    case study on 102–103
    continuous multiple Y-variables and some
        X-variables 96
    dynamic 96–97
    evaluating 88–89
    examples of 87–88
    Kryder's Law 101
    Moore's Law 101
    multiple Y-variables and no X-variables
        94–95

    multivariate 97
    product life 96
    single Y-variable and multiple X-variables
        94
    single Y-variable and no X-variables
        90–91
    single Y-variable and single X-variable
        91–93
    validation of 97–99
    variables for 2
statistical samples 36
stepwise regression
    algorithms for 516–517
    Backward 521–523, 526–527
    Forward 518–521, 525–526
    in JMP 517–518
    Mixed 523
Stinton, William 427
stock market returns
    See also exponential smoothing models
    asset variability and risk 419–421
    Capital Asset Pricing Model (CAPM)
        421–425
    case study on 444
    GE, S&P 500 Index, and Treasury Bill risk
        comparison 418–419
    introduction 417–418
    probability distribution example 123–124,
        128–129
    simple moving average of 552–554
stratification in cross-classified data 352–363
    See also independence of two nominal
        variables
    consumer preference example 353–357
    description of 352–353
    performance comparison example
        357–361
    risk analysis example 362–363
stratified samples 8, 43–44
Student's *t*-test
    critical value derived from 199
    Each Pair 260–263

regression analysis with nominal variables 494

sampling distribution from 149–150

two-way analysis of variance (ANOVA) 287–289, 304–305

Sum of Squares 293

summation symbol (Σ) 111, 162, 386

survey questionnaires 19–33

    administration of 22–24

    case study on 51

    design of 24

    introduction to 20

    layers of 21–22

    open versus closed responses in 26–33

    questions to avoid in 25–26

surveys, sampling for 33–46

    cluster 44–45

    concepts of 33–36

    multi-phase 46

    multi-stage 45

    nominal variable independence in 366–367

    non-probability samples 37–38

    overview 8–9

    probability proportional to size 45

    probability samples 36–38

    simple random 38–40

    stratified 43–44

    systematic 40–43

symmetric distributions 55, 66, 110, 133

systematic sampling 40–43

T

t-Ratio 106

    See also Student's t-test

    of hypothesis test for beta coefficient 424

    multiple regression analysis 459

    simple regression analysis 385–386

tax assessment, real estate, case study on 539

telephone interviews 22

test statistics 106, 191, 198–200

    See also means, comparing two

testing, statistics in 4

tick spacing, in graphs 71

time period mismatch, of frames 34

time plot 55, 60–61

time series analysis 96, 465–466

    See also exponential smoothing models

Time Series Analysis Forecasting and Control (Box, Jenkins, Reinsel) 548

tolerance intervals 174, 178–180

tossing fair coin example 107–108, 112, 342

Total Sum of Squares 252, 257–258, 300, 515

traffic speed limits, hypothesis testing and 221–222

transformations 83, 431–432, 471

trend component, in exponential smoothing models 550

triangular distribution 146

true values 36

Tukey HSD test 261–262, 304–305

two sample *t*-test 225–233

two-sided alternative hypothesis 188, 195–197

two-sided confidence intervals 171

two-sided hypothesis testing 202–203

two-way analysis of variance (ANOVA), with equally replicated data 292–305

    example of 296–298

    hypotheses 294

    interaction effect in 295–296

    JMP analysis example 298–305

    model and variation sources 292–294

    overview 278–279

two-way analysis of variance (ANOVA), with unequal replications 279, 306

two-way analysis of variance (ANOVA), without replications 278–292

    example of 282–284

    Fit Model Platform, in JMP, for 284–292

    hypotheses 282

    model and variation sources 280–282

    overview 278–279

Type I and Type II errors 203

U

unbounded distributions 110–111
uncertainty, in decision making 4
unequal variance, tests for 270–271
uniform distributions 135, 147
uniformly weighted moving average (UWMA)
      551
unimodal distributions 110, 133
unimodal symmetric distributions 66
unitary demand 434–438
U.S. Bureau of Labor Statistics 405, 570
U.S. Census Bureau 8
U.S. Constitution 8
U.S. Department of Commerce, Department of
      Economic Analysis 77
U.S. Department of Energy 537
U.S. Environmental Protection Agency (EPA) 4
U.S. Treasury Bills
      correlation of financial indices example
         410
      exponential smoothing example 543–549
      risk of 418–419
      stock closing prices versus 441

V

validity, in questionnaires 24
Vanguard Windsor mutual funds 441
variability, asset 419–421
variables 53–79
      See also independence of two nominal
         variables
      See also multiple regression with nominal
         variables
      See also random variables
      See also simple regression analysis
      See also statistical models
      box plots 65–66, 74–76
      case study on 79–80
      CDF (cumulative distribution function) plots
         67–69
      constants versus 11
      continuous 55–56, 58–61

distribution of two discrete random
         118–124
      fixed-effect one-way analysis of variance
         (ANOVA) 249
      graphs of 54–55, 71–72
      histograms 64–65, 72–76
      in statistical models 2
      input 449
      lurking 352
      Moments list 70–71
      noise 449
      nominal 56, 62–64
      numerical summaries of 55–56
      ordinal 56, 316
      performance 84
      presentation example of 57–58
      process 449
      quantiles 69–70
      relationships between 4
variance (σ^2)
      See also standard deviation (σ)
      calculation of 112
      confidence interval for 150–151
      definition of 56
      measure of spread 134
      residual examination for 465
      residuals 266
      σ^2 symbol for 149, 163
      standard deviation from 109
      stock portfolio example 123–124
      tests for unequal 270–271
variance inflation factors (VIFs) 473–474, 508,
         524–525
variation
      constant 92
      in decision making 4–5
      sources of 252, 280–282, 292–294
      volatility, stock
      See stock market returns
volunteer samples 38
voting, popularity 38
voting irregularities, curvilinear regression to
         test 427–430